AF615901

Low Dimensional Sigma Models

Low Dimensional Sigma Models

W J Zakrzewski

Department of Mathematical Sciences,
University of Durham

Adam Hilger, Bristol and Philadelphia

British Library Cataloguing in Publication Data

Zakrzewski, Wojciech J.
Low dimensional sigma models.
1. Elementary particles. Mathematical models
I. Title
539.7′21′0724

ISBN 0-85274-231-2

Library of Congress Cataloging-in-Publication Data

Zakrzewski, W. J.
Low dimensional sigma models

Bibliography: 2p.
Includes index.
1. Sigma particles. 2. Integrals, Path.
3. Chirality. 4. Supersymmetry. I. Title.
QC793.5.H427Z35 1989 539.7′21 88-34764
ISBN 0-85274-231-2

Consultant Editor: Professor E J Squires, University of Durham

Published under the Adam Hilger imprint by IOP Publishing Ltd
Techno House, Redcliffe Way, Bristol BS1 6NX, England
242 Cherry Street, Philadelphia, PA 19106, USA

Printed in Great Britain by JW Arrowsmith Ltd, Bristol

To my family

Contents

Preface

Two years ago I was invited by professors J P Antoine and J Weyers to give a series of lectures on two-dimensional σ models at the Catholic University of Louvain-la-Neuve. I was told that I was free to choose the specific topics of this lecture course and that the students who would attend them would already have some knowledge of field theory, path integrals and supersymmetry.

This book is based on these lectures, although it contains a couple of chapters on solutions of the $U(N)$ chiral models and on models with the so-called Wess–Zumino term, which were not discussed at Louvain-la-Neuve. They are included here because they fall naturally into the area covered by the book and also because some of the work in these chapters was performed during my stay at Louvain-la-Neuve and (which is more important) this work was done in collaboration with one of the students there, Bernard Piette.

The book is intended for both graduate students in theoretical particle physics who are interested in σ models and those working in the general area of harmonic maps in pure mathematics. The book does not aim to be a comprehesive review of the work in these areas; rather, it presents a view of the area seen from an angle of the potential applications of path integral techniques with emphasis on the work which has particularly interested me or in which I have been involved. No attempt is made to compile a comprehensive bibliography or establish priorities; these can be found elsewhere (a list of recent books and some excellent review articles, which contain such bibliographies is included at the end of the book). Instead, references to papers are, for the most part, given as I used them in preparing my lecture course. I hope the numerous authors whose work is not mentioned here will not be offended by being missed out. I feel that my rather free-and-easy approach to references will make the book less heavy and more readable, and the interested reader will find further references from the list of books and articles at the end of the book.

As is easy to see, a large part of the book reports on the work in which I have been involved, in particular in collaboration with A M Din. I would like to use this opportunity to thank him for many years of working together in a pleasant and fruitful atmosphere. Clearly, he is a moral co-author of large parts of this book.

I would also like to thank my other collaborators and, in particular, the above-mentioned Bernard Piette. My stay at Louvain-la-Neuve was enjoyable and I was very impressed by the scientific atmosphere and the hospitality of the people there. At the same time I was very pleased to meet Bernard; he has turned out to be a particularly good collaborator.

I also wish to express my gratitude to my colleagues at the University of Durham who, through their discussions with me, have greatly improved my understanding of many topics covered in this book; in particular I want to thank professor E J Squires, who, in addition, has encouraged me to write this book and who has made constructive comments on a preliminary version.

Finally, I want want to thank Dr D K Campbell and other members of the Center for Non-Linear Studies at Los Alamos National Laboratory for their warm hospitality. The final touches to the manuscript were made during my stay at the Center.

The book starts with two introductory chapters. Many readers can probably skip them; they are included for completeness and to set up notations and to provide some motivation for what is discussed in the rest of the book. The discussion of path integrals and functional techniques given at the beginning of chapter 1 is particulary sketchy; however, its material is so well known now that its inclusion here may be slightly redundant. At the same time, as I see it, quantisation via the path integral formalism provides the main *raison d'etre* for studying classical solutions in Euclidean space, and so not to mention them at all would seem to amount to sacrilege.

In the second chapter we discuss some applications of path integral techniques and also introduce Witten's supersymmetric quantum mechanics. Again, the material is probably well known to most potential readers. The next two chapters delve into classical solutions (called harmonic maps in mathematical literature) of the Euclidean sigma models, for which the target manifold is the $U(N)/(U(k) \times U(N-k))$ coset space, the best known example of which is the $\mathbb{C}P^{N-1}$ manifold (also called $\mathbb{P}C(N)$ by mathematicians). Chapter 3 discusses the general structure of these solutions and some of their properties; chapter 4 reports on attempts to use path integral methods to investigate quantum properties.

In chapter 5 we discuss various generalisations of these models involving fermions and in chapter 6 we study some $1/N$ expansions of these models (with and without fermions). We return to purely bosonic models in chapter 7, where we discuss classical solutions of σ models, for which the target manifold is the group $U(N)$ (or $SU(N)$). Most of the work reported here is quite new and so the discussion is more detailed. In chapter 8 we briefly investigate the concept of the integrability of our σ models, concentrating on properties of the associated Lax pair problem for the model. We observe that the solutions of the equations of this problem provide us with solutions of another problem; namely, they solve the classical equations of sigma models with the Wess–Zumino term. We discuss briefly some properties of such solutions.

In chapter 9, following closely some work of Braaten *et al*, we look at the general formulation of σ models with and without the Wess–Zumino term. We discuss these models from a geometrical point of view, pointing out that the appearance of the Wess–Zumino term can be interpreted as the imposition of torsion on the target manifold. In chapter 10 we make a short excursion to four dimensions, where we discuss briefly the Skyrme model. We then report on some similar studies in three dimensions. The work here centres on the so-called $\mathbb{C}P^1$ skyrmions, for which some of their properties can be studied in full generality. We also discuss their spin; as emphasised by Wilczek and Zee, these $\mathbb{C}P^1$ skyrmions can be quantised in any way which is intermediate between bosons and fermions, and the term responsible for their spin is provided by an expression which corresponds to the Hopf term of the theory.

We finish the book with a chapter on the boson fermion equivalence and the Kac–Moody and Virasoro algebras. We also discuss briefly various conserved local and nonlocal charges in our models.

1 Introduction to Path Integrals

1.1 Quantum Mechanics

1.1.1 Canonical Quantisation

Consider a one-dimensional mechanical system governed by the Hamiltonian $H(p, q)$, where q denotes the generalised coordinate and p the corresponding generalised momentum. When the system is quantised canonically, p and q become operators $\hat{p}$ and $\hat{q}$, which satisfy the well known commutation relation

$$[\hat{q}, \hat{p}] = i, \tag{1.1}$$

where we have set $\hbar = 1$. In the position representation they act in the Hilbert space of square integrable functions $\Psi(q, t)$, and are given by

$$\hat{q} = q, \quad \hat{p} = -i\frac{\partial}{\partial q}. \tag{1.2}$$

The Schrödinger equation

$$i\frac{\partial \Psi}{\partial t} = \hat{H}\Psi, \tag{1.3}$$

where $\hat{H}$ is the operator Hamiltonian obtained from the classical Hamiltonian by replacing p and q by operators $\hat{p}$ and $\hat{q}$, has the formal solution

$$\Psi(t_1) = \hat{U}(t_1, t_0)\Psi(t_0), \tag{1.4}$$

where $\hat{U}(t_1, t_0)$ is the *evolution operator* and is given by

$$\hat{U}(t_1, t_0) = \exp\left[-i(t_1 - t_0)\hat{H}\right]. \tag{1.5}$$

We shall be interested in calculating

$$\langle q'|\, \hat{U}(t_1, t_0)\, |q\rangle\,, \tag{1.6}$$

i.e. the matrix elements of the evolution operator taken between position eigenstates, since all physically relevant quantities can be expressed in terms of them. These matrix elements were shown by Feynman [1] to satisfy

$$\langle q' | \hat{U}(t_1, t_0) | q \rangle = \langle \exp(iS) \rangle , \tag{1.7}$$

where $\langle \exp(iS) \rangle$ denotes an average (in a sense to be made more precise below) over all trajectories $q(\tau)$, $p(\tau)$ in phase space, subject to the condition that $q(t_0) = q$ and $q(t_1) = q'$. Here S is the action, which is given by

$$S = \int_{t_0}^{t_1} [p(\tau)\dot{q}(\tau) - H(q(\tau), p(\tau))] \, \mathrm{d}\tau, \qquad \dot{q}(\tau) = \frac{\mathrm{d}q(\tau)}{\mathrm{d}\tau}. \tag{1.8}$$

Such averages are integrals over all the 'possible' classical paths that the system can follow and so are called path integrals.

1.1.2 Path Integrals

The most convenient way to proceed is to define our path integral as a limit of a finite-dimensional integral $J_N(q, q', t)$ $(t = t_1 - t_0)$ [2]. Various approximations (limiting procedures) are possible. One such approximation involves the division of $[t_0, t_1]$ into N equal parts with intermediate points denoted by $\tau_1, \tau_2, \ldots, \tau_{N-1}$. To determine our phase space 'trajectories' we have to specify $q(\tau)$ at all $N - 1$ intermediate points, i.e. $q(\tau_i)$ and $p(\tau)$ at N points for which we can take τ_i and either t_0 or t_1. In addition, we have to decide how to interpolate between these values. A possible approximation would corespond to choosing a linear interpolation between these given values, for both $q(\tau)$ and $p(\tau)$. A more 'natural' prescription would involve choosing $p(\tau)$ which are constant and $q(\tau)$ which are linear on each segment $[\tau_i, \tau_{i+1}]$ (i.e. are given by classical mechanics). An example of such a 'trajectory' is shown in figure 1.1 below.

Denoting by q_i the values $q(\tau_i)$ for $i = 1, \ldots, N - 1$ and similarly for p_i (this time with the addition of $p_N = p(t_1)$) we see that 'a trajectory' is known if $q_1, \ldots, q_{N-1}, p_1 \ldots, p_N$ are given, and so S is a function of q, q' and of the $N - 1$ variables defined above. Then we *define*

$$J_N(q, q', t) = \frac{1}{(2\pi)^N} \int_{-\infty}^{\infty} \mathrm{d}p_1 \ldots \mathrm{d}p_N \int_{-\infty}^{\infty} \mathrm{d}q_1 \ldots \mathrm{d}q_{N-1} \exp(iS), \tag{1.9}$$

and can show that, for most potentials in H,

$$\lim_{N \mapsto \infty} J_N(q, q', t) = \langle q' | \exp(-it\hat{H}) | q \rangle . \tag{1.10}$$

Let us check this result for two special cases:

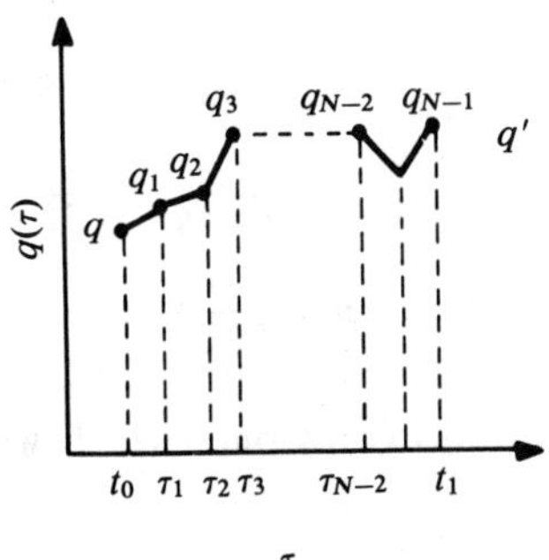

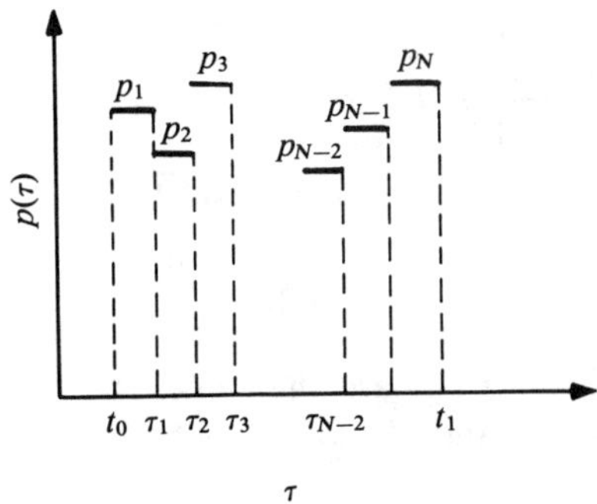

Figure 1.1

1. $H = H(q)$,
2. $H = H(p)$.

(They are both unrealistic but show the type of calculation one has to perform in more realistic cases.)

Consider first case 1, i.e. $H = H(q)$ only. Using the explicit form of our 'trajectories' we calculate S and find:

$$\begin{aligned} S &= \int_{t_0}^{t_1} (p\dot{q} - H)\,\mathrm{d}\tau \\ &= p_1(q_1 - q) + p_2(q_2 - q_1) + \ldots + p_N(q' - q_{N-1}) - \int_{t_0}^{t_1} H(q(\tau))\,\mathrm{d}\tau. \end{aligned}$$

Next we integrate over p_i. To perform these integrations we use

$$\frac{1}{2\pi}\int_{-\infty}^{\infty} \mathrm{d}p\, e^{ipx} = \delta(x), \tag{1.11}$$

and so we obtain the product of the delta functions:

$$\delta(q_1 - q)\delta(q_2 - q_1)\ldots\delta(q' - q_{N-1}).$$

Due to these δ's we can now integrate over q_i and obtain

$$J_N(q, q', t) = \delta(q - q')\exp[-i(t_1 - t_0)H(q)]. \tag{1.12}$$

There is no need now to take the limit $N \mapsto \infty$ (as nothing depends on N) and we recognise that the derived expression is $\langle q'|\exp[-i(t_1 - t_0)\hat{H}(q)]\,|q\rangle$.

Consider now the second case, i.e. when $H = H(p)$ only. We proceed exactly as before except that now we perform the integration over q_i first. We obtain

$$\begin{aligned}\frac{1}{(2\pi)^N}\int_{-\infty}^{\infty} &\mathrm{d}q_1\ldots\mathrm{d}q_{N-1}\\ &\times\exp\{i[p_1(q_1-q)+p_2(q_2-q_1)+\ldots+p_N(q'-q_{N-1})]\}\\ &=\frac{1}{2\pi}\exp[i(p_Nq'-p_1q)]\,\delta(p_1-p_2)\ldots\delta(p_{N-1}-p_N).\end{aligned} \tag{1.13}$$

Hence all p_i are the same. We integrate over all p_i except one (which we call p) and obtain

$$\frac{1}{2\pi}\int_{-\infty}^{\infty}\mathrm{d}p\,\exp[ip(q'-q)-i(t_1-t_0)H(p)],$$

which again is

$$\langle q'|\exp[-i(t_1-t_0)\hat{H}(p)]\,|q\rangle\,,$$

as can be seen by inserting $\int \mathrm{d}p\,|p\rangle\,\langle p| = 1$. It is easy to check that the other 'obvious' approximation (corresponding to a linear interpolation for $q(\tau)$ and $p(\tau)$) gives the same result in both cases.

If H depends on both p and q, the proof is much more complicated. It turns out that, in general, the limit depends on the finite-dimensional approximation. This is related to the well known problem of operator ordering in canonical quantisation. A good discussion of this point can be found in an article by Hietarinta [3].

If $H = H_1(p) + H_2(q)$ there is no problem.

There is no consistent notation for functional integrals. We use

$$\begin{aligned}\int_{\substack{q(t_0)=q\\ q(t_1)=q'}}\exp(iS)\prod_{\tau}\frac{\mathrm{d}q(\tau)\mathrm{d}p(\tau)}{2\pi} &= \int_{\substack{q(t_0)=q\\ q(t_1)=q'}}\exp(iS)\,\mathrm{D}q(\tau)\,\mathrm{D}p(\tau)\\ &= \int_{\substack{q(t_0)=q\\ q(t_1)=q'}}\exp(iS)\,\mathrm{D}[q]\,\mathrm{D}[p].\end{aligned} \tag{1.14}$$

This notation is useful, even though in the finite-dimensional approximation there are $N-1$ integrations over q and N integrations over p. If the dynamical system depends on more variables, we make the replacements

$$H(p,q) \Longrightarrow H(q_1,q_2,\ldots,q_n,p_1,p_2,\ldots,p_n)$$

and

$$(2\pi)^{-N} \Longrightarrow (2\pi)^{-Nn}, \qquad \mathrm{d}q_k \Longrightarrow \prod_{i=1}^{n}\mathrm{d}q_{ik}, \quad \mathrm{d}p_k \Longrightarrow \prod_{i=1}^{n}\mathrm{d}p_{ik}$$

and the integrals are now denoted by

$$\int_{\substack{q(t_0)=q \\ q(t_1)=q'}} \exp(iS) \prod_\tau \prod_{i=1}^{n} \left(\frac{\mathrm{d}q_i(\tau)\mathrm{d}p_i(\tau)}{2\pi} \right). \tag{1.15}$$

The discussion has so far relied on the phase space being R^{2n}. It is easy to generalise it to other cases. Let us briefly discuss the case when the phase space is given by a cylinder, i.e. is $U(1) \times R$. For simplicity, let us consider only the case when $H = H(p)$. Using ϕ and p as our variables (and remembering that $\phi \cong \phi + 2\pi$) we have

$$\langle \phi' | \exp[-i(t_1 - t_0)\hat{H}] \, |\phi\rangle = \int \exp\left(i \int (p \, \mathrm{d}\phi - H \, \mathrm{d}\tau) \right) \prod_\tau \frac{\mathrm{d}\phi \mathrm{d}p}{2\pi}. \tag{1.16}$$

In our approximation the exponent takes the form

$$i \sum_{i=1}^{N} p_i(\phi_{i+1} - \phi_i) - i \sum_{i=1}^{N} H(p_i)\Delta\tau.$$

Next we have to perform integrations over N p_i variables and $(N-1)$ ϕ_i variables. What is their range? Clearly we integrate over p_i from $-\infty$ to $+\infty$ but what about ϕ_i? As two values of ϕ which differ by a multiple of 2π are equivalent, a little thought shows that the appropriate procedure is to integrate over ϕ_i from $-\infty$ to $+\infty$ and then *sum* over all functions $\phi(t)$ which satisfy

$$\phi(t_0) = \phi, \quad \phi(t_1) = \phi' + 2\phi n,$$

where n in an *integer*. Thus we obtain

$$\frac{1}{(2\pi)^N} \sum_{n=-\infty}^{\infty} \prod_i^N \mathrm{d}p_i \mathrm{d}\phi_i$$
$$\times \exp\left(i \sum_i^N p_i(\phi_{i+1} - \phi_i) + 2\pi n p_N - \sum_i^N H(p_i)\Delta\tau \right).$$

The integration over $\phi_1, \phi_2, \ldots, \phi_{N-1}$ gives

$$\frac{1}{(2\pi)^N} \int \prod_{i=1}^{N-1} \mathrm{d}\phi_i \exp\left(i \sum_{i=1}^{N} p_i(\phi_{i+1} - \phi_i) \right)$$
$$= \frac{1}{2\pi} \left(\prod_{i=1}^{N-1} \delta(p_i - p_{i+1}) \exp[i(p_N\phi' - p_1\phi)] \right).$$

Now we integrate over all momenta but one and obtain

$$\frac{1}{2\pi}\int \mathrm{d}p \sum_{n=-\infty}^{\infty} \exp[ip(\phi' - \phi) + 2\pi inp - i(t - t_0)H(p)]. \qquad (1.17)$$

However, as

$$\sum_{n=-\infty}^{\infty} \exp(2\pi inp) = \sum_{m=-\infty}^{\infty} \delta(p - m), \qquad (1.18)$$

we finally obtain

$$\frac{1}{2\pi}\sum_{m=-\infty}^{\infty} \exp[im(\phi' - \phi)]\,\exp[-i(t_1 - t_0)H(m)], \qquad (1.19)$$

which is the sum over all possible values of the momentum variable $p = m$ conjugate to the compact coordinate ϕ.

The path integrals we have defined above correspond to sums over trajectories in phase space. Often, and in particular in field theories, we consider path integrals which involve sums over trajectories in configuration space. In that case we consider Lagrangians and to define our path integrals we need also an approximation for the $\dot{q}$ terms in the Lagrangian. The standard approximation corresponds to a consideration of the differences of $q(\tau)$ at adjacent points.

We will not discuss such path integrals here as there exist several books in which these path integrals are derived and discussed in detail (see, e.g., [4]). Moreover, we will introduce them in the section in which we discuss path integrals in field theories.

In addition, one should add that in a number of cases, in which the Hamiltonian takes a simple form, we can perform the integration over $p(\tau)$; the resultant path integral can be interpreted as corresponding to the sum over trajectories in configuration space.

1.2 Fermions

Physical particles which are fermionic obey the Pauli exclusion principle. In field theory language they are described by fields, which, when quantised, satisfy anticommutation relations. In the path integral formulation approach they are described by anticommuting quantities for which special rules of integration have to be introduced. Such anticommuting quantities are introduced by considering first a Grassmann algebra with a finite even number of generators denoted by x_i and $x_i^\star$, where $i = 1, \ldots, n$. These generators anticommute:

$$x_i x_j + x_j x_i = 0, \quad x_i^\star x_j^\star + x_j^\star x_i^\star = 0$$

and also

$$x_i x_j^\star + x_j^\star x_i = 0$$

and so we see that

$$x_i^2 = 0, \quad x_j^{\star 2} = 0.$$

When one considers an arbitrary function of x_i and $x_i^\star$ the Taylor's series expansion gives us

$$f(x, x^\star) = \sum_{a_i, b_i = 0,1} C_{a_1 \dots a_n b_1 \dots b_n} x_1^{a_1} x_2^{a_2} \dots x_n^{a_n} x_n^{\star b_n} \dots x_1^{\star b_1}, \tag{1.20}$$

where $C_{a_1 \dots a_n b_1 \dots b_n}$ are complex coefficients. We define the conjugate field $f^\dagger$

$$f \mapsto f^\dagger = \sum_{a_i, b_i = 0,1} \bar{C}_{a_1 \dots a_n b_1 \dots b_n} x_1^{b_1} \dots x_n^{b_n} x_n^{\star a_n} \dots x_1^{\star a_1}. \tag{1.21}$$

This conjugation, defined for grassmannian quantities, plays the role of the hermitian conjugation for the commuting variables and in what follows we will refer to it as to the hermitian conjugation.

We observe that we can extend our Grassmann algebra by adding to our generators $\{x_i,\ x_j^\star\}$ a further set of self-conjugate generators. However, any pair of them, say y_1, y_2, corresponds to two mutually conjugate generators: $\tilde{x} = y_1 + iy_2$, $\tilde{x}^\star = y_1 - iy_2$ and so we see that we have a genuine extension of our algebra only if the number of the self-conjugate generators is odd. Although in the rest of this section we shall, for simplicity, consider only the case of an even number of the generators the generalisation to the case of the odd number is straightforward.

Next we define the integration operation for anticommuting variables. We want it to preserve some familiar properties of our intuitive notion of integration such as linearity and invariance under a translation of the integration variable by a constant and so we postulate that

$$\begin{gathered} \int \mathrm{d}x_i = 0, \qquad \int x_i \, \mathrm{d}x_i = 1, \\ \int (c_1 f_1 + c_2 f_2) \, \mathrm{d}x^\star \mathrm{d}x = c_1 \int f_1 \, \mathrm{d}x^\star \mathrm{d}x + c_2 \int f_2 \mathrm{d}x^\star \mathrm{d}x, \end{gathered} \tag{1.22}$$

where the integrations over $x^\star$ variables satisfy analogous relations and where we defined

$$\int f(x, x^\star) \, \mathrm{d}x^\star \mathrm{d}x = \int f(x_1, \dots, x_n, x_1^\star, \dots, x_n^\star) \, \mathrm{d}x_1^\star \mathrm{d}x_1 \mathrm{d}x_2^\star \mathrm{d}x_2 \dots \mathrm{d}x_n^\star \mathrm{d}x_n. \tag{1.23}$$

In the expression above c_i are complex numbers, and $\mathrm{d}x_i$ and $\mathrm{d}x_i^\star$ are anticommuting quantities which anticommute with each other and with the generators of the algebra. These rules were first postulated by Berezin [5] and were shown by him to provide a useful definition of the integration procedure. Notice that it does not make sense to ask what the limits of the integration are, although sometimes it is convenient to think of the limits as being from grassmannian $-\infty$ to $+\infty$. Let us observe that for our function f given above we obtain

$$\int f\,\mathrm{d}x^\star\mathrm{d}x = C_{1\ldots1}. \tag{1.24}$$

A little thought shows that, in contradistinction to the usual case, the Berezin integration procedure is not the inverse of the differentiation but, in fact, it coincides with it.

We would like to finish this section by giving two formulae, which become useful in practical applications of functional integrals involving fermions. They are

$$\int \exp(-x^\star Ax)\,\mathrm{d}x^\star\,\mathrm{d}x = \det A \tag{1.25}$$

and

$$\frac{\int \exp\left(-x^\star Ax + \eta^\star x + x^\star\eta\right)\mathrm{d}x^\star\,\mathrm{d}x}{\int \exp(-x^\star Ax)\,\mathrm{d}x^\star\,\mathrm{d}x} = \exp(\eta^\star A^{-1}\eta), \tag{1.26}$$

where $x^\star Ax = \sum_{ik} a_{ik}x_i^\star x_k$ (here a_{ik} are complex numbers) and where $\eta^\star x = \sum_i \eta_i^\star x_i$ and $x^\star\eta = \sum_i x_i^\star\eta_i$. Here η_i and $\eta_i^\star$ are anticommuting quantities; they anticommute with each other and with the generators x_i, $x_i^\star$.

To prove the first of these formulae we expand $\exp(\ldots)$ in power series and observe that only the nth term contributes, and its contribution is given by the right-hand side of the formula. To prove the second formula we perform the shift $x \to x+\eta$ and $x^\star \to x^\star+\eta^\star$ of the integration variables in order to cancel the linear terms and then use the first formula.

1.3 Functional Integrals in Field Theory

Consider a self-interacting scalar field whose classical action is given by

$$S = \int \mathrm{d}^4x\,\left(\tfrac{1}{2}\partial_\mu\varphi\partial^\mu\varphi - \tfrac{1}{2}m^2\varphi^2 - \lambda\varphi^4/4!\right). \tag{1.27}$$

To define the functional integral over the field $\varphi(x_\mu)$ we proceed as follows. First we take a large cubic volume V in Minkowski spacetime and divide it into N^4 equal small cubes V_i ($i = 1, \ldots, N^4$). Then we choose to approximate $\varphi(x)$ by functions which are constant in each cube V_i, i.e.

$\varphi(x) = \varphi_i$ for $x \in V_i$. For the derivatives we use the difference approximation:

$$\partial_\mu \varphi(x_\nu) \simeq \frac{1}{\Delta l}[\varphi(x_\nu + \delta_{\mu\nu}\Delta l) - \varphi(x_\nu)].$$

Then S becomes a function of N^4 variables φ_i. After the integration over φ we have to take the limit $V \to \infty$ and also $N \to \infty$ and $V_i \to 0$. One can show that this is a possible definition of functional integration for a field theory.

Let us note that this is a slightly different approximation from the one we used in the quantum mechanical case. (There we could also have used approximations involving integration only over $q(\tau)$; they can be thought of as coming from our approximation after integrating out $p(\tau)$.) In the field theory context we can also use the previous form of the integrals and, although normally this is not so convenient, we will mention it for completeness. To do this we introduce

$$S[\varphi, \pi] = \int \left(\pi\partial_0\varphi - \tfrac{1}{2}\pi^2 - \tfrac{1}{2}(\vec{\partial}\varphi)^2 - \tfrac{1}{2}m^2 - \lambda\varphi^4/4!\right) \mathrm{d}^4x.$$

Here $\varphi(x)$ plays the role of the coordinate density, and $\pi(x)$ that of the momentum density. Notice that $S[\varphi, \pi] = S[\varphi]$ if $\pi \to \partial_0\varphi$ and that the Hamiltonian is given by

$$H = \int \mathrm{d}^3x \, (\tfrac{1}{2}\pi^2 + \tfrac{1}{2}(\vec{\partial}\varphi)^2 + \tfrac{1}{2}m^2 + \lambda\varphi^4/4!).$$

Then we can show that

$$\int \exp(iS[\varphi]) \prod_x \mathrm{d}\varphi(x) \cong \int \exp(iS[\varphi, \pi]) \prod_x \mathrm{d}\varphi(x)\mathrm{d}\pi(x). \qquad (1.28)$$

To prove the equivalence of these two formulations we perform the shift transformation $\pi(x) \to \pi(x) + \partial_0\varphi(x)$ and obtain

$$\int \exp(iS[\varphi]) \prod_x \mathrm{d}\varphi(x) \int \exp\left(-\tfrac{1}{2}i\int \pi^2(x)\mathrm{d}^4x\right) \prod_x \mathrm{d}\pi(x).$$

The last factor does not depend on any parameters of the Lagrangian and, being an overall factor, will cancel in all practical applications. Notice that we have established only the equivalence of both formulations (and not their equality). The equivalence, up to an overall factor, is the best we can hope for; the appearance of such overall factors plagues all functional calculations and, in practice, such overall factors are often infinite. In the latter case special care has to be exercised so that one does not derive meaningless answers.

1.4 Further Comments on Functional Methods

1.4.1 Functional Differentiation

So far we have discussed functional integrals and functional integration. In practice, we need functional differentiation as often as functional integration. To define functional derivatives we will assume, for simplicity, that our functions depend on only one variable and that they are square integrable. Let $\eta(x)$ be such a function. In addition we will assume that the space of such functions has an orthonormal basis, i.e. that there exists a set of functions $u_i(x)$ such that

$$\begin{aligned}\int u_i(x)\, u_j(x)\, \mathrm{d}x &= \delta_{ij} \\ \sum_i u_i(x)\, u_i(y) &= \delta(x-y).\end{aligned} \tag{1.29}$$

Then any function $\eta(x)$ can be expanded as

$$\eta(x) = \sum_i \eta_i u_i(x)$$

and the scalar product of two such functions is given by

$$(\eta, \psi) = \int \mathrm{d}x\, \eta(x)\, \psi(x) = \sum_i \eta_i \psi_i. \tag{1.30}$$

Consider now $F(\eta)$ — a functional defined on the space of functions $\eta(x)$. An example of such a functional will be

$$F(\eta) = \int \mathrm{d}x\, f(x)\, \eta(x),$$

for some fixed $f(x)$. This functional is linear; a general multilinear functional will be given by

$$F(\eta) = \sum_n F_n(\eta),$$

where

$$F_n(\eta) = \int \mathrm{d}x_1 \ldots \mathrm{d}x_n\, f(x_1, \ldots, x_n)\, \eta(x_1) \ldots \eta(x_n).$$

The functional derivative of $F(\eta)$ with respect to $\eta(x)$, which is normally denoted by $\delta F(\eta)/\delta\eta(x)$, is formally defined by

$$\int \frac{\delta F(\eta)}{\delta \eta(x)} \lambda(x)\, \mathrm{d}x = \lim_{\epsilon \to 0} \frac{F(\eta + \epsilon\lambda) - F(\eta)}{\epsilon}, \tag{1.31}$$

or, using the δ function,

$$\left.\frac{\delta F(\eta)}{\delta\eta(x)}\right|_{x=\xi} = \lim_{\epsilon\to 0}\frac{F(\eta+\epsilon\delta(x-\xi))-F(\eta)}{\epsilon}.$$

The last definition is less rigorous as $\delta(x-\xi)$ is not square integrable. The functional derivative $\delta F(\eta)/\delta\eta(x)$ is a natural extension of the *gradient* of a finite-dimensional vector space. To see this we use the basis $u_i(x)$ and expand $\eta(x) = \sum_i \eta_i u_i(x)$. Then

$$\begin{aligned} F(\eta) &\to F(\eta_i) \\ \eta(x)+\epsilon\delta(x-\xi) &\to \sum_i [\eta_i + \epsilon u_i(\xi)]u_i(x). \end{aligned} \tag{1.32}$$

Thus

$$\begin{aligned} F(\eta+\epsilon\delta(x-\xi)) &\to F(\eta_i+\epsilon u_i(\xi)) \\ &= F(\eta_i) + \epsilon\sum_j u_j(\xi)\frac{\partial F(\eta_i)}{\partial\eta_j}. \end{aligned}$$

Therefore,

$$\frac{\delta F(\eta)}{\delta\eta(x)} \to \sum_j u_j(x)\frac{\partial F(\eta_i)}{\partial\eta_j},$$

and so

$$\frac{\delta}{\delta\eta(\xi)} = \sum_j u_j(\xi)\frac{\partial}{\partial\eta_j}. \tag{1.33}$$

This last expression clearly brings out the analogy with the gradient:

$$\underline{\nabla} \Longleftrightarrow \sum_j \hat{x}_j\frac{\partial}{\partial x_j}.$$

Let us look at an example. Take $F(\eta) = \int \eta(x)f(x)\,\mathrm{d}x = \sum_i \eta_i f_i$. Then

$$\frac{\delta F(\eta)}{\delta\eta(\xi)} = \sum_j u_j(\xi)f_j = f(\xi),$$

while, if we take

$$F(\eta) = \int \mathrm{d}x_1\mathrm{d}x_2\,\eta(x_1)\eta(x_2)\,f(x_1,x_2)$$

we obtain

$$\frac{\delta F(\eta)}{\delta\eta(x)} = \int \mathrm{d}x'\,\eta(x')[f(x,x')+f(x',x)].$$

Moreover, we note that

$$\frac{\delta\eta(y)}{\delta\eta(x)} = \sum_{ij} u_i(x)\frac{\partial}{\partial\eta_i}\eta_j\, u_j(y) = \delta(x-y),$$

and also that, if $G(\eta) = G(F(\eta))$, then

$$\frac{\delta G}{\delta\eta(x)} = \frac{\partial G}{\partial F}\frac{\delta F}{\delta\eta(x)}.$$

In addition, we have a functional version of the Taylor expansion:

$$\begin{aligned} F(\eta+\lambda) &\to F(\eta_i+\lambda_i) \\ &= \exp\left(\sum_j \lambda_j \frac{\partial}{\partial\eta_j}\right) F(\eta_i) = \exp\left(\lambda, \frac{\delta}{\delta\eta}\right) F(\eta). \end{aligned} \tag{1.34}$$

Also we have the commutation relation:

$$\left[\eta(x), \frac{\delta}{\delta\eta(y)}\right] = -\delta(x-y). \tag{1.35}$$

1.4.2 *Functional Integrals Revisited*

Functional integrals, such as

$$I = \int \mathrm{d}[\eta]\, F(\eta), \tag{1.36}$$

are often defined in a different manner than that which we have discussed so far, namely they are often defined in terms of the integration over the coefficients of the function $\eta(x)$ in our basis $u_i(x)$. In this alternative definition we use

$$\eta(x) = \sum_i \eta_i u_i(x), \tag{1.37}$$

and set

$$I = \lim_{n\to\infty} \int F(\eta_1 \ldots \eta_n)\, \mathrm{d}\eta_1 \ldots \mathrm{d}\eta_n. \tag{1.38}$$

This alternative definition can be shown to be equivalent to the previous one for all square integrable functions $\eta(x)$. Let us use this new definition of the functional integration to perform the integration over the gaussian functional

$$F(\eta) = \exp(-\pi(\eta,\eta)). \tag{1.39}$$

We obtain

$$\int F(\eta)\,\mathrm{d}[\eta] = \lim_{n\to\infty}\prod_{\alpha=1}^{n}\int_{-\infty}^{\infty}\mathrm{d}\eta_\alpha\,\exp(-\pi\eta_\alpha^2)$$
$$= \lim_{n\to\infty}\prod_{\alpha=1}^{n} 1 = 1. \tag{1.40}$$

Observe that it was crucial to take π in the exponential; any other choice would have given us 0 or ∞.

It would be appropriate to add a few comments at this point about the functional integration measure. Let us observe that, had we changed η to $m\eta$ in our integration measure $\mathrm{d}\eta_i$, we would have picked up an extra factor:

$$\lim_{n\to\infty} m^n \tag{1.41}$$

and this factor differs from 0 or ∞ only for $m = 1$. Usually we set $F(\alpha) = \exp(-\frac{1}{2}(\alpha,\alpha))$ and then

$$\int F(\alpha)\,\mathrm{d}[\alpha/\sqrt{2\pi}] = 1. \tag{1.42}$$

A further comment should be made about the boundary conditions that the functions $\eta(x)$ are expected to satisfy; in most applications $\eta(x)$ are expected to vanish at ∞, i.e. $\eta(\pm\infty) = 0$. (Such conditions guarantee the existence of the basis $u_i(x)$.) Simple changes of variables in our functional integration over the functional gaussians allow us to obtain further formulae for functional integrals, which are used in many applications. Thus, performing the change $\alpha(x) \mapsto \alpha(x) + \phi(x)$ followed by the change $\phi(x) \mapsto -i\phi(x)$ gives

$$\int \exp\left(-\tfrac{1}{2}(\alpha,\alpha) + i(\alpha,\phi)\right)\mathrm{d}[\alpha/\sqrt{2\pi}] = \exp\left(-\tfrac{1}{2}(\phi,\phi)\right). \tag{1.43}$$

And from this we obtain a formal result that

$$\int \exp\left(-\tfrac{1}{2}(\alpha,\alpha) + i(\alpha,\phi)\right) F(\alpha)\,\mathrm{d}[\alpha/\sqrt{2\pi}] = F\left(\frac{1}{i}\frac{\delta}{\delta\phi}\right)\exp\left(-\tfrac{1}{2}(\phi,\phi)\right) \tag{1.44}$$

which can be proved by the Taylor series expansion. The functional delta function is given by

$$\delta[\alpha-\beta] = \int \exp(i[\alpha-\beta,\omega])\mathrm{d}[\omega/\sqrt{2\pi}]$$
$$= \lim_{n\to\infty}\prod_{i=1}^{\infty}\int \mathrm{d}[\omega_i/\sqrt{2\pi}]\,\exp(i\omega_i[\alpha_i-\beta_i]) \tag{1.45}$$
$$= \lim_{n\to\infty}\prod_{i}[\delta(\alpha_i-\beta_i)\sqrt{2\pi}],$$

and so it satisfies

$$\int \delta[\alpha - \beta] F(\alpha) \mathrm{d}[\alpha/\sqrt{2\pi}] = F(\beta). \tag{1.46}$$

Moreover, we have functional analogues of the Fourier transforms:

$$F(\alpha) = \int \hat{F}(\omega)\, e^{i(\omega,\alpha)} \mathrm{d}[\omega/\sqrt{2\pi}] \tag{1.47}$$

and

$$\hat{F}(\alpha) = \int F(\alpha)\, e^{-i(\omega,\alpha)} \mathrm{d}[\alpha/\sqrt{2\pi}]. \tag{1.48}$$

1.4.3 Changes of Variables

Consider first the linear changes of variables, i.e. changes which involve $\alpha(x) \to \alpha'(y)$ and are given by

$$\alpha(x) = \int \mathrm{d}y\, K(x,y)\, \alpha'(y), \tag{1.49}$$

with $K(x,y)$ a symmetric kernel. (Such changes are often denoted as $\alpha = K \circ \alpha'$.) To discuss such changes of variables we use the orthonormal basis $u_i(x)$ such that

$$\int K(x,y)\, u_j(y)\, \mathrm{d}y = \lambda_j u_j(x), \tag{1.50}$$

i.e. $K \circ u_j = \lambda_j u_j$. Then

$$\alpha_i = \int u_i(x)\, \alpha(x)\, \mathrm{d}x = \lambda_i \alpha_i' \tag{1.51}$$

and $\prod_i \mathrm{d}\alpha_i = \prod_i \mathrm{d}\alpha_i' \lambda_i$, and so

$$\mathrm{d}[\alpha/\sqrt{2\pi}] = \mathrm{d}[\alpha'/\sqrt{2\pi}] \prod_i \lambda_i. \tag{1.52}$$

The last product corresponds to the functional determinant. To see this let us introduce $\ln K$ as the operator with eigenvalues $\mu_i = \ln \lambda_i$, i.e. it satisfies

$$\int \ln K(x,y)\, u_j(y)\, \mathrm{d}y = \mu_j\, u_j(x). \tag{1.53}$$

Then

$$\begin{aligned}\prod_i \lambda_i &= \exp\left(\sum_j \mu_j\right) = \exp\left(\sum_j \int \ln K(x,y) u_j(x) u_j(y)\, \mathrm{d}x\, \mathrm{d}y\right)\\ &= \exp\left(\int \ln K(x,y) \delta(x-y)\, \mathrm{d}x\, \mathrm{d}y\right)\\ &= \exp\left(\int \ln K(x,x)\, \mathrm{d}x\right) = \exp\left(\mathrm{Tr}\,(\ln K)\right),\end{aligned} \tag{1.54}$$

and so provides a definition of the functional *trace*. Moreover,

$$\exp\left(\mathrm{Tr}\,(\ln K)\right) = \det K \tag{1.55}$$

provides a definition of the functional *determinant*.

What is $\ln K$? If we set $K = 1 + L$, i.e. $K(x,y) = \delta(x-y) + L(x,y)$ we see that

$$\ln K \;=\; \ln(1+L) \;=\; \sum_{n=1}^{\infty} \frac{1}{n} L^n \tag{1.56}$$

with $L^n(x,y) = \int \mathrm{d}z\, L^{n-1}(x,z)\, L(z,y)$, etc. Clearly

$$\ln K \circ u_j \to \ln \lambda_j\, u_j \;=\; \mu_j\, u_j . \tag{1.57}$$

Let us return to our change of variables ($\alpha = K \circ \alpha'$) and try to use it to perform the integral

$$I \;=\; \int \mathrm{d}[\alpha/\sqrt{2\pi}]\, \exp\left(- \int \frac{\alpha^2(x)}{g^2(x)}\, \mathrm{d}x \right) . \tag{1.58}$$

To perform this calculation we set

$$\alpha(x) \;=\alpha'(x)\, g(x) \;=\; \int \mathrm{d}y\, K(x,y)\alpha'(y), \tag{1.59}$$

where $K(x,y) = g(x)\delta(x-y)$. Then to calculate $\det K$ we set $g = 1 + f$ and call $L(x,y) = f(x)\delta(x-y)$. Then $L^n(x,y) = f^n(x)\,\delta(x-y)$, and so

$$\ln K(x,y) = \sum_{n=1}^{\infty} \frac{f^n(x)}{n} \delta(x-y) \;=\ln\, g(x)\,\delta(x-y), \tag{1.60}$$

showing that

$$\det K \;=\; \exp\left(\delta(0) \int \mathrm{d}x\, \ln g(x) \right) . \tag{1.61}$$

Hence

$$I \;=\; \det K \int \mathrm{d}[\alpha'/\sqrt{2\pi}]\, \exp\left(- \int [\alpha'(x)]^2\, \mathrm{d}x \right) \;=\; \det K,$$

and so we see that functional methods primarily provide us with a mathematical dependence on functions.

1.4.4 Nonlinear Changes of Variables

So far we have discussed only linear changes of variables. When we consider more general changes of variables, the situation is much more complicated, as such changes of variables produce not only the usual Jacobians but in addition provide further terms. As a consequence two theories, which differ from each other by a naive change of variables in the functional integral, are in fact different at the quantum level, and the difference between them appears already at the two-loop level. This fact was first observed by Edwards and Gulyaev [6] who investigated the change from Cartesian to polar coordinates in the path integral formalism, and found that additional terms arise due to the stochastic nature of path integrals.

The original results of Edwards and Gulyaev have been rederived and extended by many other authors [7]. Here we will reproduce them in the case of canonical point transformations following closely the discussion given by Gervais and Jevicki [8].

In their article Gervais and Jevicki consider a simple quantum mechanical system (with n degrees of freedom) based on the Lagrangian

$$L = L(q, \dot{q}) = \tfrac{1}{2}\sum_{a=1}^{n} \dot{q}_a^2 - V(q) \tag{1.62}$$

and study the effects of a general canonical point transformation

$$q_a(t) = F^a(Q(t)) \tag{1.63}$$

from $(q_1, \ldots, q_n)$ to $Q = (Q_1, \ldots, Q_n)$ on the evaluation of the following path integral:

$$K(q^2, t^2; q^1, t^1) = \int_{q(t^1)=q^1}^{q(t^2)=q^2} \prod_{a=1}^{n} D[q_a] \exp\left(\frac{i}{\hbar} S[q]\right), \tag{1.64}$$

where $q^i = (q_1^i, \ldots, q_n^i)$, $i = 1, 2$.

Notice that this path integral is nothing else but our kernel integral discussed in the first section, in which we have now integrated out the p variables. Moreover, although we perform our integral in Euclidean space we keep the notation of Minkowski space, and we have inserted $\hbar$ to keep track of the order of all terms we are considering.

To discuss the effects of the change of variables we treat the path integral as a limit of finite-dimensional integrations. As usual we subdivide the time $t^2 - t^1$ into N equal intervals denoting $t_k = t^1 + \epsilon k$ and set $q_a(k) = q_a(t_k)$. Then our path integral is given by

$$\lim_{N\to\infty} \int \prod_{a=1}^{n} \prod_{k=1}^{N-1} \frac{dq_a(k)}{(2\pi i \hbar \epsilon)^{1/2}} \prod_{k=1}^{N-1} \exp\left(\frac{i}{\hbar} S[q(k+1), q(k)]\right), \tag{1.65}$$

with $q_a(0) = q_a^1$ and $q_a(N) = q_a^2$. In the exponent the action is approximated by

$$S[q(k+1), q(k)] = \frac{1}{2\epsilon} \sum_{a=1}^{n} (q_a(k+1) - q_a(k))^2 - \epsilon V(q(k)). \tag{1.66}$$

Next, we perform our change of variables, equation (1.63). The new integration measure is given by

$$\prod_{a=1}^{n} \mathrm{d}q_a(k) = \left[\det g_{ij}(Q(k))\right]^{1/2} \prod_{i=1}^{n} \mathrm{d}Q_i(k), \tag{1.67}$$

where

$$g_{ij}(Q) = \sum_{a=1}^{n} \frac{\partial F^a(Q)}{\partial Q_i} \frac{\partial F^a(Q)}{\partial Q_j}, \tag{1.68}$$

and the action is given by

$$\begin{aligned} \check{S}[Q(k+1), Q(k)] &= S[F(Q(k+1)), F(Q(k))] \\ = \frac{1}{2\epsilon} \sum_{a=1}^{n} &[F^a(Q(k+1)) - F^a(Q(k))]^2 - \epsilon V(F(Q(k))). \end{aligned} \tag{1.69}$$

Now we expand $F^a(Q(k+1)) - F^a(Q(k))$ and approximate the action, keeping all terms which are of $\mathrm{O}(\epsilon)$. As Edwards and Gulyaev have observed [6], the stochastic nature of path integrals gives $(\Delta Q)^2 = \mathrm{O}(\epsilon)$, which shows that we have to keep terms of up to the fourth order in ΔQ.

To simplify the calculations, we expand the functions $F^a(Q(k+1))$ and $F^a(Q(k))$ about the midpoint:

$$\bar{Q}_i = \tfrac{1}{2}(Q_i(k+1) + Q_i(k)) \tag{1.70}$$

and so we obtain

$$\begin{aligned} \bar{S}[Q(k+1), Q(k)] = \frac{1}{2\epsilon} \Bigg(& g_{ij}(\bar{Q}(k))\, \Delta Q_i(k)\, \Delta Q_j(k) \\ + \frac{1}{12} \frac{\partial F^a}{\partial Q_i}(\bar{Q}(k)) & \frac{\partial^3 F^a}{\partial Q_j \partial Q_l \partial Q_m}(\bar{Q}(k)) \Delta Q_i(k)\, \Delta Q_j(k)\, \Delta Q_l(k)\, \Delta Q_m(k) \Bigg) \\ & - \epsilon V(F(\bar{Q}(k))), \end{aligned} \tag{1.71}$$

and so our kernel is given by

$$\begin{aligned} & \left[\det g_{ij}(Q^2) \det g_{ij}(Q^1)\right]^{-1/4} \lim_{N\to\infty} \int \prod_{i=1}^{n} \prod_{k=1}^{N-1} \frac{\mathrm{d}Q_i(k)}{(2\pi i \epsilon \hbar)^{1/2}} \\ & \times \prod_{k=1}^{N-1} J(Q(k+1), Q(k)) \exp\left(\frac{i}{\hbar} \bar{S}[Q(k+1), Q(k)]\right), \end{aligned} \tag{1.72}$$

where

$$J(Q(k+1), Q(k)) = \left[\det g_{ij}(Q(k+1)) \det g_{ij}(Q(k))\right]^{1/4}. \tag{1.73}$$

Next, we expand this Jacobian about $\bar{Q}_i(k)$ and after some lengthy calculations find

$$J(Q(k+1), Q(k)) \sim [\det g_{ij}(\bar{Q}(k))]^{1/2}[1 + \tfrac{1}{16}\{g^{ij}(\bar{Q}(k))g_{ij,lm}(\bar{Q}(k)) + g^{ij}{}_{,l}(\bar{Q}(k))g_{ij,m}(\bar{Q}(k))\}\,\Delta Q_l(k)\,\Delta Q_m(k)], \tag{1.74}$$

where $_{,i}$ denotes $\partial/\partial Q_i$.

Now we make use of the following integrals:

$$\int \prod_{i=1}^{n} \mathrm{d}x_i \,\exp\left(\frac{i}{2\hbar\epsilon}\, g_{ij}\, x_i x_j\right) x_\alpha x_\beta = (2\pi i\epsilon\hbar)^{n/2}\, g^{-1/2}\, i\hbar\epsilon g^{\alpha\beta} \tag{1.75}$$

and

$$\int \prod_{i=1}^{n} \mathrm{d}x_i \,\exp\left(\frac{i}{2\hbar\epsilon}\, g_{ij}\, x_i x_j\right) x_\alpha x_\beta x_\gamma x_\delta = (2\pi i\epsilon\hbar)^{n/2}\, g^{-1/2}\, (i\hbar\epsilon)^2 (g^{\alpha\beta} g^{\gamma\delta} + g^{\alpha\gamma} g^{\beta\delta} + g^{\alpha\delta} g^{\beta\gamma}), \tag{1.76}$$

where $g = \det(g_{ij})$.

These formulae allow us to replace the $(\Delta Q)^4/\epsilon$ terms in the action and the $(\Delta Q)^2$ terms in the Jacobian by a potential-like term of the form $\epsilon V(Q)$. Thus we obtain

$$K(q^2, t^2; q^1, t^1) = [g(Q^2)\, g(Q^1)]^{-1/4} \lim_{N\to\infty} \int \prod_{k=1}^{N-1} \prod_{i=1}^{n} \frac{\mathrm{d}Q_i(k)}{(2\pi i\hbar\epsilon)^{1/2}} \times \prod_{k=0}^{N-1} [g(\bar{Q}(k))]^{1/2} \exp\left(\frac{i}{\hbar} S_{\text{eff}}[\bar{Q}(k),\, \Delta Q(k)]\right), \tag{1.77}$$

where $g(Q)$ denotes the determinant of the corresponding matrix and where

$$S_{\text{eff}} = \frac{1}{2\epsilon} \sum_{ij} g_{ij}(\bar{Q}(k))\, \Delta Q_i(k)\, \Delta Q_j(k) \;-\; \epsilon[V(F(\bar{Q}(k))) + \Delta V(\bar{Q}(k))], \tag{1.78}$$

with

$$\Delta V(\bar{Q}) = \tfrac{1}{8} \hbar^2\, \Gamma^i{}_{jl}(\bar{Q})\, \Gamma^j{}_{im}(\bar{Q}) g^{lm}(\bar{Q}), \tag{1.79}$$

where $\Gamma^l{}_{ij}$ is defined by

$$F^a{}_{,ij}(Q) = \Gamma^l{}_{ij}(Q)\, F^a{}_{,l}(Q). \tag{1.80}$$

We observe that the effective action (1.78) differs from the 'naive' one by the additional potential term $\Delta V(Q)$ proportional to $\hbar^2$. This, then, explains why the formal change of variables breaks down starting at the two-loop level. Moreover, the explicit form of this extra term depends on the midpoint definition of our path integrals; had we chosen a different definition, the form of the additional term would have been different.

It is instructive to write down the equivalent phase space path integral, which has the form

$$[g(Q^2)\,g(Q^1)]^{-1/2} \lim_{N\to\infty} \int \prod_{i=1}^{n} \prod_{k=1}^{N-1} \mathrm{d}Q_i(k) \prod_{k=0}^{N-1} \frac{\mathrm{d}P_i(k)}{2\pi\hbar}$$
$$\times \exp\left(\frac{i}{\hbar}\sum_{k=0}^{N-1}[P(k)\,\Delta Q(k) - H(P(k),\bar{Q}(k))]\right), \tag{1.81}$$

with the following Hamiltonian:

$$H(P,Q) = \tfrac{1}{2}\sum_{i,j=1}^{N} g^{ij}(Q)P_iP_j + V(F(Q)) + \Delta V(Q). \tag{1.82}$$

Observe that the operator Hamiltonian corresponding to the above phase space path integral contains non-commuting factors. However, as Gervais and Jevicki have shown, there exists a unique ordering of non-commuting factors so that the final operator Hamiltonian describes the same quantum theory as the original one.

A further discussion of the relation between the operator ordering and the definition of the corresponding path integrals can be found in an article by Hietarinta [3].

1.4.5 Multi-dimensional Gaussian Integrals

So far we have discussed functional gaussian integrals. However, in most applications we find we have to perform integration over expressions such as $\int \eta(x)\,K(x,y)\,\eta(y)\,\mathrm{d}x\,\mathrm{d}y$, where K is often a properly defined, differential operator. (Such expressions, as we will discuss shortly, arise in the path integral quantisation of any theory with dynamics.) To discuss them we want to treat K as a limit of a finite-dimensional operator, and so we have to generalise our gaussian integrals to N-dimensional gaussian integrals. Recall that

$$G(a) = \int_{-\infty}^{\infty} \mathrm{d}x\, e^{-ax^2} = \frac{\sqrt{\pi}}{\sqrt{a}}. \tag{1.83}$$

Then generalising to N degrees of freedom we consider

$$G(A) = \int_{-\infty}^{\infty} \mathrm{d}x_1 \ldots \mathrm{d}x_N \exp\left(-\sum_{ij} x_i a_{ij} x_j\right), \tag{1.84}$$

where A is a *real symmetric* matrix ($N \times N$). In what follows we shall write $\sum_{ij} x_i a_{ij} x_j = x^\top A x$ with $A^\top = A$. A can be diagonalised by a rotation:

$$A = R^\top D\, R, \qquad R\, R^\top = R^\top R = 1,$$

where D is a diagonal matrix with entries $d_1 \ldots d_N$. Then

$$\begin{aligned} G(A) &= \int_{-\infty}^{\infty} \mathrm{d}x_1 \ldots \mathrm{d}x_N \, \exp\left(-x^\top R^\top D R x\right) \\ &= \int_{-\infty}^{\infty} \mathrm{d}y_1 \ldots \mathrm{d}y_N \, \exp\left(-y^\top D y\right), \end{aligned} \tag{1.85}$$

with $y = Rx$. As $\det R = 1$ the Jacobian is 1, and we find

$$\begin{aligned} G(A) = G(d_1) \ldots G(d_N) &= (\pi)^{N/2} \, (d_1 \ldots d_N)^{-1/2} \\ &= (\pi)^{N/2} (\det A)^{-1/2}. \end{aligned} \tag{1.86}$$

This expression is well defined, provided all $d_i > 0$. Let us note that this result is very different from the case of grassmann variables, where $\det A$ appears in the numerator. Thus we can envisage situations where both commuting and anticommuting quantities are involved, and they appear effectively with the same operator A; then we can have a cancellation of the determinants. Such situations arise in supersymmetric theories and, as we will discuss a little later, they lead to the cancellation of quantum corrections to some classical results.

1.5 Green Functions in Field Theory

Consider for simplicity a scalar field theory based on a Lagrangian (action) density:

$$L = \tfrac{1}{2}\partial_\mu \phi \, \partial^\mu \phi - \tfrac{1}{2} m^2 \phi^2 - V(\phi), \tag{1.87}$$

where $V(\phi)$ is an interaction term like $V(\phi) = \lambda \phi^4/4!$. The action is defined as

$$S = \int \mathrm{d}^4 x \, L(\phi, \partial_\mu \phi).$$

All information about the quantum mechanical properties of the theory is contained in various functional integrals (see, for example, [9]), or in the response of the vacuum state of the theory to an arbitrary driving force (a *source*). Thus we need a generating functional $\langle \omega_{\text{out}} | \omega_{\text{in}} \rangle_J$, where $|\omega\rangle$ denotes vacuum 'in' and 'out' states and J represents an arbitrary source $J(x)$. In the path integral formulation this generating functional is represented by [9]

$$W[J(x)] = N' \int \mathrm{D}[\phi] \, \exp\left(i \int \mathrm{d}^d x (\tfrac{1}{2}\partial_\mu \phi \, \partial^\mu \phi - \tfrac{1}{2} m^2 \phi^2 - V(\phi) + J\phi) \right). \tag{1.88}$$

Here N' stands for the normalisation factor and d for the number of dimensions (usually 4 or 2). Observe that the integrand of the expression above is not suppressed for large ϕ (due to the i in the exponential) but only oscillates rapidly. This shows that proper care has to be exercised when the integrals are calculated and/or any approximations applied. This problem is partially remedied by either the introduction of a convergence factor:

$$\exp\left(-\epsilon\int\phi^2\,\mathrm{d}^d x\right),$$

where $\epsilon > 0$, or by going over to the Euclidean space (performing the Wick rotation):

$$\begin{aligned} x_0 &= -i\overline{x}_0 \\ \partial_\mu\phi\,\partial^\mu\phi &\rightarrow -\overline{\partial_\mu}\phi\,\overline{\partial_\mu}\phi \\ \mathrm{d}^d x &= -i\mathrm{d}^d\overline{x}. \end{aligned} \tag{1.89}$$

The formal expansion of $W[J]$ in the power series of J defines our Green functions:

$$W[J] = \sum_{N=0}^{\infty}\frac{(i)^N}{N!}\int\prod_{i=1}^{N}\mathrm{d}^d x_i\; J(x_1)\ldots J(x_N)\,G^N(x_1\ldots x_N), \tag{1.90}$$

which shows that

$$G^N(x_1\ldots x_N) = \frac{1}{(i)^N}\frac{\delta}{\delta J(x_1)}\cdots\frac{\delta}{\delta J(x_N)}\,W[J]\,\Big|_{J=0}. \tag{1.91}$$

In the momentum space these Green functions become transition amplitudes.

Observe that as

$$\frac{1}{i}\frac{\delta}{\delta J(x)}\exp\left(i\int\mathrm{d}^d yJ(y)\phi(y)\right) = \phi(x)\exp\left(i\int\mathrm{d}^d yJ(y)\phi(y)\right), \tag{1.92}$$

these Green functions are also given by

$$\begin{aligned} G^N(x_1\ldots x_N) =& N''\int\mathrm{D}[\phi]\,\phi(x_1)\ldots\phi(x_N) \\ &\times\exp\left(i\int\mathrm{d}^d x[\tfrac{1}{2}\partial_\mu\phi\partial^\mu\phi - \tfrac{1}{2}m^2\phi^2 - V(\phi)]\right). \end{aligned} \tag{1.93}$$

How do we calculate the above given Green functions? It is clear that if $V(\phi) = 0$, i.e. the theory is free, this is quite easy; however, when $V(\phi)$ is almost anything else (including in particular $V(\phi) = \lambda\phi^4$) the calculation becomes highly nontrivial, and in fact cannot be performed analytically in a closed form. The reason for this is that only gaussian integrals can be performed analytically in a closed form; for others we need some sort of approximation scheme. Before we discuss such approximation schemes we would like to mention, for completeness, other quantities of interest. In addition to $W[J]$, another generating functional is also often introduced. This generator is $Z[J]$, which is related to $W[J]$ through

$$e^{i\,Z[J]} = W[J],$$

and its power series expansion gives all the connected Green functions. This is explained in all standard texts on field theory (see, e.g, [9]).

In addition we also introduce the *effective action*, which is defined as a Legendre transform of $Z(J)$:

$$\Gamma[\phi_{\rm cl}] = Z[J] - \int \mathrm{d}^d x\, J(x)\phi_{\rm cl}(x), \tag{1.94}$$

where

$$\phi_{\rm cl}(x) = \frac{\delta Z[J]}{\delta J(x)} \quad \text{and so} \quad J(x) = -\frac{\delta\Gamma(\phi_{\rm cl})}{\delta\phi_{\rm cl}(x)}. \tag{1.95}$$

As it is difficult to determine $Z[J]$, it is also difficult to determine $\phi_{\rm cl}$. We observe that $\phi_{\rm cl}$ is an implicit function of $J(x)$. It can be shown [9] that

$$(\Delta_x + m^2)\phi_{\rm cl}(x) = J(x) - \frac{1}{W[J]}V'\Big(-\frac{\delta}{\delta J(x)}\Big)\, W[J], \tag{1.96}$$

so that for $V(\phi) = \lambda\phi^4/4!$ we obtain

$$(\Delta_x + m^2)\,\phi_{\rm cl}(x) = J(x) - \frac{\lambda}{3!}\phi_{\rm cl}^3(x) + \frac{\lambda}{3!}\frac{\delta^2\phi_{\rm cl}(x)}{\delta J^2(x)} + \frac{i\lambda}{4}\frac{\delta\phi_{\rm cl}^2(x)}{\delta J(x)}. \tag{1.97}$$

The obtained equation is highly nonlinear. Clearly, in order to solve it, we have to apply some sort of perturbation theory. An expansion in powers of λ suggests itself, and it is easy to show that this expansion corresponds to the conventional perturbation theory. We can hope that this expansion will give us useful information if λ is small; but what should we do if this condition is not satisfied? Is there any other way to proceed? In fact, having introduced the path integral formulation of the theory, we can exploit this fact and set up expansion schemes based on the saddle point evaluation of integrals, which we will discuss next.

1.6 Saddle-Point Evaluation of Integrals

The idea of the saddle-point evaluation of integrals is quite simple. Consider

$$I = \int_{-\infty}^{\infty} \mathrm{d}x\, e^{-a(x)p}, \tag{1.98}$$

and approximate $a(x)$ by expanding it around its stationary points, i.e. the solutions of

$$\frac{da(x)}{dx} = 0.$$

Let us assume, for simplicity, that there is only one solution to this equation and that this solution is given by $x = x_0$. Then

$$a(x) \cong a(x_0) + \tfrac{1}{2}(x-x_0)^2 a^{(2)} + \tfrac{1}{3!}(x-x_0)^3 a^{(3)} + \tfrac{1}{4!}(x-x_0)^4 a^{(4)} + \ldots, \quad (1.99)$$

where $a^{(n)}$ denotes the nth derivative of a. Then

$$I \cong e^{-pa(x_0)} \int_{-\infty}^{\infty} \exp\Big(-\tfrac{1}{2}p(x-x_0)^2 a^{(2)} - \tfrac{1}{3!}p(x-x_0)^3 a^{(3)} - \tfrac{1}{4!}p(x-x_0)^4 a^{(4)} + \ldots\Big)\, dx.$$

Keeping only $a^{(2)}$, and assuming that it is positive, we obtain

$$I \cong e^{-pa(x_0)} \left[\left(\frac{2\pi}{pa^{(2)}(x_0)}\right)^{1/2} + \cdots\right]. \quad (1.100)$$

If we want to keep higher terms in our expansion of $a(x)$ we have to expand the exponential; in most applications $a^{(3)}$ vanishes, so let us derive the leading correction in this case. We expand

$$\exp(-\tfrac{1}{4!}p(x-x_0)^4 a^{(4)}(x_0)) = 1 - \frac{p}{4!}(x-x_0)^4 a^{(4)}(x_0) + \ldots \quad (1.101)$$

and use

$$\begin{aligned} \int_{-\infty}^{\infty} dx\, e^{-x^2 a} &= \sqrt{\frac{\pi}{a}} \\ \rightarrow \int_{-\infty}^{\infty} dx\, x^4\, e^{-x^2 a} &= \frac{\partial^2}{\partial a^2} \int_{-\infty}^{\infty} dx\, e^{-x^2 a} \\ &= \frac{3}{4}\sqrt{\frac{\pi}{a}}\frac{1}{a^2} \end{aligned} \quad (1.102)$$

to find

$$I \cong e^{-pa(x_0)} \left(\frac{2\pi}{pa^{(4)}(x_0)}\right)^{1/2} \left(1 - \frac{a^{(4)}(x_0)}{8p(a^{(2)}(x_0))^2} + \ldots\right). \quad (1.103)$$

This expression gets 'better' as $a^{(4)}$ gets 'smaller' (so that we can ignore the higher order terms denoted by ... above). Moreover, the expansion also gets better as $p \mapsto \infty$. (Of course, there are some small overall factors, but here we are concerned with the $1/p^n$ series.) This means that we can apply our method when $a(x)$ is of the form shown in the figure on the left below, but it also means that the application will not give reliable results if $a(x)$ is of the form shown on the right.

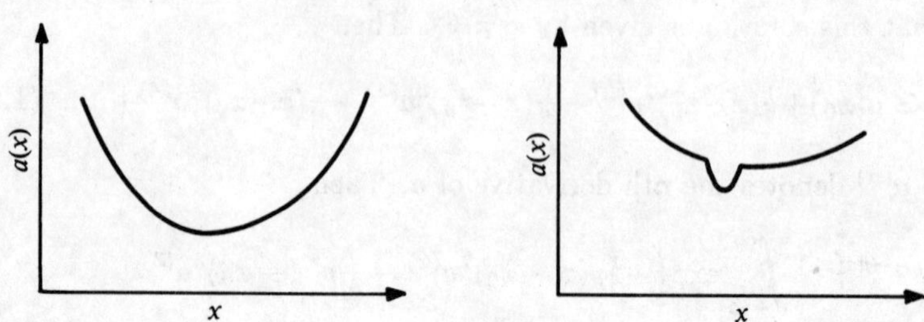

1.7 Application of the Saddle-Point Method to Functional Integrals

Now we will try to apply the saddle-point method technique to our *Euclidean* generating functional $W[J]_{\mathrm{E}}$,

$$W[J]_{\mathrm{E}} = N_{\mathrm{E}} \int \mathrm{D}[\phi] \exp\left(-\frac{1}{\hbar} S_{\mathrm{E}}[\phi, J]\right), \tag{1.104}$$

where we have reinstated $\hbar$ (i.e. $\hbar \neq 1$) and where

$$S_{\mathrm{E}}[\phi, J] = \int \mathrm{d}^d x [\tfrac{1}{2}\partial_\mu\phi\, \partial_\mu\phi + \tfrac{1}{2}m^2\phi^2 + V(\phi) - J\phi]. \tag{1.105}$$

Thus, first of all, we have to find all the stationary points of the action (i.e. solve the 'equations of motion' of ϕ in the presence of J):

$$\frac{\delta S_{\mathrm{E}}}{\delta\phi} = 0 \quad \Longrightarrow \quad \phi = \phi_0,$$

where ϕ_0 satisfies

$$-\partial_\mu\partial_\mu\phi_0 + m^2\phi_0 + V'(\phi_0) = J. \tag{1.106}$$

Observe that when $J = 0$ these equations are just the ordinary Euclidean equations of motion. Next we evaluate S for $\phi = \phi_0$, and find

$$S_{\mathrm{E}}[\phi_0, J] = \tfrac{1}{2}\int \mathrm{d}^d x\ [2V(\phi_0) - \phi_0 V'(\phi_0) - J\phi_0], \tag{1.107}$$

and so

$$S_{\mathrm{E}}[\phi, J] = S_{\mathrm{E}}[\phi_0, J] + \tfrac{1}{2}\int \mathrm{d}^d x\, \mathrm{d}^d y\, \frac{\delta^2 S_{\mathrm{E}}}{\delta\phi(x)\delta\phi(y)}\Big|_{\phi=\phi_0} \times [\phi(x) - \phi_0(x)]\,[\phi(y) - \phi_0(y)] + \ldots . \tag{1.108}$$

We see that

$$\frac{\delta^2 S_{\mathrm{E}}}{\delta\phi(x)\delta\phi(y)} = [-\partial_\mu\partial_\mu + m^2 + V''(\phi)](x)\,\delta(x-y)$$

is a differential operator, which we call the *fluctuation operator*, and so when we next perform the integration over $\phi - \phi_0$ (dropping all the higher order terms) we need to calculate its determinant. We obtain

$$\begin{aligned} W_{\mathrm{E}}[J] \cong & N_{\mathrm{E}} \exp\left(-\frac{1}{\hbar} S_{\mathrm{E}}[\phi_0, J]\right) \\ & \times \int \mathrm{D}[\phi] \exp\left(-\frac{1}{2\hbar}\int \mathrm{d}^d x\, \mathrm{d}^d y\, \phi(x) \frac{\delta^2 S_{\mathrm{E}}}{\delta\phi(x)\delta\phi(y)}\Big|_{\phi=\phi_0} \phi(y)\right) \\ \cong & N'_{\mathrm{E}} \exp\left(-\frac{1}{\hbar} S_{\mathrm{E}}[\phi_0, J]\right) \\ & \times \left[\det\left((-\partial_\mu\partial_\mu + m^2 + V''(\phi_0))\frac{\delta(x-y)}{\hbar}\right)\right]^{-1/2} . \end{aligned} \tag{1.109}$$

Thus, in principle, the procedure is very simple. We have to

1. solve for ϕ_0
2. calculate S_0
3. calculate $\det[(-\partial_\mu\partial_\mu + m^2 + V''(\phi_0))\delta(x-y)/\hbar]$
4. and finally calculate all the interesting Green functions.

However, there are very serious difficulties with finding all the ϕ_0's. (Notice that the equation for ϕ_0 is nonlinear, and that it involves an arbitrary $J(x)$.) Moreover, how do we evaluate our determinant? Presumably, not from the definition, by writing it as a product of all the relevant eigenvalues. Clearly we need some further approximation if we are to make any progress. A possible approximation is based on the neglect of $J(x)$ in the equation for ϕ_0. The main justification for this approximation is that, in fact, we do not have to calculate the generating functional $W[J]$ but only some of the Green functions:

$$G^N(x_1 \ldots x_N) = \int \mathrm{D}[\phi]\,\phi(x_1)\ldots\phi(x_N)\, e^{-S/\hbar}. \tag{1.110}$$

And, in practice, we are only interested in G^N for a few (low) values of N. In this approximation, which gets progressively worse as N gets larger, we assume that, in the equation for ϕ_0, the only terms that matter are those of the action (i.e. the inclusion of the factors $\phi(x_i)$'s in the functional integral does not alter its stationary points). A further partial justification for this comes from the smallness of $\hbar$. In addition, the theory we might be considering frequently has some further symmetry. For example we may want to consider a theory which involves N fields ϕ_i, with their interaction being a function of, say, $(\phi_i\phi_i)$. Then, if N is large enough, we may consider treating N as an additional large quantity (like p in our example from the last section), and so perform an additional expansion in powers of $1/N$. Such expansions go under the name of $1/N$ limits. Sometimes we combine both methods. We should remember that each method is only an approximation scheme, and that, in practice, we can never find all the terms in such an expansion; and that often we have no real hope of being able to face any question about the convergence of the expansion series. Moreover, each expansion scheme has its (limited) range of validity, which is often not known or easy to determine. We finish this chapter with two important questions:

1. What should we do when there is more than one solution for $\phi_0[J]$?
2. What should we do if the operator, whose determinant we have to evaluate, has zero eigenvalues?

We will be able to provide at least partial answers to these questions when we consider some explicit applications.

References

[1] Feynman R P and Hibbs A R 1965 *Quantum Mechanics and Path Integrals* (New York: McGraw-Hill)
Marinov M S 1980 Path integrals in quantum theory: an outlook of basic concepts *Phys. Rep.* **60** 1

[2] Popov V N 1987 *Functional Integrals and Collective Excitations* (Cambridge: Cambridge University Press)

[3] Hietarinta J 1982 Quantum canonical transformations as integral transformations *Phys. Rev.* D **25** 2103

[4] Bailin D and Love A 1986 *Introduction to Gauge Field Theory* (Bristol: Adam Hilger)

[5] Berezin F 1966 *The Method of Second Quantization* (New York: Academic)

[6] Edwards S F and Gulyaev Y V 1964 Path integrals in polar coordinates *Proc. R. Soc.* A **279** 229

[7] Schulman L S 1981 *Techniques and Applications of Path Integration* (New York: Wiley)

[8] Gervais J L and Jevicki A 1976 Point canonical transformations in the path integral *Nucl. Phys.* B **110** 93

[9] Ramond P 1981 *Field Theory: a Modern Primer* (Reading, MA: Benjamin-Cummings)

2 Some Simple Calculations

2.1 Harmonic Oscillator in Quantum Mechanics

The application of the functional integral techniques to the harmonic oscillator, together with other standard examples of the application of path integrals in quantum mechanics, was already given in the original book by Feynman [1]. In the harmonic oscillator case we want to calculate

$$K(q_a, t_a; q_b, t_b) = \int_{q(t_a)=q_a}^{q(t_b)=q_b} \mathrm{d}[q]\, e^{iS/\hbar}, \tag{2.1}$$

where

$$S = \frac{m}{2}\int_{t_a}^{t_b} (\dot{q}^2 - \omega^2 q^2)\,\mathrm{d}t. \tag{2.2}$$

We could evaluate this integral directly from first principles, but we can also evaluate it using the saddle-point method discussed in the previous chapter. Moreover, as the integrand is an exponential of a quadratic form in $\dot{q}$ and q, the result will be exact. Let us, first of all, determine the classical path around which we will perform our expansion.

The equation for the classical path is

$$\ddot{q} + \omega^2 q = 0, \tag{2.3}$$

and let us call $\bar{q}(t)$ the $q(t)$ which solves this equation, and which satisfies our boundary conditions:

$$\bar{q}(t_a) = q_a, \qquad \bar{q}(t_b) = q_b.$$

Now we can set $q(t) = \bar{q}(t) + \eta(t)$, where $\eta(t_a) = \eta(t_b) = 0$ and perform our expansion obtaining

$$\begin{aligned} S(q) = S(\bar{q}+\eta) &= \tfrac{1}{2}m\int_{t_a}^{t_b} [(\dot{\bar{q}} + \dot{\eta})^2 - \omega^2(\bar{q}+\eta)^2]\,\mathrm{d}t \\ &= S_{\mathrm{cl}} + \tfrac{1}{2}m\int_{t_a}^{t_b} (\dot{\eta}^2 - \omega^2\eta^2)\,\mathrm{d}t, \end{aligned} \tag{2.4}$$

as

$$\int_{t_a}^{t_b} (\dot{\bar{q}}\dot{\eta} - \omega^2\eta^2)\mathrm{d}t = -\int_{t_a}^{t_b} (\ddot{\bar{q}} + \omega^2 q)\eta\,\mathrm{d}t = 0,$$

due to the fact that $\bar{q}$ satisfies the equations of motion.

The explicit form of $\bar{q}$ is given by

$$\bar{q}(t) = q_a \cos[\omega(t - t_a)] + \frac{q_b - q_a\cos(\omega T)}{\sin(\omega T)} \sin[\omega(t - t_a)], \tag{2.5}$$

where $T = t_a - t_b$. A few lines of algebra show that

$$S_{\text{cl}} = \frac{m\omega}{2\sin(\omega T)}[(q_a^2 + q_b^2)\cos(\omega T) - 2q_a q_b]. \tag{2.6}$$

Thus we find

$$\begin{aligned} K(q_a, t_a; q_b, t_b) &= \exp\left(\frac{iS_{\text{cl}}}{\hbar}\right) \int_{\eta(t_a)=\eta(t_b)=0} \mathrm{d}[\eta] \\ &\quad \times \exp\left(\frac{mi}{2\hbar}\int_{t_a}^{t_b} (\dot{\eta}^2 - \omega^2\eta^2)\mathrm{d}t\right). \end{aligned} \tag{2.7}$$

We are, therefore, left with having to calculate the remaining path integral

$$F(T,\omega) = \int_{\eta(t_a)=\eta(t_b)=0} \mathrm{d}[\eta] \exp\left(\frac{mi}{2\hbar}\int_{t_a}^{t_b} (\dot{\eta}^2 - \omega^2\eta^2)\mathrm{d}t\right). \tag{2.8}$$

All the quantum effects are contained in this remaining path integral, which, of course, is nothing else but $(\det)^{-1/2}$ of the appropriate fluctuation operator (which in this case is, up to a constant, given by $(-\partial_t^2 + \omega^2)$). Notice the simple boundary conditions (the η function vanishes at both endpoints of this functional integral).

Next we expand $\eta(t)$, using a complete set of functions, which in this case are

$$u_n(t) = \sin[n\pi(t - t_a)/T] \tag{2.9}$$

obtaining

$$\eta(t) = \sum_n a_n\, u_n(t).$$

Then $\mathrm{d}[\eta] = J\prod_{n=1}^{\infty} \mathrm{d}a_n$, where the Jacobian J is clearly independent of ω, m and $\hbar$. Next, we substitute the expansion of $\eta(t)$ into the exponential in $F(T,\omega)$ and obtain

$$\begin{aligned} \int_{t_a}^{t_b} \dot{\eta}^2\,\mathrm{d}t &= \tfrac{1}{2}T\sum_{n=1}^{\infty}\left(\frac{n\pi}{T}\right)^2 a_n^2, \\ \int_{t_a}^{t_b} \eta^2\,\mathrm{d}t &= \tfrac{1}{2}T\sum_{n=1}^{\infty} a_n^2. \end{aligned} \tag{2.10}$$

Putting everything together, we find that

$$F(T,\omega) = \lim_{N\to\infty} J \prod_{n=1}^{N} \int_{-\infty}^{\infty} da_n \, \exp\left\{\frac{miT}{4\hbar}\sum_{n=1}^{N} a_n^2 \left[\left(\frac{n\pi}{T}\right)^2 - \omega^2\right]\right\}. \tag{2.11}$$

Next we perform the integration over a_n and obtain

$$F(T,\omega) = \lim_{N\to\infty} P \prod_{n=1}^{N} \left(1 - \frac{\omega^2 T^2}{n^2\pi^2}\right)^{-1/2},$$

where P is an overall factor proportional to J. Using the product representation of $\sin(x)$ we find that

$$F(T,\omega) = P' \left(\frac{\omega T}{\sin(\omega T)}\right)^{1/2}. \tag{2.12}$$

The new overall constant P' can be determined by comparing the result with the calculation of $F(T,\omega)$ for the free motion ($\omega \to 0$). As $F(T,0) = (m/2\pi i\hbar T)^{1/2}$ we finally obtain

$$F(T,\omega) = \left(\frac{m\omega}{2\pi i\hbar \sin(\omega T)}\right)^{1/2}. \tag{2.13}$$

This long calculation has shown the complications encountered in performing functional integrations. The calculation was difficult and we performed it only up to an overall factor which, in this case, we determined by comparing our result with a known case. However, we are often not in such a fortunate situation and we have to be satisfied with calculating functional integrals only up to an overall factor. On the other hand, functional integrals provide a convenient setting for calculating Green functions in cases where there are more than one solution to the classical equations of motion. We will make this clearer in the next sections.

2.2 Ground State Energy

One of the most important quantities describing a quantum system is the value of its ground state energy. This quantity can easily be derived from $K(q_a,t_a;q_b,t_b)$. To do this, let us generalise our discussion a little and consider an arbitrary potential $V(q)$. Recall the definition of $K(\ldots)$:

$$K(q_a,t_a;q_b,t_b) = \langle q_a| \exp[-i\hat{H}(t_b - t_a)/\hbar]\,|q_b\rangle. \tag{2.14}$$

Assume that $q = 0$ corresponds to the classical minimum of $V(q)$ and choose $q_a = q_b = 0$. Next, consider also the limit $t_b - t_a = T \to \infty$. Then inserting a complete set of the Hamiltonian ($\hat{H}$) eigenstates

$$\hat{H}\,|\phi_n\rangle = E_n\,|\phi_n\rangle \tag{2.15}$$

into the expression for $K(\ldots)$, gives

$$\begin{aligned} K(0,t_a;0,t_b) &= \sum_n \langle 0|\phi_n\rangle \exp(-iE_nT/\hbar)\langle\phi_n|0\rangle \\ &= \sum_n |\langle 0|\phi_n\rangle|^2 \exp(-iE_nT/\hbar). \end{aligned} \tag{2.16}$$

Recall that we prefer to perform all our calculations in Euclidean space. This corresponds to taking $t = -i\tau$ and $S_{\mathrm{E}} = -iS_{\mathrm{Mink}}$. Thus going over to Euclidean space results in

$$\exp\left(-\frac{iE_nT}{\hbar}\right) \to \exp\left(-\frac{E_n\tau}{\hbar}\right).$$

Next we consider again $\tau \to \infty$. Then, if the energy spectrum is discrete, we find

$$K_{\mathrm{E}} \Longrightarrow |\langle 0|\phi_0\rangle|^2 \exp(-E_0\tau/\hbar), \tag{2.17}$$

where $|\phi_0\rangle$ corresponds to the ground state of the system and E_0 is its energy. In the case of the harmonic oscillator discussed in the last section, setting $q_a = q_b = 0$ results in $\bar{q} \equiv 0$ and so $S_{\mathrm{cl}} = 0$. Then

$$\lim_{\tau\to\infty} K(\tau,0) = \lim_{\tau\to\infty} F(\tau,\omega) = \lim_{\tau\to\infty} \sqrt{\frac{m\omega}{\pi\hbar}}\;\; e^{-\omega\tau/2}. \tag{2.18}$$

Thus we see that $E_0 = \frac{1}{2}\hbar\omega$, which is, of course, the well known zero point energy. We see that this energy comes entirely from the functional determinant $F(\tau,\omega)$ (the classical energy is clearly zero). This supports our earlier claim that all the quantum corrections reside in the determinant and in further terms, if we have to apply perturbation theory; in the harmonic oscillator case there are no such further terms and so the result is exact.

2.3 Double Well Potential and Instantons

We consider the simplest example of a double well potential and so take

$$L = \tfrac{1}{2}\dot{q}^2 - \tfrac{1}{4}(q^2-1)^2, \tag{2.19}$$

in which we have set all the irrelevant quantities to 1. Classically, this potential has two minima ($q = \pm 1$). Next we go over to Euclidean space by setting $t = -i\tau$, and $S_{\mathrm{E}} = -iS_{\mathrm{Mink}}$ and let $\tau \to \infty$. We obtain

$$S_{\mathrm{E}} = \int_{-\infty}^{\infty} \mathrm{d}\tau \left[\frac{1}{2}\left(\frac{\mathrm{d}q}{\mathrm{d}\tau}\right)^2 + \frac{1}{4}\left(q^2-1\right)^2\right].$$

Notice that the continuation to Euclidean space effectively corresponds to the setting $V \to -V$, so that now V_{E} is given by

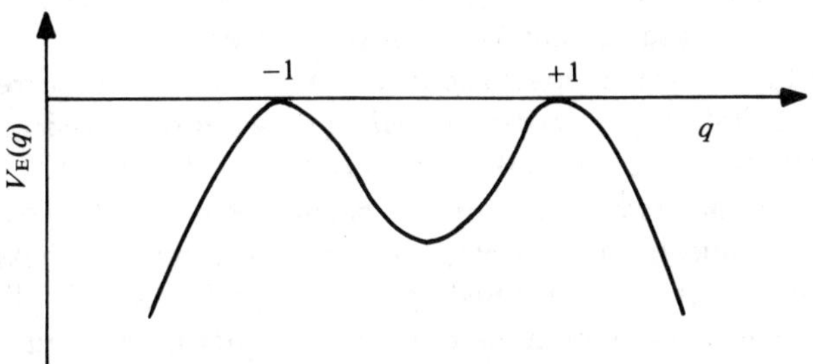

The Euclidean equations of motion (or strictly speaking, the Euler–Lagrange equations) are given by

$$\frac{\mathrm{d}^2 q}{\mathrm{d}\tau^2} = \frac{\mathrm{d}V}{\mathrm{d}q} = q^3 - q. \tag{2.20}$$

Next we consider all solutions to these equations, which at $\tau = \pm\infty$ go to the minima of the Minkowski space potential (and so the maxima of the Euclidean potential) i.e. $q = \pm 1$. We find that we have four such solutions:

$$\begin{aligned} \bar{q}_{1,2} &= \pm 1 \\ \bar{q}_{3,4} &= \pm \tanh\left(\frac{\tau - \tau_0}{\sqrt{2}}\right). \end{aligned} \tag{2.21}$$

(These four solutions are the analogues of the only solution ($\bar{q} = 0$) of the harmonic oscillator.)

Notice that of these four solutions the first two correspond to q remaining at the top of either of the two 'hills' of V_{E}, while the remaining two describe the movement from one hill to the other (in either direction) as τ varies from $-\infty$ to $+\infty$. Moreover, the action, evaluated for these solutions, is finite. For the first two it is clearly zero; to see that it is finite for the other two, we observe that as

$$\dot{\bar{q}}_{3,4} = \frac{\pm 1}{\sqrt{2}} \frac{1}{\cosh^2(\tau/\sqrt{2})} = \pm\sqrt{2V},$$

$$S_{\mathrm{E,cl}} = \int_{-\infty}^{\infty} \mathrm{d}\tau \left(\frac{\mathrm{d}q}{\mathrm{d}\tau}\right)^2 = \frac{1}{\sqrt{2}} \int_{-\infty}^{\infty} \mathrm{d}\tau \frac{1}{\cosh^4 \tau} = \int_{-1}^{1} \mathrm{d}q' \sqrt{2V(q')}. \tag{2.22}$$

As

$$\int_{-\infty}^{\infty} \mathrm{d}\tau \frac{1}{\cosh^4 \tau} < \infty,$$

the action is finite and positive (in fact it is $\frac{4}{3}$).

A solution of the classical Euclidean equations of motion, whose action is finite, is called *an instanton* [2] (or an anti-instanton — by convention $\bar{q}_3$ describes an instanton and $\bar{q}_4$ an anti-instanton).

What is the role of such solutions and what is their interpretation? It can be shown that they describe *tunnelling* between the classical minima of the potential (of course, in Minkowski space) [2]. Let us recall [3] that if the classical system has two (or more) equivalent minima of the potential then quantum mechanically, in general, there will be tunnelling between these minima. As a result of this tunnelling, the energy levels will split, with the energy of one of them being higher and of the other lower than their initial common energy. The true (quantum mechanical) minimum corresponds to the lower of the two, and the value of this energy is, of course, also modified by the addition of the zero point energy to the classical value. These effects can be easily discussed in our functional integral formulation of quantum mechanics; in fact this formulation provides a natural setting for studying such problems.

2.4 Role of Instantons

Following Coleman [2] we generalise our discussion a little, and so consider the case of an even potential $V(q) = V(-q)$ and denote its two (assuming there are only two) minima by $\pm q_0$. Let also us denote $V''(\pm q_0) = \omega^2$.

We would like to calculate

$$\langle -q_0|\exp(-\hat{H}T/\hbar)\,|-q_0\rangle \;=\; \langle q_0|\exp(-\hat{H}T/\hbar)\,|q_0\rangle \qquad (2.23)$$

and

$$\langle q_0|\exp(-\hat{H}T/\hbar)\,|-q_0\rangle \;=\; \langle -q_0|\exp(-\hat{H}T/\hbar)\,|q_0\rangle\,. \qquad (2.24)$$

We proceed as before by approximating the functional integral by its semiclassical limit (i.e. by performing the saddle-point calculation with the expansion terminated at the first nontrivial term, which is the determinant of the fluctuation operator). We consider T large, but strictly speaking not infinite. As in the case discussed in the previous section, we have four possible solutions, which describe either remaining at rest at $\pm q_0$ or the motion from $-q_0$ at $\tau = -\frac{1}{2}T$ to q_0 at $\tau = \frac{1}{2}T$ (or the other way round). The last two cases correspond to our previous instanton and anti-instanton solutions. As before we have

$$\frac{\mathrm{d}q}{\mathrm{d}\tau} \;=\; \sqrt{2V},$$

the instanton solutions of which are given by $q = \bar{q}(\tau)$, where

$$\tau \;=\; \tau_0 + \int_0^{\bar{q}} \mathrm{d}q' \sqrt{\frac{1}{2V}}. \qquad (2.25)$$

Here τ_0 is a free parameter; it corresponds to the value of τ at which $q = 0$.

We look at the graph of $q(\tau)$ given below:

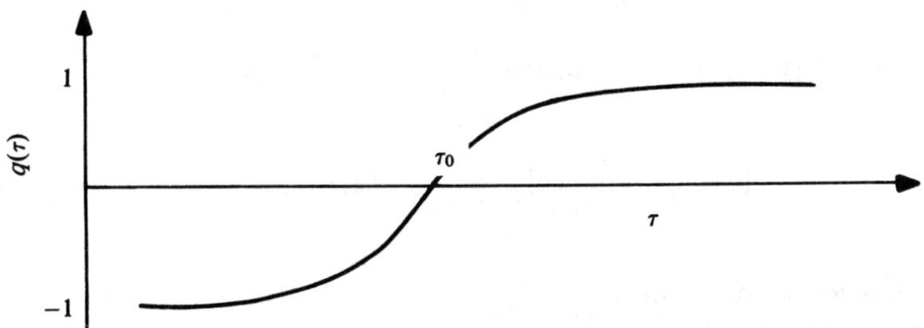

Referring also to the expression for $dq/d\tau$ above, we see that $q(\tau)$ changes little with τ for most values of τ, except those close to τ_0, (where $dq/d\tau$ has a maximum); hence we will talk about the instanton 'at τ_0'. The value of the action, evaluated for the instanton solution is, by analogy with the previous case, given by

$$S_0 = \int_{-T/2}^{T/2} \left(\frac{\mathrm{d}q}{\mathrm{d}\tau}\right)^2 \mathrm{d}\tau = \int_{-q_0}^{q_0} \mathrm{d}q' \sqrt{2V}. \tag{2.26}$$

Next, we have to evaluate the fluctuation determinant. To do this we put

$$q(\tau) = q_{\mathrm{cl}}(\tau - \tau_0) + \eta(\tau), \tag{2.27}$$

where we have written $q_{\mathrm{cl}}(\tau - \tau_0)$ for $\bar{q}(\tau)$ to emphasise that the classical solution depends on τ_0 only through $\tau - \tau_0$. Expanding $S_{\mathrm{E}}(q(\tau))$, we find

$$S_{\mathrm{E}}(q(\tau)) = S_{\mathrm{E}}(q_{\mathrm{cl}}(\tau - \tau_0)) + \tfrac{1}{2} \int \eta(\tau) \hat{O} \eta(\tau)\, \mathrm{d}\tau + \ldots, \tag{2.28}$$

where

$$\hat{O} = -\frac{\partial^2}{\partial \tau^2} + \frac{\mathrm{d}^2 V}{\mathrm{d}q^2}\bigg|_{q_{\mathrm{cl}}(\tau - \tau_0)}. \tag{2.29}$$

In the limit $(T \mapsto \infty)$, which we ultimately want to consider,

$$S_{\mathrm{E}}(q_{\mathrm{cl}}(\tau - \tau_0)) \to \int_{-\infty}^{\infty} \mathrm{d}\tau \left(\frac{\mathrm{d}q_{\mathrm{cl}}}{\mathrm{d}\tau}\right)^2 = S_0 \tag{2.30}$$

and is independent of τ_0.

Let us next consider

$$\frac{1}{\sqrt{S_0}} \frac{\mathrm{d}q_{\mathrm{cl}}(\tau - \tau_0)}{\mathrm{d}\tau} = \eta_0(\tau - \tau_0). \tag{2.31}$$

Since

$$\ddot{q} = \frac{\partial V}{\partial q} \qquad \frac{\mathrm{d}}{\mathrm{d}\tau}\ddot{q} = \frac{\mathrm{d}^2}{\mathrm{d}\tau^2}\dot{q} = \frac{\partial^2 V}{\partial q^2}\dot{q}$$

it is evident that $\eta_0(\tau - \tau_0)$ satisfies

$$\left(-\frac{\partial^2}{\partial \tau^2} + \frac{\mathrm{d}^2 V}{\mathrm{d}q^2}\bigg|_{q_{\mathrm{cl}}(\tau-\tau_0)}\right)\eta_0(\tau - \tau_0) = 0, \tag{2.32}$$

and so is a zero mode of $\hat{O}$.

The problem of the appearance of zero modes of the fluctuation operator arises because our instanton solution breaks a symmetry of the action. The action is translationally invariant, but our explicit solution depends on τ, and so breaks this invariance. In fact, one can show that this is an example of a general situation; each time a solution breaks an invariance of the action, the fluctuation operator exhibits a corresponding zero mode (associated with the translation in the direction of this breaking).

How do we deal with the situation when the fluctuation operator possesses zero modes? Clearly the determinant vanishes, and we have to find a way of dealing with the resultant infinity. In fact, this problem was studied by various people [4]. These studies resulted in the construction of a general method of dealing with such cases; the method used is the so-called *method of the collective coordinates.* This method relies on restricting the space of fluctuations by eliminating η_0 from it, and replacing it by the integration over τ_0. The method of the collective coordinates is an extremely useful tool, and so we discuss it in some detail in appendix A.

In this appendix we show also that the introduction of the collective coordinates leads to the following result for $\langle q_0|\exp(-\hat{H}T/\hbar)\,|-q_0\rangle$:

$$\langle q_0|\exp(-\hat{H}T/\hbar)\,|-q_0\rangle_{1\,\mathrm{inst}} = NT\sqrt{\frac{S_0}{2\pi\hbar}}e^{-S_0/\hbar}[\det{}'(-\partial_t^2 + V''(\bar{q}))]^{-1/2}. \tag{2.33}$$

Here, as we explain in detail in appendix A, the prime over $\det A$ denotes that the zero mode of A is omitted. Moreover, we have put the subscript '1 inst' to emphasise that this is an (approximate) contribution of one instanton. In addition, we have also put in the factor T in place of $\int_{-\infty}^{\infty}\mathrm{d}\tau_0$, which is clearly divergent, bearing in mind, however, that we are only interested in the limit $T \mapsto \infty$ (only in this case, strictly speaking, we have a zero mode!).

Now let us quote the result of Coleman [2], who showed that

$$\left(\det{}'(-\partial_t^2 + V''(\bar{q}))\right)^{-1/2} = A\left(\det(-\partial_t^2 + \omega^2)\right)^{-1/2}, \tag{2.34}$$

where A is some constant independent of T. Thus for large T, using the previously derived expression (2.18) for $\det(-\partial_t^2 + \omega^2)$, we have

$$\langle q_0 | \exp(-\hat{H}T/\hbar) | -q_0 \rangle_{1\text{ inst}} = \sqrt{\frac{\omega}{\pi\hbar}} \sqrt{\frac{S_0}{2\pi\hbar}}\, BT \exp\left(-\tfrac{1}{2}\omega T - \frac{S_0}{\hbar}\right) \tag{2.35}$$

where B is some unknown constant.

So far we have included only the solution of the equations of motion, and estimated our functional integral by performing the expansion only up to and including the determinant. However, there is no guarantee that this will give a correct approximation to the exact answer; strictly speaking we should estimate the contribution of the higher order terms. On the other hand, it has been suggested [2] that a better and more practical approximation is to keep the expansion only up to the determinant, and to include also the contribution due to the approximate solutions of the equations of motion corresponding to alternating instantons and anti-instantons. Such approximate solutions are described by functions $q(\tau)$ as shown below:

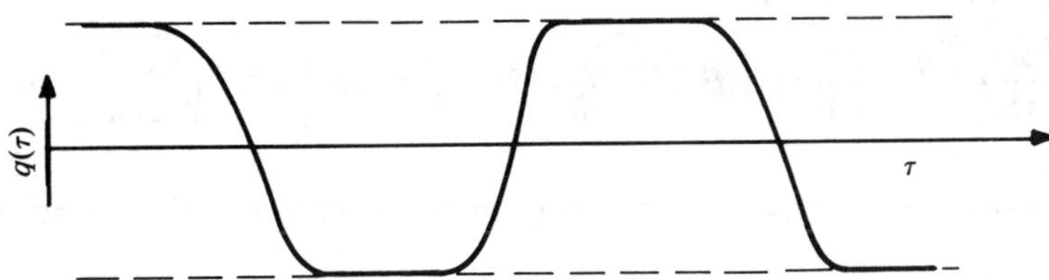

We should add further words of justification of our procedure. First of all, it is clear that the approximate solutions become correct solutions in the limit of $T \mapsto \infty$. In addition, the errors involved in using these functions are not very large as each of them is close to $\bar{q}_3$ or $\bar{q}_4$ (for some value of τ_0) for most values of their argument. A further justification can be given by distorting the 'functional integration contour' into the complex plane and so picking up the contribution of complex solutions. This was studied in some detail by the Marseille group [5], who found that the contribution of complex solutions resembles the contribution of the approximate solutions we are going to discuss. The last justification is the result itself; we will find that the method reproduces well the familiar tunnelling result.

Let us observe that, when we include approximate solutions, we can only alternate instantons and anti-instantons; that each of them can be located at any time τ, but that due to the ordering the zero mode factor T of one instanton should be replaced by

$$\int_{-T/2}^{T/2} d\tau_1 \int_{-T/2}^{\tau_1} d\tau_2 \cdots \int_{-T/2}^{\tau_{N-1}} d\tau_N = \frac{T^N}{N!}. \tag{2.36}$$

Moreover, other coefficients also factorise and so will appear to a power given by the instanton number. Thus we find

$$\langle -q_0|\exp(-\hat{H}T/\hbar)\,|-q_0\rangle$$
$$= \sqrt{\frac{\omega}{\pi\hbar}}e^{-\omega T/2}\sum_{\text{even }N}\frac{\left[BT\sqrt{(S_0/2\pi\hbar)}\right]^N}{N!}\exp\left(-\frac{NS_0}{\hbar}\right), \tag{2.37}$$

and also

$$\langle q_0|\exp(-\hat{H}T/\hbar)\,|-q_0\rangle$$
$$= \sqrt{\frac{\omega}{\pi\hbar}}e^{-\omega T/2}\sum_{\text{odd }N}\frac{\left[BT\sqrt{(S_0/2\pi\hbar)}\right]^N}{N!}\exp\left(-\frac{NS_0}{\hbar}\right). \tag{2.38}$$

Thus we have

$$\langle \pm q_0|\exp(-\hat{H}T/\hbar)\,|-q_0\rangle$$
$$= \sqrt{\frac{\omega}{\pi\hbar}}\,e^{-\omega T/2}\frac{1}{2}\left[\exp\left(BT\sqrt{\frac{S_0}{2\pi\hbar}}e^{-S_0/\hbar}\right)\mp\exp\left(-BT\sqrt{\frac{S_0}{2\pi\hbar}}e^{-S_0/\hbar}\right)\right] \tag{2.39}$$

and we see that we have two low-lying energy eigenstates with energies:

$$E_\pm = \tfrac{1}{2}\hbar\omega \pm \sqrt{\frac{\hbar S_0}{\pi}}\,Be^{-S_0/\hbar}. \tag{2.40}$$

In this expression we have an unknown constant B. Clearly this constant depends on the details of the potential.

To conclude this section, we note that the classical degeneracy of the energy eigenstates is broken by the tunnelling — the barrier penetration. The difference between the new energies is proportional to the barrier penetration factor $\exp(-S_0/\hbar)$. The state of the lower energy E_- corresponds to the spatially even combination of harmonic oscillator states centred at the bottoms of the two wells. The exact value of the energy shift depends on the details of the potential through the constant B. Thus, in this simple example, we see some very important quantum effects at work: the conventional quantum effects raise the classical ground state energy level by giving it a zero point energy; the effects of instantons is opposite — they lower this energy, and they lower it by an amount which is nonperturbative (i.e. not given by a power series in $\hbar$).

The effects discussed here are not restricted to the double well potential in quantum mechanics; similar effects arise in many field theories and in many models which possess nontrivial solutions to their Euclidean equations of motion. In quantum mechanics we often consider periodic potentials (they are good candidates to describe some solid state systems). An example of such a potential is

$$V(q) = 1 + \cos(\omega q).$$

This potential clearly has an infinite number of minima and, as is well known, the corresponding quantum system exhibits bands of energy called 'Brillouin zones'. Within each of them the energy is given by

$$E(\theta) = \tfrac{1}{2}\hbar\omega + A\cos\theta\, e^{-S_0/\hbar}, \tag{2.41}$$

where A is some constant dependent on the details of the potential. Similar conclusions can be reached in some field theory models ($\mathbb{C}P^{N-1}$ models in two dimensions or $SU(N)$ QCD theory in four dimensions). We will return to this point later on.

2.5 Supersymmetric Quantum Mechanics

So far we have discussed only bosonic models. However, in most physical applications, fermions also appear and often play an important role. Since in field theory fermions are described by anticommuting fields, we may wonder what their analogue would be in quantum mechanics. Moreover, we can further ask what the analogue would be in a supersymmetric theory. Such a system (supersymmetric quantum mechanics, SUSYQM) was first introduced by Berezin and Marinov [6] and by Brink *et al* [7] to discuss the first quantization of a relativistic spin $\frac{1}{2}$ particle. Later, it was reintroduced and popularised by Witten [8], who used it to obtain some insight into the nature of the spontaneous breakdown of supersymmetry.

Witten's SUSYQM Lagrangian is given by (one dimension)

$$L = \tfrac{1}{2}\dot{x}^2 - \tfrac{1}{2}W^2(x) + i\psi^*\dot{\psi} - \tfrac{1}{2}[\psi^*, \psi]\, W'(x), \tag{2.42}$$

where

$$W'(x) = \frac{\partial W}{\partial x}, \qquad \{\psi^*, \psi\} = \{\psi, \psi\} = \{\psi^*, \psi^*\} = 0. \tag{2.43}$$

It is easy to check that this Lagrangian transforms as a total derivative (and so the action is invariant) under the following transformation:

$$\begin{aligned} \delta x &= (\psi\epsilon + \epsilon^*\psi^*), \\ \delta\psi &= \epsilon\dot{x} - i\epsilon^* W, \\ \delta\psi^* &= \epsilon\dot{x} + i\epsilon W, \end{aligned} \tag{2.44}$$

where ϵ and ϵ^* are constant anticommuting parameters of the transformation.

Witten's SUSYQM Lagrangian can be derived by dimensional reduction from a supersymmetric field theory (treating SUSYQM as a SUSY field theory, which depends only on one variable — time), or by formulating the theory directly in superspace and integrating out the anticommuting components of this superspace. For pedagogical reasons we will discuss this second approach.

The superspace which we want to consider is a space which, in addition to the usual commuting variable t, contains also two anticommuting quantities (θ and θ^*). Thus they satisfy

$$\{\theta, \theta^*\} = \{\theta, \theta\} = \{\theta^*, \theta^*\} = [\theta, t] = [\theta^*, t] = 0. \tag{2.45}$$

Supersymmetry transformations are grassmannian even translations in superspace, which mix both types of variables, i.e.

$$\begin{aligned} t &\to t' = t - i(\theta^* \epsilon - \epsilon^* \theta) \\ \theta &\to \theta' = \theta + \epsilon \\ \theta^* &\to \theta^{*\prime} = \theta^* + \epsilon^*. \end{aligned} \tag{2.46}$$

Such finite transformations are generated by

$$G = \exp(i[\epsilon^* Q^* + Q\epsilon]), \tag{2.47}$$

where Q and Q^* are supersymmetry generators. Under supersymmetry transformations all quantities A are transformed as

$$A \to A' = GAG^*, \tag{2.48}$$

which, for infinitesimal transformations, reduces to

$$\delta A = i[\epsilon^* Q^* + Q\epsilon, A]. \tag{2.49}$$

A little thought shows that the transformation laws for t, θ and θ^* are reproduced if the generators Q and Q^* are given by

$$\begin{aligned} Q &= i\partial_\theta - \theta^* \partial_t \\ Q^* &= -\, i\partial_{\theta^*} + \theta\partial_t. \end{aligned} \tag{2.50}$$

Then, as is very easy to check,

$$\begin{aligned} \{Q, Q^*\} &= 2i\partial_t = 2H, \\ [Q, H] &= [Q^* H] = 0. \end{aligned} \tag{2.51}$$

Moreover, it is also easy to check that two further operators, which we will call covariant derivatives,

$$D_\theta = \partial_\theta - i\theta^*\partial_t \quad \text{and} \quad D_{\theta^*} = \partial_{\theta^*} - i\theta\partial_t \tag{2.52}$$

are invariant under our supersymmetry transformations, i.e. they anticommute with Q and Q^*.

Next we consider superfields, i.e. functions defined on superspace, which are a natural generalisation of fields of ordinary field theory. The superfields are defined as functions on superspace which have a definite grassmann parity, i.e. they commute or anticommute with each other. In what follows, for simplicity, we shall consider a real scalar superfield, i.e. a grassmann-even superfield which satisfies

$$\begin{aligned} \phi(t,\theta,\theta^*) &= \phi^*(t,\theta,\theta^*), \\ \phi'(t',\theta',\theta^{*\prime}) &= \phi(t,\theta,\theta^*). \end{aligned} \tag{2.53}$$

We can now expand $\phi(t,\theta,\theta^*)$ in the power series of θ and θ^* and, due to their anticommuting nature, we obtain only a few terms of this expansion, which we can write as

$$\phi(t,\theta,\theta^*) = x(t) + i\theta\psi(t) - i\psi^*(t)\theta^* + \theta^*\theta D(t). \tag{2.54}$$

This expansion serves as a definition of $x(t)$, $\psi(t)$, $\psi^*(t)$ and $D(t)$.

To see how these functions transform under our supersymmetry transformation, we use the transformation property of the field ϕ:

$$\delta\phi \;=\; i[\epsilon^*Q^* + Q\epsilon\,,\,\phi]. \tag{2.55}$$

This, combined with

$$\delta\phi = \delta x + i\theta\delta\psi - i\delta\psi^*\theta^* + \theta^*\theta\delta D,$$

shows that

$$\begin{aligned} \delta x(t) &= -\epsilon^*\psi^*(t) + \psi(t)\epsilon, \\ \delta\psi(t) &= \epsilon^*\dot{x}(t) - i\epsilon^* D(t), \\ \delta\psi^*(t) &= \epsilon\dot{x}(t) + i\epsilon D(t), \\ \delta D(t) &= \epsilon\dot{\psi} + \dot{\psi}^*\epsilon^* = (\mathrm{d}/\mathrm{d}t)(\epsilon\psi + \psi^*\epsilon^*). \end{aligned} \tag{2.56}$$

Moreover, our covariant derivatives are given by

$$D_\theta\phi = i\psi - \theta^* D - i\theta^*\dot{x} + \theta^*\theta\dot{\psi} \quad \text{etc.} \tag{2.57}$$

A general action invariant under SUSY transformations is given by

$$S \;=\; \int \mathrm{d}t\, \mathrm{d}\theta^*\, \mathrm{d}\theta[\tfrac{1}{2}|D_\theta\phi|^2 - f(\phi)], \tag{2.58}$$

where

$$f(\phi) = \sum a_n \phi^n .$$

Due to the rules of the integration over anticommuting variables, we see that only the coefficient of $\theta\theta^*$ in $[\frac{1}{2}|D_\theta\phi|^2 - f(\phi)]$ contributes to S. Moreover, as $(D_\theta\phi)^* = -D_{\theta^*}\phi$, it is easy to find that

$$(D_\theta\phi)^*(D_\theta\phi) = \theta\theta^*[\dot{x}^2 + D^2 + i(\psi^*\dot{\psi} - \dot{\psi}^*\psi)] + \ldots, \qquad (2.59)$$

and also

$$f(\phi) = \theta\theta^*(-D\dot{f} - \tfrac{1}{2}[\psi^* \, \psi]\,\ddot{f})\,\Big|_{\theta^*=\theta=0} + \ldots, \qquad (2.60)$$

where, in both the expressions above, ... stands for terms of lower order in θ and θ^*. As a result we see that, after the integration over θ and θ^*, we get

$$S = \int \mathrm{d}t\{\tfrac{1}{2}\dot{x}^2 + \tfrac{1}{2}D^2 + \tfrac{1}{2}i(\psi^*\dot{\psi} - \dot{\psi}^*\psi) + D\dot{f} + \tfrac{1}{2}[\psi^* \, , \psi]\ddot{f}\}. \qquad (2.61)$$

As we are interested in the quantum formulation of the theory we consider the functional integration over all the fields ($x(t)$, $\psi(t)$, $\psi^*(t)$ and $D(t)$). Of these the D integration can be performed straight away:

$$\begin{aligned} &\int \mathrm{D}[D] \exp\left(-i\int \mathrm{d}t(\tfrac{1}{2}D^2 + D\dot{f})\right) \\ &\qquad = \int \mathrm{D}[D+f] \exp\left(-i\int \mathrm{d}t[\tfrac{1}{2}\{D+\dot{f}\}^2 - \tfrac{1}{2}\dot{f}^2]\right) \qquad (2.62) \\ &\qquad = N \exp\left(\tfrac{1}{2}i\int \mathrm{d}t\, \dot{f}^2\right), \end{aligned}$$

where N is some irrelevant constant. So we see that in the functional formulation of the quantum theory we can keep only the 'dynamical' variables of the theory (x, ψ and ψ^*), and so consider

$$\int \mathrm{D}[x]\,\mathrm{D}[\psi]\,\mathrm{D}[\psi^*]\, e^{iS_1}, \qquad (2.63)$$

where S_1 denotes the effective action for the 'dynamical' fields, and is given by

$$S_1 = \int \mathrm{d}t\,\{\tfrac{1}{2}\dot{x}^2 + \tfrac{1}{2}i(\psi^*\dot{\psi} - \dot{\psi}^*\psi) - \tfrac{1}{2}W^2 - \tfrac{1}{2}[\psi^* \, , \psi]\,\dot{W}\}, \qquad (2.64)$$

where $W = -\dot{f}$. The function f, which appears in our expressions, is often called the *superpotential*.

The interested reader may ask what was the real purpose of introducing the nondynamical variable D. After all, having introduced it, we eliminated it by integrating it out. A little thought shows that with the nondynamical variables present supersymmetry is implemented linearly; this can be important quantum mechanically as it ensures that the supersymmetric Ward identities are also linear. Without the nondynamical variables supersymmetry is realised in a nonlinear way, which may lead to some problems with the quantum formulation of the theory. The introduction of the nondynamical variables (like the D field in our case) avoids these problems.

Having rewritten our supersymmetric theory as a theory based on the effective action S_1, we can now consider various interesting cases (like a harmonic oscillator, an anharmonic oscillator, etc) and attempt to calculate the Green functions of the theory by calculating all the required determinants, etc. On the other hand, it is instructive to consider also the conventional quantisation procedure for the theory based on the action S_1.

In such a procedure we quantise the theory by imposing the following commutation and anticommutation relations:

$$[x\,,p] \;=\; i, \qquad \{\psi^*\,,\psi\} \;=\; 1. \tag{2.65}$$

Here, p is a momentum conjugate to x and ψ^* is conjugate to ψ; moreover, we have set $\hbar = 1$.

What are the SUSY generators in this formulation? Clearly, they should perform the SUSY transformations and reproduce the Hamiltonian. To decide what they are, we should, first of all, choose a convenient basis in which to address this problem. To do this let us observe that, if $\{\psi^*, \psi\} = 1$, we can choose

$$\psi \;=\; \sigma^- \;=\; \begin{pmatrix} 0 & 0 \\ 1 & 0 \end{pmatrix}, \quad \psi^* \;=\; \sigma^+ \;=\; \begin{pmatrix} 0 & 1 \\ 0 & 0 \end{pmatrix}. \tag{2.66}$$

Then, as can be easily checked,

$$\{\psi^*\,,\psi\} = 1, \quad [\psi^*\,,\psi] = \sigma_3 = \begin{pmatrix} 1 & 0 \\ 0 & -1 \end{pmatrix}. \tag{2.67}$$

In this basis the Hamiltonian takes the form originally written by Witten:

$$H = \tfrac{1}{2}[p^2 + W^2(x)] + \tfrac{1}{2}\sigma_3\, W'(x). \tag{2.68}$$

It is very convenient to define a fermion-number like operator $\hat{n}$:

$$\hat{n} \;=\; \psi^*\psi \;=\; \begin{pmatrix} 1 & 0 \\ 0 & 0 \end{pmatrix}. \tag{2.69}$$

This operator commutes with the Hamiltonian and its eigenstates correspond to the upper and to the lower components of the total wavefunction. Here we write

$$|\phi\rangle = \phi_u(x)\,|\uparrow\rangle + \phi_d(x)\,|\downarrow\rangle \tag{2.70}$$

and so represent the state $|\phi\rangle$ by a two-component wavefunction:

$$|\phi\rangle \sim \phi(x) = \begin{pmatrix} \phi_u(x) \\ \phi_d(x) \end{pmatrix}.$$

What are the symmetry generators Q and Q^*? Clearly, they must satisfy

$$\{Q^*\,, Q\} = 2H, \quad [Q\,, H] = [Q^*\,, H] = 0 \tag{2.71}$$

and have the property that

$$Q\,|\downarrow\rangle = A\,|\uparrow\rangle \quad Q^*\,|\uparrow\rangle = B\,|\downarrow\rangle\,, \tag{2.72}$$

and similarly for Q^*. It is easy to check that

$$\begin{aligned} Q^* &= (p - iW)\psi = \left(\frac{1}{i}\frac{\partial}{\partial x} - iW\right)\sigma^- \\ Q &= (p + iW)\psi^* = \left(\frac{1}{i}\frac{\partial}{\partial x} + iW\right)\sigma^+ \end{aligned} \tag{2.73}$$

satisfy all our requirements, and so can be considered our required SUSY generators.

Let us observe, in passing, that if we introduce $\bar{Q} = Q + Q^*$ then $2H = (\bar{Q})^2$. Since

$$\bar{Q} = \frac{1}{i}\frac{\partial}{\partial x}\begin{pmatrix} 0 & 1 \\ 1 & 0 \end{pmatrix} + W\begin{pmatrix} 0 & i \\ -i & 0 \end{pmatrix} \tag{2.74}$$

we see that $\bar{Q}$ is hermitian and so all eigenvalues of H are non-negative. Consequently, the lowest eigenstate of the Hamiltonian (the ground state) must have its energy $E_{\mathrm{g}} \geq 0$. Can it be zero?

To discuss this point let us call this eigenstate $|g\rangle$. It is easy to show, using the commutation and anticommutation relations satisfied by Q, Q^* and H, that $E_{\mathrm{g}} = 0$ if and only if both $Q\,|g\rangle = 0$ and $Q^*\,|g\rangle = 0$. We consider the case when $E_{\mathrm{g}} \neq 0$ as describing a system in which SUSY is spontaneously broken, and so we see that we have a convenient criterion to decide whether the supersymmetry is spontaneously broken or not. To translate this into the conditions on the wavefunction of the ground state, we represent $|g\rangle$ by

$$|g\rangle \sim \begin{pmatrix} \phi_{\mathrm{u}}^g \\ \phi_{\mathrm{d}}^g \end{pmatrix}$$

and so find that the condition for no spontaneous breakdown is

$$\begin{aligned} \left(\frac{1}{i}\frac{\partial}{\partial x} - iW\right)\phi_{\mathrm{u}}^{g} &= 0 \\ \left(\frac{1}{i}\frac{\partial}{\partial x} + iW\right)\phi_{\mathrm{d}}^{g} &= 0. \end{aligned} \tag{2.75}$$

These equations can easily be solved, and we find

$$\begin{aligned} \phi_{\mathrm{u}}^{g}(x) &= \phi_{\mathrm{u}}^{g}(x_0)\exp\left(\int_{x_0}^{x} W(x')\,\mathrm{d}x'\right) \\ \phi_{\mathrm{d}}^{g}(x) &= \phi_{\mathrm{d}}^{g}(x_0)\exp\left(-\int_{x_0}^{x} W(x')\,\mathrm{d}x'\right). \end{aligned} \tag{2.76}$$

The fact that we have found solutions to (2.75) might suggest that spontaneous symmetry breakdown can never occur. This suggestion is, however, incorrect. To see this let us recall that, in order that $\phi(x)$ be a wavefunction of a physical state, it must be normalisable. So it remains to check this normalisability. Looking at our expressions above, we observe that, if the leading term of $W(x)$ (leading as $x \to \infty$) is an even function of x, the leading term of $\int_{x_0}^{x} W(x')\,\mathrm{d}x'$ is an odd function of x, and so neither ϕ_d or ϕ_u are normalisable. This observation, originally made by Witten, shows that, if the leading term of $W(x)$ is an even function of x, the system has no zero energy ground state, and so exhibits spontaneous symmetry breakdown. Of course, since

$$H = \tfrac{1}{2}[p^2 + W^2(x)] + \tfrac{1}{2}\sigma_3 W'(x), \tag{2.77}$$

we see that, when supersymmetry is not broken, it is the $\phi_{\mathrm{d}}(x)$ which describes the ground state (the other component is zero). Moreover, the equation which ϕ_{d} satisfies is

$$H_{\mathrm{g}} = \tfrac{1}{2}[p^2 + W^2(x) - W'(x)].$$

Let us consider now a couple of examples. First of all let us take the harmonic oscillator, which is described by $W = \omega x$. Then

$$H = \tfrac{1}{2}(p^2 + \omega^2 x^2 \mp \omega), \tag{2.78}$$

and

$$\begin{aligned} \phi_{\mathrm{u}} &= C_{\mathrm{u}} \exp(\tfrac{1}{2}\omega x^2) \\ \phi_{\mathrm{d}} &= C_{\mathrm{d}} \exp(-\tfrac{1}{2}\omega x^2), \end{aligned} \tag{2.79}$$

and we see that $\phi_{\mathrm{d}}(x)$ is the usual harmonic oscillator wavefunction. The effective potential for $\phi_{\mathrm{d}}(x)$ is

$$V_{\mathrm{d}} = \tfrac{1}{2}\omega^2 x^2 - \tfrac{1}{2}\omega. \tag{2.80}$$

The first term in the expression for $V_{\rm d}$ can be thought of as representing the bosonic contribution (it comes from $W^2(x)$), while the second represents the fermionic contribution. If we disregarded this fermionic contribution we would find that the quantum effects modify the vanishing ($E = 0$) classical value of the energy ($x = 0 \rightarrow W^2 = 0$), and change it to $E = \frac{1}{2}\omega$, due to the zero point energy (recall we have set $\hbar = 1$). However, the fermionic contribution reduces this value back to zero. This, as was pointed out by Witten, is a simple example of the claim, often made in the literature, that in supersymmetric theories the bosonic and fermionic determinants can cancel. We see (considering also other examples which correspond to no supersymmetry breakdown) that these effects persist to all orders, if SUSY is not spontaneously broken. In fact, it is useful to look at this point from a different angle.

Consider

$$\phi_{\rm d} = A \exp\left(-\frac{ax^{2n}}{2n}\right). \tag{2.81}$$

This function has no nodes and would appear a reasonable candidate for being a wavefunction of an appropriate ground state. Since

$$\frac{\rm d}{{\rm d}x}\phi_{\rm d} = -ax^{2n-1}\phi_{\rm d}$$

and

$$\frac{{\rm d}^2\phi_{\rm d}}{{\rm d}x^2} = a^2\,x^{4n-2}\phi_{\rm d} - a(2n-1)x^{2n-2}\phi_{\rm d}$$

we see that our $\phi_{\rm d}$ is the ground state wavefunction of a system with $W = ax^{2n-1}$. It would be quite difficult to prove this point from the point of view of the functional determinants (especially if we took a more complicated form of $\phi_{\rm d}$).

On the other hand, when the leading term in $W(x)$ is an even function, e.g. $W(x) = \frac{1}{2}x^2$, the ground state is degenerate, and so its energy is positive. In the example above the effective potentials for the two lowest energy states (corresponding to $|\uparrow\rangle$ and $|\downarrow\rangle$) are now given by

$$V(x) = \tfrac{1}{2}(W^2(x) \pm W'(x)) = \tfrac{1}{8}x^4 \pm \tfrac{1}{2}x. \tag{2.82}$$

Hence, in the purely anharmonic potential case, the fermionic contribution does not bring the quantum value of the energy down to zero (the energy remains positive).

Finally let us add that SUSYQM has turned out to be a very useful tool, not only for checking various ideas about supersymmetry, but also for proving some mathematical theorems and for solving various equations. Thus, exploiting the similarity of the chirally decomposed Dirac operator and the Q and Q^* operators which appear in SUSYQM, Alvarez Gaume [9] has managed to construct a simple proof of the Atiyah–Singer index theorem on compact spaces. At the same time it has been shown [10] that the solutions of all the exactly solvable potentials of ordinary quantum mechanics, and of some other related problems, can be obtained by exploiting the existence of a supersymmetric extension of the Hamiltonian and, consequently, its factorisation.

References

[1] Feynman R P and Hibbs A R 1965 *Quantum Mechanics and Path Integrals* (New York: McGraw-Hill)
Marinov M S 1980 Path integrals in quantum theory: an outlook of basic concepts *Phys. Rep.* **60** 1

[2] Coleman S 1978 *The Uses of Instantons. Lectures given at the Erice Summer School of Subnuclear Physics, 1977* ed Zichichi (New York: Academic)

[3] Sudbery A 1986 *Quantum Mechanics and the Particles of Nature* (Cambridge: Cambridge University Press)

[4] Gervais J L and Sakita B 1975 Extended particles in quantum field theories *Phys. Rev.* D **11** 2943
Tomboulis E 1975 Canonical quantization of nonlinear waves *Phys. Rev.* D **12** 1678

[5] Richard J L and Rouet A 1981 Complex saddle points in the double well oscillator *Nucl. Phys.* B **183** 251; 1981 Complex saddle points versus dilute gas approximation in the double well anharmonic oscillator *Nucl. Phys.* B **185** 47

[6] Berezin F A and Marinov M S 1975 Classical spin and Grassmann algebra *JETP Lett.* **21** 320

[7] Brink L, Deser S, Zumino B, Di Vecchia P and Howe P S 1976 Local supersymmetry for spinning particles *Phys. Lett.* **64B** 435
Brink L, Di Vecchia P and Howe P S 1977 A Lagrangian formulation of the classical and quantum dynamics of spinning particles *Nucl. Phys.* B **118** 76

[8] Witten E 1981 Dynamical breaking of supersymmetry *Nucl. Phys.* B **188** 513
Cooper F and Freedman B 1983 Aspects of supersymmetric quantum mechanics *Ann. Phys., NY* **146** 262

[9] Alvarez-Gaumé L 1983 Supersymmetry and the Atiyah–Singer index theorem *Commun. Math. Phys.* **90** 161

[10] Cooper F, Ginocchio J N and Wipf A 1987 Supersymmetry, hidden shape invariance and exactly solvable potentials *Los Alamos preprint LA-UR-87-4101*
Cooper F, Khare A, Musto R and Wipf A 1988 Supersymmetry and the Dirac equation *Los Alamos preprint LA-UR-88-169*

3 Grassmannian Models and Their Classical Solutions

3.1 Introduction: A few Comments about Yang–Mills Theories

Nonabelian gauge theories are likely to play an important role in any final field theoretic description of the theory of elementary particles. In fact our present theory — the standard model — is based on such theories. Nonabelian gauge theories, for example in the case of an $SU(2)$ theory, are defined in terms of a Lagrangian density:

$$L = \frac{1}{2g^2} \operatorname{Tr} F_{\mu\nu} F_{\mu\nu}, \tag{3.1}$$

where

$$F_{\mu\nu} = \partial_\mu A_\nu - \partial_\nu A_\mu + [A_\mu , A_\nu], \tag{3.2}$$

and where A_μ is an $SU(2)$ valued vector function of Euclidean four-dimensional spacetime. (In fact, from now on and unless said to the contrary, we will always assume that the spacetime has been analytically continued to be Euclidean.)

In the previous chapter we said that we did not know how to calculate general functional integrals, and that, in practice, the only procedure we have available for their calculation consists of expansion around stationary points of the action and then perturbation theory of the resultant effective theory. So, if we were to quantise the theory based on the Lagrangian (3.1), we would, first of all, have to know the stationary points of the action. These, as is well known, are given by the solutions of the equations

$$D_\mu F_{\mu\nu} = \partial_\mu F_{\mu\nu} - [A_\mu , F_{\mu\nu}] = 0. \tag{3.3}$$

These equations, when written in terms of A_μ, are highly nonlinear. However, due to the Bianchi identity:

$$D_\mu {}^*F_{\mu\nu} = 0, \tag{3.4}$$

where

$${}^*F_{\mu\nu} = \tfrac{1}{2}\epsilon_{\mu\nu\alpha\beta} F_{\alpha\beta}, \tag{3.5}$$

a subclass of solutions of (3.3) is provided by the solutions of

$$F_{\mu\nu} = \pm {}^*F_{\mu\nu}. \tag{3.6}$$

These equations, being the first-order equations for A_μ, are called 'self-duality' and 'antiself-duality' equations, and should, in principle, be easier to solve. They can be thought of as having come from the requirement that the Lagrangian density L is given by

$$L = \pm q,$$

where $q(x) = (1/2g^2)\,\mathrm{Tr}\, F_{\mu\nu}\,{}^*F_{\mu\nu}$ is the density of the topological charge.

In fact, it is easy to show that

$$Q = \int \mathrm{d}x^4\, q(x) \tag{3.7}$$

is a topological quantity which characterises A_μ, treated as maps of S^3 at spacetime infinity into $SU(2)$. As such, Q takes discrete values ($32\pi^2\times$ integer), and it divides all maps A_μ into different classes, characterised by the value of this integer. As $q(x)$ is a density of a topological quantity, it is easy to check that it can be written locally as a total divergence:

$$q(x) = \partial_\mu J_\mu, \tag{3.8}$$

where

$$J_\mu(x) = 2\epsilon_{\mu\nu\lambda\sigma}\,\mathrm{Tr}\, A_\nu(\partial_\lambda A_\sigma + \tfrac{2}{3}A_\lambda A_\sigma). \tag{3.9}$$

Hence, $q(x)$ divides all field configurations into sets corresponding to different values of Q.

On the other hand (3.8) shows that Q is a topological quantity. To see this vary A_μ. Then as $A_\mu \to A_\mu + \delta A_\mu$ $Q \to Q + \delta Q$, where

$$\begin{aligned}\delta Q = \int \mathrm{d}^4x\, \partial_\mu \{ 2\epsilon_{\mu\nu\lambda\sigma}\,\mathrm{Tr}\,[\delta A_\nu(\partial_\lambda A_\sigma + \tfrac{2}{3}A_\lambda A_\sigma) \\ + A_\nu(\partial_\lambda \delta A_\sigma + \tfrac{2}{3}\delta A_\lambda A_\sigma + \tfrac{2}{3}A_\lambda \delta A_\sigma)]\}.\end{aligned} \tag{3.10}$$

But as $\epsilon_{\mu\nu\lambda\sigma}\partial_\mu(A_\nu\partial_\lambda\delta A_\sigma) = -\epsilon_{\mu\nu\lambda\sigma}\partial_\mu(\partial_\lambda A_\nu\delta A_\sigma)$ we see that

$$\delta Q = \int \mathrm{d}^4x\, \partial_\mu(\delta A_\nu R_{\mu\nu}) \tag{3.11}$$

for some $R_{\mu\nu}$ and so is determined only by the behaviour of δA_μ at infinity. Hence any two fields A_μ which have the same behaviour at infinity would give the same contribution to Q. This establishes the topological nature of Q.

In addition, we have a Bogomolnyi-like bound [1]. To derive it we observe that, as

$$\mathrm{Tr}\, F_{\mu\nu}\, F_{\mu\nu} \;=\; \mathrm{Tr}\, {}^*F_{\mu\nu}\, {}^*F_{\mu\nu}, \tag{3.12}$$

we have

$$\begin{aligned} 0 &\le (F_{\mu\nu} \pm^* F_{\mu\nu})(F_{\mu\nu} \pm^* F_{\mu\nu}) \\ &= 2F_{\mu\nu}F_{\mu\nu} \pm 2F^*_{\mu\nu}F_{\mu\nu}, \end{aligned} \tag{3.13}$$

and so

$$S - |Q| \ge 0. \tag{3.14}$$

Thus we see that all field configurations which satisfy the 'self-duality' or 'antiself-duality' equations (3.6) correspond to the local minima of the action. Hence, such field configurations are stable (modulo zero modes) under small fluctuations.

All finite action solutions of the self-duality (and anti-selfduality) equations have been implicitly, though unfortunately not explicitly, determined by Atiyah *et al* [2]. This work builds on the work of many people; it was started by Belavin *et al* [3] and 't Hooft [4]. All these finite action solutions are called *instantons* and *anti-instantons*. However, the question then arises as to whether there are further solutions (of finite action) to the full equations of motion, i.e. (3.3), which are not self-dual or antiself-dual. Despite a large amount of work in this area the answer to this question is still unknown. Moreover, the problem of calculating all the corrections due to the fluctuations about the instanton solutions has turned out to be a hard mathematical problem [5].

Are there simpler field theories which bear some resemblance to four-dimensional nonabelian gauge theories? To answer this question, we must decide which properties of four-dimensional nonabelian gauge theories we want to keep. In order to do this we have to look at gauge theories in more detail, a topic which lies outside the scope of this book. We refer all interested readers to any standard book on field theory, e.g. the book by Ramond [6], and here we simply state that we would like to preserve:

1. the conformal invariance of the action,
2. the existence of nontrivial solutions to the equations of motion (related to the point above),

3. the existence of a convenient parameter (like N in $SU(N)$ gauge theories) to allow a possible $1/N$ expansion,
4. asymptotic freedom,
5. the confinement of fermions if they are coupled in.

Of these five properties, in this study we are particularily concerned with properties 2 and 3. Properties 1 and 4, of less relevance to us here, are discussed in most standard texts dealing with gauge theories, see e.g. the above-mentioned book by Ramond [6], and property 5 refers to the quark-field confinement in QCD which, though not proved to hold, is widely believed to be true. Moreover, it appears to be true experimentally.

A *two-dimensional* model, which possesses similar properties (but being two-dimensional, is hopefully simpler to study), was first proposed by Golo and Perelomov [7] and Eichenherr [8]. This model, the so-called $\mathbb{C}P^{N-1}$ model, has very quickly become a very popular 'toy model' for investigating various questions of physical interest. As a consequence, soon after its invention, it was used in 1979 by D'Adda *et al* [9], and later by others, in order to study some of the questions mentioned above. In fact, the $\mathbb{C}P^{N-1}$ model belongs to a whole class of such models (called grassmannian models) which recently have become interesting in their own right.

At this stage, to avoid possible confusion, we would like to stress that the word 'grassmannian' is used here in a sense which is different from the way we have used it before. In the previous chapters we spoke of 'grassmann variables' , i.e. anticommuting quantities, used when fermions or fermion-like quantities are considered in the path integral formulation framework. In this chapter the word 'grassmannian' is associated with 'grassmannian models'. These models are described by matrix valued bosonic fields (i.e. involving commuting quantities). In fact, later on, the fields themselves will be often referred to as 'grassmannians'. In the next section we introduce these models and discuss their instanton solutions.

3.2 Two-dimensional Grassmannian σ Models and their Instanton Solutions

Before we introduce our grassmannian models let us mention briefly the so-called $O(N)$ σ models, which are, perhaps, more familiar to most readers and are a natural generalisation of the original $O(4)$ σ model, first introduced by Gell-Mann and Levy [10]. The most common formulation of these models uses as basic fields the N-dimensional real unit vectors ϕ^a (i.e. ϕ^a satisfies $\phi^a \cdot \phi^a = 1$), which are, of course, functions of our two-dimensional space locally parametrised by x_1 and x_2. The Lagrangian density is given by

$$L = \partial_\mu \phi^a \cdot \partial_\mu \phi^a, \tag{3.15}$$

taken, of course, together with the constraint $\phi^a \cdot \phi^a = 1$.

In this formulation, the target manifold (the manifold where the fields ϕ^a lie) is the unit sphere in N dimensions— S^N, and the model is invariant under global $O(N)$ transformations (i.e. transformations independent of x_1 and x_2):

$$\phi^a \quad \rightarrow \phi^{a'} = O^{a'a}\phi^a, \tag{3.16}$$

where $O^{a'a} \in O(N)$.

However, there is another way of thinking about these models which brings out the coset space nature of the target manifold. In this new formulation we choose a standard vector ϕ_0^a, and then observe that any vector ϕ^a can be obtained from our standard vector by some local orthogonal transformation $O(N)$, i.e. $O(N) = O(N)(x_1, x_2)$. However, as the little group of the standard vector ϕ_0^a (i.e. a group of transformations which leave this vector invariant) is $O(N-1)$, we see that we can identify the target manifold with the coset space $O(N)/O(N-1)$ (i.e. transformations $O(N)$ modulo $O(N-1)$). Thus, strictly speaking, the $O(N)$ σ models should be called either S^N or $O(N)/O(N-1)$ models.

Let us now introduce our grassmannian models. We will do this by considering first the target manifold which we take in the form of a coset space, this time the matrices of the transformations being unitary. Thus we will denote by $G(M, N)$ the complex Grassmann manifold which can be written as a quotient space:

$$G(M, N) = \frac{U(N)}{U(M) \times U(N-M)}, \tag{3.17}$$

where $U(K)$ denotes the group of $K \times K$ unitary matrices. It can be shown that, geometrically, $G(M, N)$ corresponds to the manifold of complex M-planes in C^N. Now, let $g = g(x)$, where $x = (x_1, x_2) \in R^2$, be an element of $U(N)$, which we decompose as

$$g = (z, y), \tag{3.18}$$

where

$$z = (z_1 \ldots z_M), \qquad y = (z_{M+1} \cdots z_N), \tag{3.19}$$

in which each z_i $(i = 1 \ldots N)$ is an N-component column vector. Then from the unitarity of g we find that the vectors z_i are orthonormal:

$$z_i^\dagger z_k = \delta_{ik}. \tag{3.20}$$

The grassmannian models can be defined in terms of the $N \times M$ matrix z (the analogue of our ϕ^a vector of the $O(N)$ sigma models), considered as a dynamical variable, together with the constraint

$$z^\dagger z = 1_M, \tag{3.21}$$

where 1_M denotes the $M \times M$ unit matrix and † the hermitian conjugation. The Lagrangian density and the action of the model are defined as

$$L = \mathrm{Tr}\,(D_\mu z)^\dagger D_\mu z, \quad S = \int \mathrm{d}^2 x L, \tag{3.22}$$

where the covariant derivative is defined by

$$D_\mu \psi = \partial_\mu \psi - \psi A_\mu, \tag{3.23}$$

in which A_μ is a composite gauge field defined by

$$A_\mu = z^\dagger \partial_\mu z, \quad A_\mu^\dagger = -A_\mu. \tag{3.24}$$

The above Lagrangian density is invariant under global $U(N)$ transformations:

$$z \to z' = Kz, \quad K \in U(N), \quad K = \text{constant} \tag{3.25}$$

and also under local $U(M)$ transformations:

$$z \to z' = zh \text{ where } h = h(x) \in U(M), \tag{3.26}$$

since then

$$\begin{aligned} D_\mu z \to (D_\mu z)' &= (D_\mu z)h, \\ (D_\mu z)'^\dagger &= h^\dagger (D_\mu z)^\dagger. \end{aligned} \tag{3.27}$$

In the special case when $M = 1$, the above model is the $\mathbb{C}P^{N-1}$ model, as the complex Grassmann manifold in this case is equivalent to the complex projective space. The $\mathbb{C}P^{N-1}$ model is therefore described by a complex N-component column vector, together with the normalisation condition $z^\dagger z = |z|^2 = 1$.

The $\mathbb{C}P^{N-1}$ model possesses abelian $U(1)$ symmetry, and its composite gauge field A_μ is a pure imaginary function of our (Euclidean) two-dimensional x. The simplest $\mathbb{C}P^{N-1}$ model corresponds to $N = 2$. Moreover, it can be shown that this model ($\mathbb{C}P^1$) is equivalent to the $O(3)$ σ model mentioned above.

To exhibit this equivalence, it suffices to take

$$\phi^i = z^\dagger \sigma_i z, \tag{3.28}$$

where z is now a two-component $\mathbb{C}P^1$ field, and σ_i are the familiar Pauli matrices. Then

$$\begin{aligned} (D_\mu z)^\dagger D_\mu z &\to \partial_\mu \phi_i \, \partial_\mu \phi_i, \\ \text{and} \qquad |z|^2 = 1 &\to \phi_i \, \phi_i = 1. \end{aligned} \tag{3.29}$$

However, it can be shown that no similar relation holds for $N \neq 2$. It can be proved that our grassmannian model, to which a fermionic contribution can be added, possesses all the properties mentioned at the end of the last section.

Let us now consider the problem of finding finite action solutions of the Euler–Lagrange equations of our grassmannian model. These equations are given by

$$D_\mu D_\mu z + z\,(D_\mu z)^\dagger\, D_\mu z \;=\; 0, \tag{3.30}$$

together with the constraint $z^\dagger z = 1$, and we see that (by analogy with the gauge theory case) these equations, considered as equations for z, are highly nonlinear. However, as in the gauge theory case, we can obtain a class of solutions of these equations by seeking solutions of the simpler equations, namely:

$$D_\mu z \;=\; \pm i\epsilon_{\mu\nu}\, D_\nu z, \tag{3.31}$$

(together with the constraint $z^\dagger z = 1$).

It is easy to show that all the solutions of (3.31) are also automatically solutions of (3.30) but, being first order in derivatives, are easier to solve. In fact, as in nonabelian gauge theories, our first-order equations (3.31) are associated with the existence of a topological charge, and can be considered as having come from the requirement that the Lagrangian density L satisfies

$$L \;=\; \pm q, \tag{3.32}$$

where

$$q \;=\; i\,\epsilon_{\mu\nu}\,\partial_\mu\,\mathrm{Tr}\,(z^\dagger \partial_\nu z) \tag{3.33}$$

is again the topological charge density.

The topological charge $Q = \int \mathrm{d}^2x\, q$ divides all fields z into different classes, each characterised by a discrete value associated with the topological properties of the overall phase of z at spatial infinity. In fact, as $x^2 + y^2 \to \infty$, $x/y = \tan\phi$ =constant, $z \to z_0 e^{i\psi(\phi)}$, and Q provides a topological characterisation of the map

$$S^1 \ni \phi \to e^{i\psi(\phi)} \in U(1) \sim S^1. \tag{3.34}$$

Thus, by analogy with

$$F_{\mu\nu} = \pm^* F_{\mu\nu}$$

of nonabelian gauge theories to which they correspond, the equations (3.31) are called 'self-duality' (and 'antiself-duality') equations. Their finite action solutions are, once again, called instantons and anti-instantons respectively.

To obtain a better insight into the properties of grassmannian models it is convenient to change the Euclidean variables (x_1, x_2) to holomorphic and antiholomorphic variables:

$$x_\pm \;=\; x_1 \pm x_2, \tag{3.35}$$

and then rewrite all the expressions in terms of these variables. Thus we find

$$\begin{aligned} L &= 2\,\mathrm{Tr}\,[(D_+z)^\dagger(D_+z) + (D_-z)^\dagger(D_-z)] \\ q &= 2\,\mathrm{Tr}\,[(D_+z)^\dagger(D_+z) - (D_-z)^\dagger(D_-z)], \end{aligned} \tag{3.36}$$

where $D_\pm$ denotes the covariant derivatives in which the differentiation is performed with respect to the $x_\pm$ variables. Since

$$[D_+\,,\,D_-]\psi = \tfrac{1}{2}\psi q, \tag{3.37}$$

we see that the equations of motion can be rewritten as

$$D_-D_+z + z(D_+z)^\dagger(D_+z) = 0, \tag{3.38}$$

or equivalently

$$D_+D_-z + z(D_-z)^\dagger(D_-z) = 0, \tag{3.39}$$

and the equations of self-duality (antiself-duality) are given by

$$D_\mp z = 0. \tag{3.40}$$

The solutions of the self-duality equations in the $\mathbb{C}P^{N-1}$ case have already been given in the original paper by D'Adda *et al* [9]; for the general grassmannian models they were given by Macfarlane [11]. In the $\mathbb{C}P^{N-1}$ model they are given by

$$z = \frac{f}{|f|}, \tag{3.41}$$

where, in the instanton case, the f vector is a function of only x_+ (for anti-instantons, of x_-). The finiteness of the action imposes conditions on the components of f — they have to be rational functions of their argument. However, due to the invariance under an overall factor multiplication, it is sufficient to consider only polynomial components of f (with no overall factors). This was discussed by D'Adda *et al* in their original paper.

By analogy, instanton solutions of the general $G(M,N)$ model are obtained from a set of M linearly independent holomorphic vectors $f_1 \ldots f_M$, properly orthonormalised. This orthonormalisation can be performed as follows: let us denote by $\hat{z}$ the $N \times M$ matrix consisting of $f_1 \ldots f_M$, i.e.

$$\hat{z} = (f_1 \ldots f_M) \tag{3.42}$$

and define an $M \times M$ hermitian matrix $M = \hat{z}^\dagger \hat{z}$. Because of the linear independence of the vectors $f_1 \ldots f_M$, the matrix M is positive definite and invertible. Hence the hermitian square root matrices $M^{1/2}$ and $M^{-1/2}$ exist and are unique. Then it is easy to see that

$$z = \hat{z}(M^{-1/2}) \tag{3.43}$$

satisfies both the instanton equations of motion and the constraint. (For the anti-instantons we interchange x_+ with x_-.)

It is worthwhile to pause at this stage and observe that as work in two-dimensional Euclidean space and our solutions depend only on $x_+ = x + iy$, we can think of them as being defined over a complex plane coordinated by x_+. We can now exploit the conformal invariance of our model. As the conformal group in two dimensions consists of all holomorphic transformations

$$x_+ \rightarrow \zeta = \zeta(x_+), \tag{3.44}$$

we observe that the application of this transformation to any solution of the equations of motion gives a new solution. Thus we can generate new solutions this way. However, calculating the value of the action for these new solutions we can use (3.44) to perform the change of the variables of integration, and find that the resultant expression for the action density of the new solution has the same dependence on ζ as the original solution had on x_+. This may suggest equal values of the actions for both solutions. This suggestion is, however, incorrect. To see this observe that, if the degree of the map (3.44) is, say, k, then this transformation maps the complex plane (and hence our two-dimensional space) onto itself k times and, in consequence, the action of the new solution is k times larger than the original one. Thus we see that, if we want to restrict ourselves to finite action solutions, the degree of the map (3.44) must be finite, and any solution generated by the application of our conformal transformation (3.44) will already be included in our set of solutions, as it will correspond to some choice of our holomorphic vectors $f_1 \ldots f_M$ mentioned above.

It is worth adding at this stage that even though we have argued that the condition of the finiteness of the action turns our two-dimensional space into being effectively equivalent to S^2, we relax this condition here and so think of our two-dimensional space as corresponding to a complex plane. This interpretation allows us to consider transformations (3.44) characterised by the integer values of k, and to connect solutions corresponding to different values of the action.

What are the properties of these instanton solutions? What is the value of their action, and how does this value depend on the properties of $\hat{z}$? We shall not discuss this at this stage; it is more convenient to delay this discussion until we have presented noninstanton solutions as well. However, let us simply state, without any justification at this stage, that, in the $\mathbb{C}P^{N-1}$ case, the value of this action is given by

$$S = 2\pi\alpha_0, \tag{3.45}$$

where α_0 is the highest degree of the polynomial components of f.

We see that, in contradistinction to the gauge theory case, the form of all the solutions to the self-duality equations is very simple and explicit.

This suggests that it may be not too difficult to find finite action solutions of the full equations of motion, which are not solutions of the self-duality equations. This will be shown in the following sections.

Before finishing this section, we present another formulation of the model based on the introduction of an $N \times N$ projection matrix $\mathbb{P}$, defined by

$$\mathbb{P} = zz^{\dagger} = \sum_{i=1}^{M} z_i z_i^{\dagger}, \tag{3.46}$$

where $\mathbb{P}$ is a hermitian projector, and so satisfies

$$\mathbb{P} = \mathbb{P}^{\dagger} = \mathbb{P}^2.$$

In this formulation the Lagrangian density takes the form:

$$L = \mathrm{Tr}\,(\partial_\mu \mathbb{P}\, \partial_\mu \mathbb{P}) \tag{3.47}$$

together with the constraint $\mathbb{P} = \mathbb{P}^2$, and the equations of motion are given by

$$[\partial_\mu \partial_\mu \mathbb{P}\, ,\, \mathbb{P}] = 0. \tag{3.48}$$

Observe that the only difference between the different grassmannian models resides in the rank of $\mathbb{P}$. Finally, the first-order (self-dual) equations, when written in terms of $\mathbb{P}$, are given by

$$\partial_- \mathbb{P}\, \mathbb{P} = 0 \quad \text{and} \quad \mathbb{P}\, \partial_- \mathbb{P} = 0, \tag{3.49}$$

or equivalently

$$\mathbb{P}\, \partial_+ \mathbb{P} = 0 \quad \text{and} \quad \partial_+ \mathbb{P}\, \mathbb{P} = 0. \tag{3.50}$$

This (gauge-invariant) formulation is interesting in its own right, but it is also extremely useful when one tries to prove that a particular field configuration solves the equations of motion. We will use it extensively in the next sections. However, before we do this let us complete this section by pointing out that, as

$$\begin{aligned} \mathrm{Tr}\,(D_+ z)^{\dagger}(D_+ z) &= \mathrm{Tr}\, \mathbb{P}_- \mathbb{P}_+ \mathbb{P} \\ \mathrm{Tr}(D_- z)^{\dagger}(D_- z) &= \mathrm{Tr}\, \mathbb{P}_+ \mathbb{P}_- \mathbb{P}, \end{aligned} \tag{3.51}$$

on a field configuration which satisfies the equations of motion, the Lagrangian and the topological charge densities are given by

$$\begin{aligned} L &= 2\, \mathrm{Tr}\, \mathbb{P}_+ \mathbb{P}_- \\ q &= 2(\mathrm{Tr}\, \mathbb{P}\mathbb{P}_- \mathbb{P}_+ - \mathrm{Tr}\, \mathbb{P}\mathbb{P}_+ \mathbb{P}_-). \end{aligned} \tag{3.52}$$

3.3 General Solutions of the $\mathbb{C}P^{N-1}$ Model

Let us, first of all, consider the $\mathbb{C}P^{N-1}$ model, and derive the general form of the solutions to its equations of motion. Thus we want to solve

$$D_\mu D_\mu z + z[(D_\mu z)^\dagger \cdot D_\mu z] = 0 \tag{3.53}$$

for a vector z_α, $\alpha = 1, \ldots, N$, which is subject to the condition that $|z|^2 = 1$. Moreover, we are interested only in the finite action solutions of these equations, i.e. such z_α that

$$\int [|D_+ z|^2 + |D_- z|^2] \, \mathrm{d}^2 x < \infty. \tag{3.54}$$

To derive the general form of such solutions, first of all, we investigate some of the properties they must possess. Assume that $z_\alpha(x_+, x_-)$ is a finite action solution of the equations of motion. Then we will show that

$$A^m_{i,j} \equiv (D^i_- z_\alpha)^\dagger \cdot D^j_+ z_\alpha = 0 \ \text{ for } \ m \equiv i + j \geq 1. \tag{3.55}$$

To prove this we observe that $A^1_{0,1} = A^1_{1,0} = 0$. Moreover, it is easy to show [12] that the finiteness of the action and the conservation of energy momentum gives $A^2_{1,1} = 0$. To do this we observe that, due to the conformal invariance, the energy–momentum tensor is traceless and conserved. When expressed in terms of our x_+ and x_- variables this means that

$$\partial_-((D_- z)^\dagger D_+ z) = 0, \tag{3.56}$$

which shows that $(D_- z)^\dagger D_+ z$ must be analytic. However, when we impose, in addition, the requirement of the finiteness of the action, we find [13] that this analytic function must vanish.

To prove the vanishing of $A^k_{i,j}$, for other values of i and j we use induction. Thus, we assume that $A^m_{i,j} = 0$ for $i \leq i_0, j \leq j_0, m \leq m_0$. Then, as

$$\partial_+(a^\dagger_\alpha \cdot b_\alpha) = (D_- a^\dagger_\alpha \cdot b_\alpha) + (a^\dagger_\alpha \cdot D_+ b_\alpha) \tag{3.57}$$

we see that

$$\begin{aligned} A^{m+1}_{i+1,j} &= (D^{i+1}_- z_\alpha)^\dagger \cdot D^j_+ z_\alpha \\ = \partial_+[(D^i_- z_\alpha)^\dagger \cdot D^j_+ z_\alpha] &- (D^i_- z_\alpha)^\dagger \cdot D^{j+1}_+ z_\alpha = - A^{m+1}_{i,j+1}, \end{aligned} \tag{3.58}$$

and so it is enough to prove that $A^{m+1}_{0,m+1} = 0$. First we show that $A^{m+1}_{0,m+1}$ is analytic, i.e. that it satisfies $\partial_- A^{m+1}_{0,m+1} = 0$. To do this, we write

$$\partial_- A^{m+1}_{0,m+1} = (D_+ z_\alpha)^\dagger \cdot D^{m+1}_+ z_\alpha + z^\dagger_\alpha \cdot D_- D^{m+1}_+ z_\alpha, \tag{3.59}$$

and then observe that the first term can be written as

$$(D_+ z_\alpha)^\dagger \cdot D_+^{m+1} z_\alpha = \partial_+^m A_+. \tag{3.60}$$

To prove this last result we also use induction; the result is clearly correct for $m = 0$ and, in general,

$$\begin{aligned}(D_+ z_\alpha)^\dagger D_+^{m+1} z_\alpha &= \partial_+[(D_+ z_\alpha)^\dagger \cdot D_+^m z_\alpha] - (D_- D_+ z_\alpha)^\dagger \cdot D_+^m z_\alpha \\ &= \partial_+(\partial_+^{m-1} A_+) + A_+ z_\alpha^\dagger \cdot D_+^m z_\alpha = \partial_+^m A_+.\end{aligned} \tag{3.61}$$

For the second term we have

$$\begin{aligned} z_\alpha^\dagger \cdot D_- D_+^{m+1} z_\alpha &= z_\alpha^\dagger \cdot (D_+ D_- - \tfrac{1}{2} q) D_+^m z_\alpha \\ = z_\alpha^\dagger \cdot D_+ D_- D_+^m z_\alpha &= z_\alpha^\dagger \cdot D_+^m D_- D_+ z_\alpha \\ = -z_\alpha^\dagger \cdot D_+^m A_+ z_\alpha &= -\partial_+^m A_+. \end{aligned} \tag{3.62}$$

Hence, $A_{0,m+1}^{m+1}$ is analytic. To show that it vanishes, we have to use a variant of Liouville's theorem. Since

$$|A_{0,m}^m|^2 \leq |D_+^m z_\alpha|^2, \tag{3.63}$$

we see that is suffices to show that $|D_+^m z_\alpha^2|$ belongs to the space of square-integrable functions over R^2 ($L^2(R)$). Of course, from the finiteness of the action, we already know that $|D_+ z_\alpha| \in L^2(R^2)$. To show that $I_m = \int d^2x |D_+^m z_\alpha|^2$ is finite, we observe that, if $z_\alpha(x_+, x_-)$ is a solution, so is $z'_\alpha(x_+, x_-) = z_\alpha(x'_+, x'_-)$, where $x'_+ = 1/x_+$ etc. Then the proof can be completed by repeating the discussion of Borchers and Garber [14] who showed that I_m is the sum of integrals over the insides of the unit circles $|x_+| < 1$ and $|x'_+| < 1$ of real analytic function and so is finite.

Thus, for a given solution z_α, we can construct two subspaces of $\mathbb{C}^N$ defined by

$$\begin{aligned} H &= \{D_-^i z_\alpha, \quad i = 1, 2 \ldots\} \\ H' &= \{D_+^i z_\alpha, \quad i = 1, 2 \ldots\}. \end{aligned} \tag{3.64}$$

Let the dimensions of H and H' be k and m respectively. From the previous discussion we know that these two subspaces are mutually orthogonal, and both are orthogonal to z. Clearly $k + m \leq N$-1. Moreover, it can be shown that the case of $k + m < N - 1$ corresponds to an embedding, so we can restrict ourselves to considering only the case of $k + m = N - 1$. The spaces H_k and H'_m are spanned respectively by the first k and m vectors in the expression above, where we have put subscripts on H and H' to remind us of the dimensions of these spaces.

Next we will find an analytical vector $f \in \mathbb{C}^N$, such that $f \in \hat{H}_k$, where $\hat{H}_k = H_k \cup z$. We define f_α by

$$f_\alpha^\dagger \cdot D_-^i z_\alpha = \omega \delta^{ik}, \qquad i = 0, \ldots k, \tag{3.65}$$

where

$$\partial_+\omega - (z_\alpha^\dagger \cdot \partial_+ z_\alpha)\omega = 0. \tag{3.66}$$

From the definition of f we have

$$f_\alpha = \sum_{i=0}^{k} \mu_i D_-^i z_\alpha. \tag{3.67}$$

On the other hand, we know that, for $i > 0$, we have

$$(\partial_- f_\alpha)^\dagger \cdot D_-^i z_\alpha = \partial_+(f_\alpha^\dagger \cdot D_-^i z_\alpha) - f_\alpha^\dagger \cdot \partial_+(D_-^i z_\alpha). \tag{3.68}$$

Thus, as $D_+ D_-^i z_\alpha$ is a linear combination of $z_\alpha, \ldots D_-^{i-1} z_\alpha$, we see that we can rewrite this expression as

$$(\partial_- f_\alpha)^\dagger \cdot D_-^i z_\alpha = \delta^{ik}[\partial_+\omega - z_\alpha^\dagger \cdot \partial_+ z_\alpha \,\omega] = 0. \tag{3.69}$$

From this we see that, for $i = 0$, we have $f_\alpha^\dagger \cdot z_\alpha = 0$. This implies that $f_\alpha \in H_k$, and so that $\partial_- f_\alpha \in H_k$, thus showing that $(\partial_- f_\alpha)^\dagger \cdot z_\alpha = 0$. Hence, $\partial_- f_\alpha$ is orthogonal to H_k, and so must vanish:

$$\partial_- f_\alpha = 0. \tag{3.70}$$

It is not difficult to show that f_α must be meromorphic; it can be expressed as a rational function of z_α and its derivatives, which are real analytic, and since the solutions are invariant under the conformal transformation $x_+ \to x_+^{-1}$, we see that the components of f must be rational functions of x_+. Having shown that $\hat{H}_k$ is spanned by z and its covariant derivatives, and that it contains f, we will now show that $\hat{H}_k$ is also spanned by $f_\alpha, \partial_+ f_\alpha, \ldots, \partial_+^k f_\alpha$, and that, in consequence, it is possible to express z_α in this basis. This will provide us with an explicit expression for z_α.

To this end, we first show that

$$(\partial_+^l f_\alpha)^\dagger \cdot D_-^i z_\alpha = (-1)^l \omega \delta^{i+l,k}, \qquad 0 \le i + l \le k. \tag{3.71}$$

For $i = l = 0$, this is just $f_\alpha^\dagger \cdot z_\alpha = 0$. Using induction on $i + l$, we find

$$\begin{aligned}(\partial_+^l f_\alpha)^\dagger \cdot D_-^i z_\alpha =& \partial_-[(\partial_+^{l-1} f_\alpha)^\dagger \cdot D_-^i z_\alpha] - (\partial_+^{l-1} f_\alpha)^\dagger \cdot \partial_- D_-^i z_\alpha \\ = -(\partial_+^{l-1} f_\alpha)^\dagger \cdot D_-^{i+1} z_\alpha =& (-1)^l f_\alpha^\dagger \cdot D_-^{i+l} z_\alpha = (-1)^l \omega \delta^{i+l,k}.\end{aligned} \tag{3.72}$$

From this result we see that

$$(\partial_+^l f_\alpha)^\dagger \cdot z_\alpha = (-1)^l \omega \delta^{l,k}, \qquad \text{for } 0 \le l \le k, \tag{3.73}$$

and so

$$\begin{aligned}(\partial_+^l f_\alpha)^\dagger \cdot D_+ z_\alpha &= \partial_+[(\partial_+^l f_\alpha)^\dagger \cdot z_\alpha] - [z_\alpha^\dagger \cdot \partial_+ z_\alpha][(\partial_+^l f_\beta)^\dagger \cdot z_\beta] \\ &= (-1)^l \delta^{l,k}[\partial_+\omega - (z_\alpha^\dagger \cdot \partial_+ z_\alpha)\omega] = 0.\end{aligned} \tag{3.74}$$

Hence, by induction,

$$(\partial_+^l f_\alpha)^\dagger \cdot D_+^i z_\alpha = \partial_+[(\partial_+^l f_\alpha)^\dagger \cdot D_+^{i-1} z_\alpha] - [z_\alpha^\dagger \cdot \partial_+ z_\alpha][(\partial_+^l f_\beta)^\dagger \cdot D_+^{i-1} z_\beta] = 0. \tag{3.75}$$

Thus we have shown that $\partial_+^l f_\alpha$ is orthogonal to H_m', and so that $\partial_+^l f_\alpha \in \hat{H}_k$, for $l = 0, 1, \ldots, k$. As $f_\alpha, \partial_+ f_\alpha, \ldots, \partial_+^k f_\alpha$ are all independent, we see that $\hat{H}_k$ is spanned by them.

We are now ready to express z_α in terms of $f_\alpha, \partial_+ f_\alpha, \ldots, \partial_+^k f_\alpha$. To do this we define a matrix M by

$$M_{l,i} = (\partial_+^l f_\alpha)^\dagger \cdot \partial_+^i f_\alpha, \qquad i, l = 0, \ldots, k-1, \tag{3.76}$$

and then define a vector $\hat{z}_\alpha$ of $\hat{H}_k$ by

$$\hat{z}_\alpha^{(k)} = (-1)^k \left(\partial_+^k f_\alpha - \sum_{i,l=0}^{k-1} M_{i,l}^{-1}\, \partial_+ M_{l,k-1}\, \partial_+^i f_\alpha \right). \tag{3.77}$$

Next, we show that $\hat{z}_\alpha$ is orthogonal to $\partial_+^j f_\alpha$ for $0 \le j \le k-1$. This follows as

$$\begin{aligned}(\partial_+^j f_\alpha)^\dagger \cdot (-1)^k \hat{z}_\alpha^{(k)} &= (\partial_+^j f_\alpha)^\dagger \cdot \partial_+^k f_\alpha - \partial_+ M_{j,k-1} \\ = (\partial_+^2 f_\alpha)^\dagger \cdot \partial_+^k f_\alpha - \partial_+[(\partial_+^j f_\alpha)^\dagger \cdot \partial_+^{k-1} f_\alpha] &= -\,(\partial_- \partial_+^j f_\alpha)^\dagger \cdot \partial_+^{k-1} f_\alpha = 0.\end{aligned} \tag{3.78}$$

Consequently z_α and $\hat{z}_\alpha$ must by proportional to each other, and, since z_α is a unit vector defined up to a phase factor, this means that

$$z_\alpha = \frac{\hat{z}_\alpha}{|\hat{z}|}. \tag{3.79}$$

So far we have not specified ω, which appeared in our definition of f_α. To do this we recall that the gauge freedom $z_\alpha \to z_\alpha e^{i\phi}$ implies an arbitrariness $\omega \to a(x_-)e^{i\phi}\omega$. Choosing $e^{i\phi} = \bar{a}(x_+)/a(x_-)$, we see that this corresponds to $\omega \to \bar{a}(x_+)\omega$, which in turn corresponds to $f_\alpha \to \bar{a}(x_+) f_\alpha$, which, of course, preserves analyticity. Thus, with this particular choice of gauge, we see that

$$(\partial_+^k f_\alpha)^\dagger \cdot z_\alpha = (-1)^k |\hat{z}_\alpha|, \tag{3.80}$$

and so

$$\omega = |\hat{z}_\alpha|, \tag{3.81}$$

which, as is easy to check, satisfies the equation for ω.

Of course we should now check that our z_α does, indeed, satisfy the equations of motion. This can be done in many different ways. The original method [15] relied on proving that $D_+D_-z_\alpha$ was proportional to z_α. This was done by showing that both these vectors belong to $\hat{H}_k$ and both are orthogonal to $\hat{H}_{k-1}$; thus they must be proportional, this statement being equivalent to z_α satisfying the equation of motion.

However, there is another way of thinking about the solutions of the equations of motion. This alternative formulation is based on the Gramm–Schmidt orthogonalisation of various analytical vectors in $\mathbb{C}^N$. Though the completeness of the construction is not so clear in this formulation, most of the calculations which involve either proving that our field configurations solve the equations of motion or determining the properties of our solutions are considerably simpler in this formulation. Moreover, this new formulation generalises to more general grassmannian (i.e. non $\mathbb{C}P^{N-1}$) cases. In this new formulation we observe that, if we consider a general vector $g \in \mathbb{C}^N$, which is nonzero, then we can define a convenient operator P_+ by

$$(P_+g)_\alpha = \partial_+ g_\alpha - \frac{g_\alpha (g_\beta^\dagger \cdot \partial_+ g_\beta)}{|g|^2}. \tag{3.82}$$

Moreover, we can define its repeated action by

$$(P_+^k g)_\alpha = P_+(P_+^{k-1} g_\alpha). \tag{3.83}$$

Then it is a matter of arithmetic to show that our previously defined $\hat{z}_\alpha$ is given by

$$\hat{z}_\alpha^{(k)} = (-1)^k (P_+^k f)_\alpha. \tag{3.84}$$

Hence the P_+ operator can be considered as being a generator of a transformation which takes us from one solution of the equation of motion to another one. Notice that this transformation does not introduce any new parameters. All parameters of the new solutions are determined in terms of parameters of the analytic vector f_α.

To proceed further, we note the following useful properties of $P_+^k f$ (when f is an analytic vector):

$$\begin{aligned} (P_+^k f)^\dagger \cdot P_+^l f &= 0 \qquad \text{if } l \neq k \\ \partial_-(P_+^k f) &= -P_+^{k-1} f \frac{|P_+^k f|^2}{|P_+^{k-1} f|^2} \\ \partial_+ \left(\frac{P_+^{k-1} f}{|P_+^{k-1} f|^2} \right) &= \frac{P_+^k f}{|P_+^{k-1} f|^2} \\ P_+^N f &= 0. \end{aligned} \tag{3.85}$$

These properties either follow directly from the definition or are very easy to prove. In proving them, it is useful to introduce wedge products of f and its derivatives. Thus we define

$$h^{(i)} = f \wedge \partial_+ f \wedge \ldots \wedge \partial_+^i f \qquad i = 0, \ldots N-1, \tag{3.86}$$

and then it is easy to check that

$$P_+^k f \sim (h^{k-1})^\dagger \cdot h^k, \tag{3.87}$$

where the dot product denotes the summation over all the indices of h^{k-1} and all but one of h^k. Thus, if f is analytic, so is $h^{(i)}$, although it is an element of a larger-dimensional space. So we see that we have a possible interpretation of the noninstanton ($k \neq 0$) solutions — they correspond to mixtures of instantons and anti-instantons. Both instantons and anti-instantons are elements of larger-dimensional spaces and are of the special form (3.86). This special form allows their mixture (3.87) to be elements of $\mathbb{C}^N$, corresponding to the solutions of the equations of motion of the $\mathbb{C}P^{N-1}$ model. Thus, (3.87) shows that all nonlinearities of the $\mathbb{C}P^{N-1}$ model are associated with the dimension of its manifold, and, when properly reinterpreted, the equations of motion are simply the Cauchy–Riemann relations for vectors $h^{(i)}$, just as in the instanton case. However, the vectors $h^{(i)}$ have to have a rather specific form (3.86).

The orthogonality properties (3.85) of the $P_+^k f$ vectors show that we can think of them as having been obtained from the sequence of holomorphic vectors

$$f,\, \partial_+ f,\, \partial_+^2 f,\, \ldots, \partial_+^l f, \ldots \tag{3.88}$$

through the Gramm–Schmidt orthogonalisation procedure. Moreover, when these vectors are normalised, they provide solutions of the $\mathbb{C}P^{N-1}$ model equations of motion. Let us denote by

$$e_1,\, e_2,\, \ldots, e_N \tag{3.89}$$

the vectors obtained from the above sequence by the process of the Gramm–Schmidt orthonormalisation. Then, as Sasaki [16] has shown, it is easy to prove that any element of the sequence (3.89) solves the $\mathbb{C}P^{N-1}$ model equations of motion. To do this, it is convenient to use the projector operator formulation of the model discussed before.

Let us take then, say, the jth element of the sequence and consider the corresponding projector $\mathbb{P}$:

$$\mathbb{P} \;=\; e_j\, e_j^\dagger. \tag{3.90}$$

We consider also another projector defined by

$$Q \;=\; \sum_{k=1}^{j-1} e_k\, e_k^\dagger. \tag{3.91}$$

Then, using the relations

$$\partial_- e_l = e_{l-1}\,(e_{l-1}^\dagger\,\partial_- e_l) + e_l(e_l^\dagger\,\partial_- e_l) \tag{3.92}$$

and

$$\partial_+ e_l = e_{l+1}\,(e_{l+1}^\dagger\,\partial_+ e_l) + e_l(e_l^\dagger\,\partial_+ e_l), \tag{3.93}$$

which follow from (3.85), we can establish several identities satisfied by $\mathbb{P}$ and Q. Considering Q as a projector describing an instanton solution of $G(j-1, N)$, we have

$$\partial_- Q \cdot Q = 0. \tag{3.94}$$

In the same way $\mathbb{P} + Q$ describes an instanton solution of $G(j, N)$ and so it satisfies

$$\partial_-(\mathbb{P} + Q)\cdot(\mathbb{P} + Q) = 0. \tag{3.95}$$

However, due to (3.92) and (3.93), we have

$$\partial_- \mathbb{P} \cdot Q = 0, \tag{3.96}$$

which allows us to derive (from the two previous expressions)

$$\partial_- \mathbb{P}\cdot\mathbb{P} + \partial_- Q\cdot\mathbb{P} = 0. \tag{3.97}$$

Moreover, from (3.92) and (3.93), it also follows that

$$\mathbb{P}\cdot\partial_+ Q = \partial_+ Q \tag{3.98}$$

and by hermitian conjugation that

$$\partial_- Q\cdot\mathbb{P} = \partial_- Q, \tag{3.99}$$

which permits us to rewrite (3.97) as

$$\partial_- \mathbb{P}\cdot\mathbb{P} + \partial_- Q = 0. \tag{3.100}$$

If we now take the hermitian conjugate of (3.100)

$$\mathbb{P}\cdot\partial_+ \mathbb{P} + \partial_+ Q = 0 \tag{3.101}$$

and consider the combination of ∂_+ of (3.100) and ∂_- of (3.101), we obtain the desired equation:

$$[\partial_+\partial_- \mathbb{P},\,\mathbb{P}] = 0, \tag{3.102}$$

which completes the proof that the equations of motion are satisfied.

This way of checking whether a given vector satisfies the equations of motion may not appear to be the most convenient choice when $\mathbb{C}P^{N-1}$

models are studied; it is, however, very useful for the more general grassmannian models.

3.4 General Grassmannian Models

When we consider a more general grassmannian model, say $G(M, N)$, we find several classes of solutions [17,18]. In particular, we expect the most generic solutions to depend on at least M arbitrary analytic vectors $f_1, f_2, \ldots, f_M$. From them, by the generalisation of the construction from the previous section, we can construct a large class of solutions. We proceed as follows: we choose integers $k_i, i = 1, \ldots M$, such that $k_1 \geq k_2 \geq \ldots \geq k_M \geq 1$, and such that $\sum_{i=1}^{M} k_i = N$. Then, we consider the N vectors:

$$\partial_+^{l_i} f_i, \qquad l_i = 0, \ldots k_i - 1, \quad i = 1, \ldots M, \tag{3.103}$$

which, except in special cases, will span $\mathbb{C}^N$. We denote these vectors by g_β, $\beta = 1, \ldots N$ choosing a certain order.

Next, we define the subspaces H_β; $H_\beta = \{g_1, \ldots g_\beta\}$ (with $H_0 = \emptyset$ and $H_\beta = \mathbb{C}^N$ for $\beta \geq N$) and perform the Gramm–Schmidt orthonormalisation

$$\begin{aligned} \hat{e}_\beta &= g_\beta - g_\beta \downarrow H_\beta, \qquad \beta = 1, \ldots N \\ e_\beta &= \frac{\hat{e}_\beta}{|\hat{e}_\beta|}, \end{aligned} \tag{3.104}$$

where $\downarrow H$ denotes the projection onto H. Then it is not too difficult to show that the grassmannians $z(\beta)$, $\beta = 1, \ldots N - M + 1$, defined by the orthonormal vectors $e_\beta, \ldots e_{\beta+M-1}$, are solutions of the equations of motion, provided that the vectors $\{e_\beta\}$, and therefore also $\{g_\beta\}$, have been chosen in such an order that, for all β,

$$\partial_+ H_\beta \subset H_{\beta+M}. \tag{3.105}$$

Notice that this last condition allows us, straight away, to generalise our construction a little, as, instead of our vectors of the basis g_β, we can also take their linear combinations (with analytical coefficients), provided that (3.105) is satisfied. Keeping the vectors of the basis g_β, we see that this condition is fulfilled, for example, by the choice:

$$\{g_\beta\} = \{f_1, \ldots f_M, \partial_+ f_1, \partial_+ f_2, \ldots \partial_+ f_M, \partial_+^2 f_1, \ldots\}.$$

Of course, other permutations of this particular order also fulfil our condition (3.105), such as starting the sequence $\{g_\beta\}$ with a certain number of derivative vectors $\partial_+^i f_1$, and then continuing with $f_2, f_3, \ldots$, and further on with $\partial_+^j f_1$, etc. To show that the grassmannian $z(\beta)$ satisfies the equations of motion, we introduce $\mathbb{P}$, the projector on $H_{\beta+M-1}/H_{\beta-1}$, and Q, the projector on $H_{\beta-1}$. Then all we have to do is to repeat all the arguments of the last part of the last section (observing on the way that Q describes an instanton of $G(\beta-1,N)$, while $\mathbb{P}+Q$ describes an instanton of $G(\beta+M-1,N)$).

Having found these solutions of the equations of motion (based on M arbitrary analytic vector functions), we may ask whether there are further solutions of the $G(M,N)$ models. It turns out that there are, and that they come in several different classes. First of all there exist solutions which are determined in terms of a smaller number of arbitrary holomorphic vector functions (with polynomial components) $f_1, \ldots f_K$, with $K < M$. The vectors of the basis $\{g_\beta\}$ are so chosen that $\partial_+ H_\beta \subset H_{\beta+K}$ for any β. Then we can form an ordered set of vectors e_i, constructed in the same way as before, and then partition this ordered set into a sequence of subsets, each of length at least K (with the exception of perhaps the first one). Then, if we take for our grassmannian M vectors, which are given by a collection of such subsets, we find that this grassmannian satisfies the equations of motion of the $G(M,N)$ model. It is very easy to give examples of such special solutions. For example, if we take $K=1$, then all vectors of the basis are given in terms of one vector f. Then $\partial_+ H_\beta \subset H_{\beta+1}$ and any M distinct vectors from the sequence $\{e_\beta\}$ provide a solution. In the case of two holomorphic vectors f_1, f_2 with, say, $\{g_\beta\} = \{f_1, f_2, \partial_+ f_1, \partial_+ f_2, \ldots\}$, i.e. such that $\partial_+ H_\beta \subset H_{\beta+2}$, the sets $\{e_1, e_2, e_6, e_7\}$ or $\{e_3, e_4, e_7, e_8\}$ are solutions corresponding to $M=4$.

To prove that the equations of motion are satisfied, we observe that, if we denote the corresponding projector by $\mathbb{P}$, and the complementary projector by Q (i.e., such that $\mathbb{P}+Q$ is a projector on the subspace spanned by all the vectors of the basis up to the last vector in the sequence defining the grassmannian), we have

$$\begin{aligned} \partial_-(\mathbb{P}+Q)\cdot(\mathbb{P}+Q) &= 0 \\ Q\cdot\mathbb{P} = \mathbb{P}\cdot Q &= 0 \end{aligned} \tag{3.106}$$

and from $\partial_+ H_\beta \subset H_{\beta+K}$ also

$$\begin{aligned} (\mathbb{P}+Q)\cdot\partial_+ Q &= \partial_+ Q \\ \partial_-(\mathbb{P}+Q)\cdot\mathbb{P} &= 0. \end{aligned} \tag{3.107}$$

Thus,

$$\partial_-\mathbb{P}\cdot\mathbb{P} + \partial_- Q - \partial_- Q\cdot Q = 0 \tag{3.108}$$

and we see that

$$[\partial_+\partial_-\mathbb{P}, \mathbb{P}] = [\partial_+\partial_- Q, Q] \tag{3.109}$$

thus showing that $\mathbb{P}$ is a solution, if Q is also one.

To prove that Q is a solution we repeat this procedure, this time treating Q as our previous $\mathbb{P}$ and introducing a complementary Q'. We find

$$[\partial_+\partial_-Q,\ Q] = [\partial_+\partial_-Q',\ Q'], \tag{3.110}$$

thus showing that the problem is reduced to proving that Q' satisfies the equations of motion (of the corresponding grassmannian). It is quite easy to convince oneself that the final Q in this chain of steps describes an instanton of some grassmannian, thus proving that all the intermediate Q's (and $\mathbb{P}$'s) are also solutions of the corresponding equations of motion (notice that Q's and $\mathbb{P}$'s, in general, describe grassmannians corresponding to different M's).

Are there further solutions of the equations of motion? Can we repeat the proof of completeness (given before for the $\mathbb{C}P^{N-1}$ model)? In that proof, we first of all showed that the spaces spanned by $D_+^j z$ and $D_-^i z$ were orthogonal. A few lines of algebra show that, as the covariant derivatives involve nonabelian gauge fields, the previous proof does not go through. Moreover, a glance at the solutions we have found shows that

$$A_{i,j}^m = (D_-^i z)^\dagger \cdot D_+^j z \neq 0 \qquad \text{for } m = i + j \geq 1 \tag{3.111}$$

for some of them. In fact, $A_{i,j}^m$ vanishes for all generic solutions (those depending on M functions), but does not do so for some special solutions. This is true, for example, for solutions based on one vector f (where the size of the 'gap' between subsets is related to m in the equation above, for which $A_{i,j}^m \neq 0$). However, it is easy to check that, although $A_{i,j}^m \neq 0$, their trace vanishes:

$$\operatorname{Tr} A_{i,j}^m = 0 \ \text{ for } m = i + j \geq 1. \tag{3.112}$$

Can we show that this is a consequence of the finiteness of the action? It is easy to show that

$$\operatorname{Tr} A_{i+1,j}^{m+1} = -\operatorname{Tr} A_{i,j+1}^{m+1}, \tag{3.113}$$

but so far we have not succeeded in showing that $\operatorname{Tr} A_{i,j}^m = 0$. Thus it is quite possible that this condition does not have to be satisfied, in which case there may exist further, completely different solutions. In fact, as Wood has recently shown [18], when $M > 2$ there exist solutions which lie outside the class of those discussed above; to obtain them, one has to Gramm–Schmidt orthogonalise our sequence of vectors and then perform some analytic integrations. This last step is quite cumbersome, and so very few properties of these solutions are known. Although it is not very likely, it is still not excluded that these further solutions correspond to some less conventional orderings of our vectors of basis $\{e_\beta\}$. We hope that this question can be resolved in the not-too-distant future.

3.5 Properties of the Solutions of the $G(M,N)$ Models

Let us now discuss some properties of our solutions. We start by evaluating the action given by the integral of

$$L = 2\,\mathrm{Tr}\,[(D_+z)^\dagger D_+z + (D_-z)^\dagger D_-z] \tag{3.114}$$

and the topological charge given by the integral of

$$q = 2\,\mathrm{Tr}\,[(D_+z)^\dagger D_+z - (D_-z)^\dagger D_-z]. \tag{3.115}$$

First of all, we shall prove that

$$\mathrm{Tr}\,[(D_+z(\beta))^\dagger D_+z(\beta)] = \mathrm{Tr}\,[(D_-z(\beta+M))^\dagger D_-z(\beta+M)]. \tag{3.116}$$

To show this we consider

$$\begin{aligned} \mathbb{P} &= z(\beta)z(\beta)^\dagger = \sum_{i=\beta}^{\beta+M-1} e_i e_i^\dagger, \\ \mathbb{P}' &= z(\beta+M)z(\beta+M)^\dagger = \sum_{i=\beta+M}^{\beta+2M-1} e_i e_i^\dagger, \end{aligned} \tag{3.117}$$

and

$$Q = \sum_{i=1}^{\beta-1} e_i e_i^\dagger.$$

Then, it is easy to check

$$\begin{aligned} D_+z &= \partial_+\mathbb{P}\cdot z \\ (D_-z)' &= \partial_-\mathbb{P}'\cdot z', \end{aligned} \tag{3.118}$$

and so we find that the left-hand side of (3.116) is given by

$$l = \mathrm{Tr}\,[(\partial_+\mathbb{P}z)^\dagger\partial_+\mathbb{P}z] = \mathrm{Tr}\,[\mathbb{P}\cdot\partial_-\mathbb{P}\cdot\partial_+\mathbb{P}] = \mathrm{Tr}\,[(\partial_+\mathbb{P}\cdot\mathbb{P})^\dagger\partial_+\mathbb{P}\cdot\mathbb{P}], \tag{3.119}$$

while the right-hand side is given by

$$\begin{aligned} r &= \mathrm{Tr}\,[(\partial_-\mathbb{P}'z')^\dagger\partial_-\mathbb{P}'z'] \\ &= \mathrm{Tr}\,[\mathbb{P}'\cdot\partial_+\mathbb{P}'\cdot\partial_-\mathbb{P}'] = \mathrm{Tr}\,[(\partial_-\mathbb{P}'\cdot\mathbb{P}')^\dagger\partial_-\mathbb{P}'\cdot\mathbb{P}']. \end{aligned} \tag{3.120}$$

However, using the expression (3.100) first applied to $\partial_-\mathbb{P}'\mathbb{P}'$, and then to $\partial_-\mathbb{P}\mathbb{P}$,

$$\begin{aligned}\partial_-\mathbb{P}'\cdot\mathbb{P}' &= -\,\partial_-(\mathbb{P}+Q) = -\partial_-\mathbb{P} - \partial_-Q \\ &= -\,\partial_-\mathbb{P} + \partial_-\mathbb{P}\mathbb{P} = -\mathbb{P}\partial_-\mathbb{P}\end{aligned} \tag{3.121}$$

we find that

$$\begin{aligned}r = \mathrm{Tr}\,[(\mathbb{P}\cdot\partial_-\mathbb{P})^\dagger\mathbb{P}\cdot\partial_-\mathbb{P}] &= \mathrm{Tr}\,[\partial_+\mathbb{P}\cdot\mathbb{P}\cdot\partial_-\mathbb{P}] \\ = \mathrm{Tr}\,[\mathbb{P}\cdot\partial_-\mathbb{P}\cdot\partial_+\mathbb{P}] &= l,\end{aligned} \tag{3.122}$$

thus proving (3.116).

Let us observe that (3.116) in the $\mathbb{C}P^{N-1}$ case is given by

$$\left|D_-\left(\frac{P_+^{k+1}f}{|P_+^{k+1}|}\right)\right|^2 = \left|D_+\left(\frac{P_+^k f}{|P_+^k f|}\right)\right|^2, \tag{3.123}$$

which is easy to prove directly from the properties of the $P_+^k f$. Having proved (3.116), we now show that the problem of calculating the action has been reduced to that of determining the topological charge.

To see this we look first at the $M = 2$ case. Then, introducing the notation

$$|D_\pm z(\beta)|^2 = \mathrm{Tr}\,\{[D_\pm z(\beta)]^\dagger[D_\pm z(\beta)]\} \tag{3.124}$$

we find

$$\begin{aligned}L = |D_+z(\beta)|^2 + |D_-z(\beta)|^2 &= |D_+z(\beta)|^2 - |D_-z(\beta)|^2 \\ +2|D_+z(\beta-2)|^2 - 2|D_-z(\beta-2)|^2 &+ 2|D_+z(\beta-4)|^2 - 2|D_-z(\beta-4)|^2 \\ &+\ldots+2|D_+z(p)|^2,\end{aligned} \tag{3.125}$$

where $p = 2$, if β is even, or $p = 1$, if β is odd. However, as

$$|D_-z(p)|^2 = 0 \tag{3.126}$$

for $p = 1$ or 2, we see that we can rewrite (3.125) as

$$L(\beta) = q(\beta) + 2q(\beta-2) + 2q(\beta-4) + \ldots + 2q(p), \tag{3.127}$$

and in the general case (for arbitrary M)

$$L(\beta) = q(\beta) + 2q(\beta-M) + 2q(\beta-2M) + \ldots + 2q(p), \tag{3.128}$$

where $p = \beta(\mathrm{mod}\ M)$, or M, if $\beta(\mathrm{mod}\ M) = 1$. Of course, in the $\mathbb{C}P^{N-1}$ case we have

$$L(k) = q(k) + 2\sum_{i=1}^{k-1} q(i). \tag{3.129}$$

To calculate the topological charge of the configuration:

$$z = (e_\beta, e_{\beta+1}, e_{\beta+2}, \ldots, e_{\beta+M-1}), \tag{3.130}$$

we observe that, using the cyclic permutation property of the trace, we can write

$$q = 2\,\mathrm{Tr}\,[(D_+z)^\dagger D_+z - (D_-z)^\dagger D_-z] = 2\,\mathrm{Tr}\,[(\partial_+z)^\dagger \partial_+z - (\partial_-z)^\dagger \partial_-z]. \tag{3.131}$$

Then it is a matter of simple (but tedious) algebra to show that

$$q = \sum_{i=\beta}^{\beta+M-1} q_i, \qquad q_i = 2\,\partial_+\partial_-\, \ln|\hat{e}_i|^2. \tag{3.132}$$

As expected, the topological charge density is completely additive. This property allows us to rewrite the Lagrangian density corresponding to the general case as

$$L(\beta) = 2q_1 + 2q_2 + \ldots + 2q_{\beta-1} + q_\beta + q_{\beta+1} + \ldots + q_{\beta+M-1}. \tag{3.133}$$

Next, we calculate the integrated values of the topological charge and of the action. As

$$\partial_+\partial_-\, \ln|\psi|^2 = \tfrac{1}{4}\partial_\mu\partial_\mu\, \ln|\psi|^2, \tag{3.134}$$

the use of the divergence theorem in two dimensions shows that

$$\int \mathrm{d}^2x\, \partial_-\partial_+\, \ln|\psi|^2 = \pi\alpha, \tag{3.135}$$

where $|\psi| \to |x|^\alpha$ as $|x| \to \infty$, and where we have assumed that $|\psi|$ has no singularities (and no zeros) except those at ∞.

To calculate the asymptotic behaviour of $|\hat{e}_i|^2$, it is convenient to use the wedge product formalism discussed before. Using vectors of the basis, we define

$$h^{(i)} = g_1 \wedge g_2 \wedge \ldots \wedge g_i, \tag{3.136}$$

and then

$$\hat{e}_\beta \sim (h^{\beta-1})^\dagger \cdot h^\beta. \tag{3.137}$$

Moreover, it is easy to see that, up to an overall constant factor,

$$|\hat{e}_\beta|^2 \cong \frac{|h^\beta|^2}{|h^{\beta-1}|^2}. \tag{3.138}$$

Thus, we see that if we define

$$\begin{aligned} \alpha_0 &= 0 \\ \alpha_\beta &= \deg\, h^\beta \bmod \text{(overall factors)} \end{aligned} \tag{3.139}$$

we find that

$$Q_i = \int d^2x\, q_i = 2\pi(\alpha_i - \alpha_{i-1}) \tag{3.140}$$

thus showing that the topological charge and the action of our field $z(\beta)$ are given by

$$\begin{aligned} Q &= 2\pi(\alpha_{\beta+M-1} - \alpha_{\beta-1}) \\ S &= 2\pi(\alpha_{\beta+M-1} + \alpha_{\beta-1}). \end{aligned} \tag{3.141}$$

In the $\mathbb{C}P^{N-1}$ case $M = 1$ and we have

$$\begin{aligned} Q &= 2\pi(\alpha_\beta - \alpha_{\beta-1}) \\ S &= 2\pi(\alpha_\beta + \alpha_{\beta-1}). \end{aligned} \tag{3.142}$$

Observe that in this case the vector $h^{(N)}$ (if it has not vanished earlier) is uniquely determined, and so all its x_+ dependence is in the form of an overall factor. Thus $\alpha_N = 0$ and so we see that, for $\beta = N$, we have

$$Q = -S = -2\pi\alpha_{N-1}, \tag{3.143}$$

and so

$$z_\alpha^N = \frac{P_+^{N-1} f_\alpha}{|P_+^{N-1} f|} \tag{3.144}$$

corresponds to anti-instantons. Also it is clear that in the $\mathbb{C}P^1$ case ($N = 2$) there are only two levels — instantons and anti-instantons — and the P_+ operator turns instantons into anti-instantons. For $N > 2$ we have also further levels corresponding to noninstanton solutions. Looking at the values of Q and S we see that the interpretation of the noninstanton solutions as corresponding to mixture states of instantons and anti-instantons is further reinforced by their values of the action and of the topological charge. We see that the mixture configurations are very special; they not only satisfy the equations of motion but also their interaction vanishes.

Moreover, it is easy to see that, given the intepretation of our solutions as mixtures of instantons and anti-instantons, there are some selection rules in the space of all the solutions which prevent certain combinations of instantons and anti-instantons to correspond to a solution. To see this observe that, due to (3.139), $\alpha_1 \leq 2\alpha_0 - 2$ which, in turn, guarantees that we cannot have a solution corresponding to two instantons and one anti-instanton or, for that matter, one of each. The simplest noninstanton solution (i.e. the one of the smallest action) corresponds to two instantons and two anti-instantons. In a similar way one can find other 'not allowed' configurations.

Returning to more general grassmannian models, we recall that our discussion so far has been concerned with generic solutions. What about the other solutions? For the special solutions the situation is very similar, except that we have to pay special attention as to whether the vectors which form z are all adjacent or fall into groups of adjacent vectors separated by 'gaps' . Each group of adjacent vectors (separated by 'gaps') is treated as a separate unit. For its vectors, the results of (3.141) (with appropriately modified indices β and M) still hold. Then the total topological charge and the action for each grassmannian are just sums of such expressions for each group. For the additional solutions of Wood [18] the results are still unknown.

The noninstanton (and nonanti-instanton) solutions are characterised by $|Q| \neq S$, and so for them the usual topological arguments guaranteeing their stability do not apply. Thus they may correspond, not to local minima of the action, but only to its saddle points, and this is indeed the case. They are unstable; there exist modes of fluctuations around them for which the action decreases.

To see this, we take a general solution z, and consider a small fluctuation ϕ around z. As a result of this fluctuation, the z field is modified to

$$z' = z\sqrt{1 - \phi^\dagger \cdot \phi} + \phi, \tag{3.145}$$

where $\phi^\dagger \cdot z = 0$. (We can always take the general fluctuation in this form due to the invariance of the theory under $U(M)$ transformations.) The action for the modified field is given by

$$S' = 2\,\mathrm{Tr} \int \mathrm{d}^2x [(D'_+ z')^\dagger \cdot D'_+ z' + (D'_- z')^\dagger \cdot D'_- z'], \tag{3.146}$$

where $D'_\pm$ are the usual covariant derivatives written in terms of z'. Since the topological charge is invariant under small fluctuations, we can rewrite this as

$$\begin{aligned} S' &= 2\int \mathrm{d}^2x\, q(x) + 4\int \mathrm{d}^2x\ \mathrm{Tr}\,[(D'_- z')^\dagger \cdot D'_- z'] \\ &= S + 4\int \mathrm{d}^2x\, V(\phi), \end{aligned} \tag{3.147}$$

where $V(\phi)$, calculated to second order in the small fluctuation ϕ, is given by

$$\begin{aligned} V(\phi) = {} & \mathrm{Tr}\,[(D_-\phi)^\dagger \cdot D_-\phi] - \mathrm{Tr}\,[\phi^\dagger \cdot \phi (D_- z)^\dagger \cdot D_- z] \\ & - \mathrm{Tr}\,[(z^\dagger \cdot D_-\phi + \phi^\dagger \cdot D_- z)^\dagger \cdot (z^\dagger \cdot D_-\phi + \phi \cdot D_- z)] \end{aligned} \tag{3.148}$$

(or we can rewrite $V(\phi)$ in terms of expressions involving D_+ only).

Let us consider, first of all, the case when z corresponds to an instanton solution. Then $D_- z = 0$, and so we see that

$$\begin{aligned} V(\phi) &= \mathrm{Tr}\,[(D_-\phi)^\dagger \cdot D_-\phi] - \mathrm{Tr}\,[(z^\dagger \cdot D_-\phi)^\dagger \cdot (z^\dagger \cdot D_-\phi)] \\ &= \mathrm{Tr}\,[(H \cdot D_-\phi)^\dagger \cdot H \cdot D_-\phi], \end{aligned} \tag{3.149}$$

where H is a projector $(1 - zz^\dagger)$. Thus,

$$\int \mathrm{d}^2x\, V(\phi) \tag{3.150}$$

represents a non-negative quantity and we see that the instanton solutions correspond to the local minima of the action.

Next we consider a z that describes a noninstanton solution, i.e. such that $D_\pm z \neq 0$ and choose

$$\phi = \epsilon D_+ z, \tag{3.151}$$

where ϵ is a constant complex number. Then we find that

$$\begin{aligned} V = &- |\epsilon|^2\, \mathrm{Tr}\,[(D_+z)^\dagger \cdot D_+z \cdot (D_-z)^\dagger \cdot D_-z] \\ &- |\epsilon|^2\, \mathrm{Tr}\,[(D_-z)^\dagger \cdot D_+z \cdot (D_+z)^\dagger \cdot D_-z], \end{aligned} \tag{3.152}$$

and so is negative definite if neither D_-z nor D_+z vanish. In deriving this result we used the equations of motion and also the property that

$$\mathrm{Tr}\,[(D_-z)^\dagger \cdot D_+z \cdot (D_+z)^\dagger \cdot D_+z] = 0, \tag{3.153}$$

which holds for our solutions. This last result is trivial for the $\mathbb{C}P^{N-1}$ case and for the generic solutions (where not only the trace but also the whole matrix vanishes). For the special cases it is true as well, although this time this result follows from the fact that for our special solutions the following property holds; if an element of $(D_-z)^\dagger D_+z$ is nonzero, the corresponding element of the hermitian matrix $(D_+z)^\dagger D_+z$ vanishes. This guarantees that although $(D_-z)^\dagger \cdot D_+z \cdot (D_+z)^\dagger \cdot D_-z \neq 0$, its trace vanishes. It is interesting that had these matrices vanished, we would have been able to extend the proof of completeness derived for the $\mathbb{C}P^{N-1}$ model to more general grassmannian models.

We see that as S' can be smaller than S, all noninstanton solutions are unstable, i.e. they correspond to the saddle points of the action. Of course, the fluctuation ϕ of (3.151) is not the only negative mode of the action. It is easy to see that, in general, ϵ does not have to be a constant and we can take fluctuations in the direction of the next terms in our basis g_β. Given the polynomial form of f_i, such fluctuations do not take the fields outside their topological sector. Moreover, we can also take the fluctuations in the directions of the vectors of the basis, which precede the vector corresponding to our solution z (then the corresponding ϵ's cannot be constant). However, what is unclear at the moment is the existence of the superposition principle for these fluctuations and their independence. We will have more to say on this, in a general setting, after we have discussed the solutions of the associated Dirac-like background problem. At this stage, however, we can say that there is one case, albeit very special, for which we have complete understanding of not only the negative modes but also the whole fluctuation spectrum. This is the case of the $\mathbb{C}P^2$ embedded $O(3)$ instanton, which we will discuss in more detail in the next chapter.

Let us close this section with the observation that, in the $\mathbb{C}P^{N-1}$ case, in addition to the P_+ operator, we can also define the P_- operator:

$$(P_- g)_\alpha = \partial_- g_\alpha - g_\alpha \frac{g_\beta^\dagger \partial_- g_\beta}{|g|^2}, \tag{3.154}$$

and that the role of this operator is exactly opposite to that of the P_+ operator. As is easy to check,

$$(P_- P_+ g)_\alpha \sim g_\alpha, \tag{3.155}$$

and so, instead of starting with analytic vectors and the use of the P_+ operator, we could have used anti-analytic vectors and applied to them the P_- operator. Thus we see that the emerging picture is that of N levels of solutions (for $\mathbb{C}P^{N-1}$ model) characterised by different values of the topological charge, but also by the same number of zero modes of the fluctuation operator (number of parameters on which each solution depends).

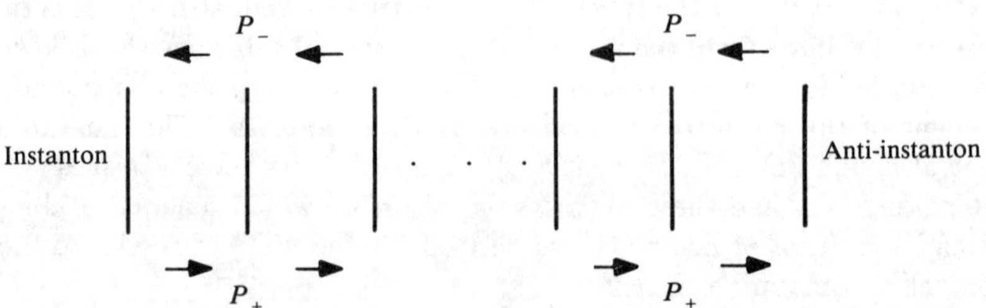

The two extreme solutions correspond to instantons and anti-instantons respectively and the operators P_+ and P_- allow us to move from one solution to the other. If we take a general f_α, given by, say, polynomials of degree N in x_+ (in the case of $\mathbb{C}P^{N-1}$), the coefficients of which are sufficiently arbitrary that no degeneracies occur, then it is easy to see that the topological charge decreases by 2 with each application of the P_+ operator.

Moreover, the solution in the middle of the sequence for N odd, or the two solutions in the middle for N even, have the largest value of the action and correspond to the largest number of instantons and anti-instantons. This suggests that there may exist a group theoretical explanation for these properties.

References

[1] Bogomolnyi E 1976 On stability of classical solutions *Sov. J. Nucl. Phys.* **24** 449

[2] Atiyah M F, Drinfeld V G, Hitchin N G and Manin Yu I 1978 Construction of instantons *Phys. Lett.* **65A** 185

[3] Belavin A, Polyakov A, Schwartz A and Tyupkin Y 1975 Pseudoparticle solutions of the Yang–Mills equations *Phys. Lett.* **59B** 85

[4] 't Hooft G 1976 Computation of the quantum effects due to a four dimensional pseudoparticle *Phys. Rev.* D **14** 3432; Erratum 1978 *Phys. Rev.* D **18** 2199

Ore F 1977 How to compute determinants exactly *Phys. Rev.* D **16** 2577

[5] Berg B and Luscher L 1979 Computations of quantum fluctuations of quark fields in an arbitrary Yang–Mills instanton background *Nucl. Phys.* B **160** 281

[6] Ramond P 1981 *Field Theory: a Modern Primer* (New York: Benjamin-Cummings)

[7] Golo V and Perelomov A 1978 Solution of the duality equations for the two-dimensional $SU(N)$ invariant chiral model *Phys. Lett.* **79B** 112

[8] Eichenherr H 1978 $SU(N)$-invariant non-linear σ models *Nucl. Phys.* B **146** 215

Eichenherr H and Forger M 1979 On the dual symmetry of the non-linear σ models *Nucl. Phys.* B **155** 381

[9] D'Adda A, Di Vecchia P and Luscher M 1978 A $1/N$ expandable series of non-linear σ models with instantons *Nucl. Phys.* B **146** 63

[10] Gell-Mann M and Levy M 1960 The axial vector current in beta decay *Nuovo Cimento* **16** 705

[11] Macfarlane A J 1979 Generalization of σ models and $\mathbb{C}P^N$ models, and instantons *Phys. Lett.* **82B** 239

[12] Din A M and Zakrzewski W J 1980 Embeddings of classical solutions of O_{2p+1} nonlinear σ models in $\mathbb{C}P^{N-1}$ model *Lett. Nuovo Cimento* **28** 121

[13] Woo G 1977 Pseudoparticle configurations in two dimensional ferromagnets *J. Math. Phys.* **18** 1264

[14] Borchers M J and Garber W D 1980 Local theory of solutions for the $O(2k+1)\,\sigma$ model *Commun. Math. Phys.* **72** 77

[15] Din A M and Zakrzewski W J 1980 General classical solutions of the $\mathbb{C}P^{N-1}$ model *Nucl. Phys.* B **174** 397; 1980 Properties of the general classical solutions of the $\mathbb{C}P^{N-1}$ model *Phys. Lett.* **95B** 419

[16] Sasaki R 1983 General classical solutions of the complex Grassmannian and $\mathbb{C}P^{N-1}$ σ models *Phys. Lett.* **130B** 69

[17] Din A M and Zakrzewski W J 1984 Further properties of classical solutions in Grassmannian σ models *Nucl. Phys.* B **237** 461

[18] Wood J C 1988 The explicit construction and parametrization of all harmonic maps from the two-sphere to a complex Grassmannian *J. Reine Angew. Math.* **386** 1

4 Quantum Fluctuations

4.1 The Case of Instantons

Having found all the finite action solutions of the $\mathbb{C}P^{N-1}$ model, we should now proceed with our programme and calculate the effects of quantum fluctuations around these solutions, i.e. perform the appropriate functional integrations. However, this programme turns to be very difficult to implement, and it is only in the case of instanton solutions that such calculations have been performed, and even then, only of the fluctuation determinant. These calculations have been done by many people, with the most complete work being that of Berg and Lüscher [1] and that of Fateev *et al* [2]. Their work, though very beautiful, is complicated due to the generality of the problem. However, if we restrict ourselves to fluctuations around one instanton in the $\mathbb{C}P^{N-1}$ model, or multi-instantons in the $\mathbb{C}P^1$ case, the problem simplifies and the calculations are more tractable. To calculate the effects of one instanton in $\mathbb{C}P^{N-1}$ we could follow the original procedure of 't Hooft [3], but more insight is gained by looking at the fluctuations in the $\mathbb{C}P^1$ model. We shall therefore discuss the effects of the fluctuations around the instanton solutions of the $\mathbb{C}P^1$ model, and in particular we shall follow closely the discussion given by Corrigan and Goddard in their Montreal lectures [4].

As we consider the $\mathbb{C}P^1$ model, f_α has only two components. However, using an overall gauge transformation, we can set one of them to 1, and so we can rewrite this field configuration in an equivalent way given by

$$z = \frac{1}{\sqrt{1+|f|^2}}\begin{pmatrix}1\\ f\end{pmatrix}. \tag{4.1}$$

We are interested in calculating various Green functions of the theory, i.e. quantities like

$$Z = \frac{1}{Z_0}\int \mathrm{D}[\phi_i]\Phi(\phi_i)\,\exp\left(-\frac{1}{g^2}S(\phi_i)\right), \tag{4.2}$$

where we have introduced the arbitrary coupling constant g, and where we have temporarily changed to the more familiar $O(3)$ notation, in which

$$\phi_1 + i\phi_2 = \frac{2f}{1+|f|^2}, \qquad \phi_3 = \frac{1-|f|^2}{1+|f|^2}. \tag{4.3}$$

In this formulation the Lagrangian density is given by

$$L = \tfrac{1}{2}(\partial_\mu \phi_i)^2 = 2\frac{\partial_\mu f^\dagger \partial_\mu f}{[1+|f|^2]^2} = 4\frac{[|\partial_+ f|^2 + |\partial_- f|^2]}{[1+|f|^2]^2}. \tag{4.4}$$

A classical solution corresponding to k instantons is given by

$$f = f_0 = c\,\frac{\prod_{i=1}^k (x_+ - a_i)}{\prod_{i=1}^k (x_+ - b_i)}, \tag{4.5}$$

where a_i, b_i and c are constant complex numbers. We see that a general k-instanton configuration is parametrised by $4k+2$ real parameters.

Next we consider functional integration over fields near this configuration. Using the topological properties of Q, we observe that, near the k-instanton configuration, $S = \int \mathrm{d}^2x\, L$ is given by

$$S = 2\pi k + 8\int \mathrm{d}^2x\, \frac{|\partial_- f|^2}{[1+|f|^2]^2}, \tag{4.6}$$

as Q is a topological invariant. Moreover, the integration measure becomes

$$\mathrm{D}[\phi_i] = \prod_x \mathrm{d}^3\phi_i\, \delta[(\underline{\phi})^2 - 1] = \prod_x \frac{\mathrm{d}^2 f}{[1+|f|^2]^2} \equiv \mathrm{d}[f], \tag{4.7}$$

and so we get $Z = \sum_k Z_k$, where

$$Z_k = \frac{1}{Z_0}\sum_k \int \mathrm{d}[f]_k \Phi(f) \exp\left(-\frac{2\pi k}{g^2} + \frac{8}{g^2}\int \mathrm{d}^2x\, \frac{|\partial_- f|^2}{[1+|f|^2]^2}\right), \tag{4.8}$$

where the subscript k on $\mathrm{d}[f]$ denotes that the integration is to be performed around the solution (4.5).

Next we consider the measure of our integration, and look at the collective coordinates associated with the zero modes of the fluctuation operator. This, as we recall from the previous chapters, leads to the change of the integration over these modes to the integration over the appropriate collective coordinates, which in our case correspond to our parameters a_i, b_i and c, which we generically call p_i. Thus, repeating the procedure discussed in

Appendix A, we find that the Jacobian of the changes of variables is given by $\det N$, where the matrix N is given by

$$N_{i,j} = \int d^2x \left(\frac{\partial f_0}{\partial p_i}^{\dagger} \cdot \frac{\partial f_0}{\partial p_j} \right) \frac{1}{[1+|f_0|^2]^2}, \tag{4.9}$$

and where f_0 describes the k-instanton solution (4.5). To calculate $N_{i,j}$ we can proceed directly or indirectly. We choose the latter method and observe that, by writing

$$f_0 = c \frac{\prod_{i=1}^{k}(x_+ - a_i)}{\prod_{i=1}^{k}(x_+ - b_i)} = \frac{n}{d}, \tag{4.10}$$

where, of course, each n and d are polynomials of degree k in x_+, we have

$$d^2 \frac{\partial f_0}{\partial p_i} = \sum_{l=1}^{2k+1} R_{i,l}\, x_+^{l-1} \tag{4.11}$$

for some coefficients $R_{i,l}$. In fact this last expression serves as a definition of the matrix R. Then a fair amount of relatively simple (but rather tedious) algebra shows that

$$\det R = \text{constant} \times c^{2k} \prod_{i<j}(b_i - b_j) \prod_{i<j}(a_i - a_j) \prod_{i,j}(a_i - b_j). \tag{4.12}$$

Instead of giving a proof of this result, let us check that this result is correct in the one-instanton case ($k = 1$). Then

$$f_0 = c\, \frac{x_+ - a}{x_+ - b}, \tag{4.13}$$

and we find $d^2 \partial f_0 / \partial a = -c(x_+ - b)$, $d^2 \partial f_0 / \partial b = c(x_+ - a)$ and $d^2 \partial f_0 / \partial c = (x_+ - a)(x_+ - b)$. As a result

$$R = \begin{pmatrix} bc & -c & 0 \\ -ac & c & 0 \\ ab & -a-b & 1 \end{pmatrix} \tag{4.14}$$

and so $\det R = c^2(b-a)$, in agreement with our general result.

Returning to the general case, we see that

$$\det N = |\det R|^2 \det M, \tag{4.15}$$

where

$$M_{i,j} = \int d^2x \frac{(x_-)^{i-1}(x_+)^{j-1}}{[|n|^2 + |d|^2]^2}. \tag{4.16}$$

Fortunately we do not have to evaluate $M_{i,j}$, as det M will cancel exactly with a contribution from the functional integral over fluctuations away from the instantons (i.e. nonzero modes). Next we calculate the determinant of the fluctuations over these nonzero modes. To do this we introduce

$$\varrho = |d|^2 + |n|^2, \tag{4.17}$$

and write

$$f = f_0 + \frac{\varrho}{d^2}\varphi. \tag{4.18}$$

Then

$$\frac{df\, df^\dagger}{[1+|f|^2]^2} = d\varphi d\varphi^\dagger, \tag{4.19}$$

and so we see that the measure has become quite simple and that the fluctuation operator, whose determinant we have to compute, is given by

$$Q = \varrho\partial_+\varrho^{-2}\partial_-\varrho. \tag{4.20}$$

To see this last point, we put the expression for f (4.18) into the expression for the action density, and obtain

$$\begin{aligned}\frac{|(1/d^2)\partial_-(\varrho\varphi)|^2}{[1+|n/d+\varrho\varphi/d^2|^2]^2} &= \frac{|\partial_-(\varrho\varphi)|^2}{[|d|^2+|n|^2]^2} + \text{higher order in } \varphi \\ &= \varrho\varphi^\dagger\partial_+\,\varrho^{-2}\,\partial_-\,\varrho\varphi,\end{aligned}$$

where we have kept the quadratic terms in φ. Thus we see that Z_k is given by

$$Z_k = \frac{1}{Z_0}\left(\frac{8\mu^2}{\pi g^2}\right)^{N(k)}\frac{1}{(k!)^2}\int\prod_i \mathrm{d}^2 p_i(\det\, N)\Phi e^{-2\pi k/g^2}[\det'(-Q/\mu^2)]^{-1}, \tag{4.21}$$

where we have inserted a mass-like parameter μ in order to soak up the dimension of Q, where $N(k) = 4k+2$, and where $1/(k!)^2$ appears due to the identity of all the a_i and b_i. So we are left, finally, with having to calculate

$$[\det'(-Q/\mu^2)]^{-1}. \tag{4.22}$$

We shall do this indirectly using the ζ function regularisation method. The reason we cannot perform our calculations directly and need some regularisation scheme comes from the fact that the functional determinant, thought of as being given by the product of all the eigenvalues of the fluctuation operator, diverges (there is an infinite number of arbitrarily large eigenvalues). Thus some regularisation has to be performed. Various regularisation schemes have been proposed and used; of these we will use the scheme first proposed by Hawking [5] and Dowker and Critchley [6].

This scheme has become quite popular recently, and a good introduction to its use can be found in Ramond's book on field theory [7]. However, for completeness, we repeat some definitions and quote some useful results in Appendix B. In this appendix we show that the determinant of any operator Q can be derived from the appropriate zeta function ζ_Q, and that the form of this zeta function is given by

$$\zeta_Q(s) = \frac{1}{\Gamma(s)} \int_0^\infty \mathrm{d}t\, t^{s-1} \int \mathrm{d}^2x\, G(x,x,t), \tag{4.23}$$

where $G(x,y,t)$ is the appropriate Green function of the operator Q, and is the generalisation of $\exp(Qt)$, when Q is a matrix operator. We are, of course, dealing with the differential operator Q (given by (4.20)), but will use the matrix notation $\exp(-Qt)$. In addition, in our case we are interested only in $\det' Q$, and so we want to project out the zero modes of Q. Thus we want to study

$$\zeta_Q(s) = \frac{1}{\Gamma(s)} \int_0^\infty \mathrm{d}t\, t^{s-1}\, \mathrm{Tr}\,[e^{Qt}(1-\Pi)], \tag{4.24}$$

where Π is the projector onto the zero modes of Q.

The direct calculation of $\zeta_Q(s)$ turns out to be quite complicated, so we adopt a different procedure. Instead of calculating $\zeta_Q(s)$ directly, let us, first of all, calculate its variation with respect to the variation of the parameters of the instantons. Thus we have to consider the variation of Q. We have

$$\delta Q = \delta(\varrho\partial_+\varrho^{-2}\partial_-\varrho) = \delta\varrho\partial_+\varrho^{-2}\partial_-\varrho + \varrho\partial_+\varrho^{-2}\partial_-\delta\varrho - 2\varrho\partial_+\varrho^{-3}\delta\varrho\partial_-\varrho. \tag{4.25}$$

Hence

$$\begin{aligned}\mathrm{Tr}\,[\delta Q e^{Qt}] &= \mathrm{Tr}\,[\delta\ln\varrho Q e^{Qt} + Q\delta\ln\varrho e^{Qt} - 2\varrho\partial_+\varrho^{-1}\delta\ln\varrho\varrho^{-1}\partial_-\varrho e^{Qt}]\\ &= 2\,\mathrm{Tr}\,[\delta\ln\varrho Q e^{Qt}] - 2\,\mathrm{Tr}\,[\delta\ln\varrho\varrho^{-1}\partial_-\varrho e^{Qt}\varrho\partial_+\varrho^{-1}].\end{aligned} \tag{4.26}$$

But

$$\begin{aligned}Q\varrho\partial_+\varrho^{-1} &= \varrho\partial_+\varrho^{-2}\partial_-\varrho^2\partial_+\varrho^{-1}\\ &= \varrho\partial_+\varrho^{-1}[\varrho^{-1}\partial_-\varrho^2\partial_+\varrho^{-1}] = \varrho\partial_+\varrho^{-1}\hat{Q},\end{aligned} \tag{4.27}$$

where

$$\hat{Q} = \varrho^{-1}\partial_-\varrho^2\partial_+\varrho^{-1}, \tag{4.28}$$

so we see that we get

$$2\,\mathrm{Tr}\,[\delta\ln\varrho\, Q\, e^{Qt} - \delta\,\ln\varrho\,\hat{Q}\, e^{\hat{Q}t}]. \tag{4.29}$$

In the discussion above we have ignored the projector $(1-\Pi)$ — projecting out the zero modes. If we reinsert it, we find that the variation of our zeta function is given by

$$\delta\zeta_Q(s) = \frac{2}{\Gamma(s)}\int_0^\infty \mathrm{d}t\, t^s \frac{\mathrm{d}}{\mathrm{d}t}\,\mathrm{Tr}\,[\delta \ln \varrho\{e^{Qt}(1-\Pi) - e^{\hat{Q}t}\}]. \tag{4.30}$$

We can now integrate by parts, and obtain

$$\delta\zeta_Q(s) = -\frac{2s}{\Gamma(s)}\int_0^\infty \mathrm{d}t\, t^{s-1}\,\mathrm{Tr}\,[\delta \ln \varrho(e^{Qt}\{1-\Pi\} - e^{\hat{Q}t})]. \tag{4.31}$$

Looking at the definition of $\zeta_Q(s)$ and the expression above, we see that

$$\zeta_Q(0) = \mathrm{Res}_{s=0}\int_0^\infty \mathrm{d}t\, t^{s-1}\,\mathrm{Tr}\,[e^{Qt}(1-\Pi)], \tag{4.32}$$

and

$$\delta\zeta_Q'(0) = -2\mathrm{Res}_{s=0}\int_0^\infty \mathrm{d}t\, t^{s-1}\,\mathrm{Tr}\,[\delta \ln \varrho(e^{Qt}\{1-\Pi\} - e^{\hat{Q}t})]. \tag{4.33}$$

What are our Green functions e^{Qt} and $e^{\hat{Q}t}$? Of course, we do not know them explicitly, but we can always represent them as

$$e^{Qt} = G(x,y,t) = \frac{e^{-|x-y|^2/t}}{\pi t}[a_0(x,y) + a_1(x,y)t + \ldots] \tag{4.34}$$

and

$$e^{\hat{Q}t} = \hat{G}(x,y,t) = \frac{e^{-|x-y|^2/t}}{\pi t}[\hat{a}_0(x,y) + \hat{a}_1(x,y)t + \ldots]. \tag{4.35}$$

This representation follows from the fact that when $\varrho = 1$ our Green functions reduce to each other and satisfy

$$\frac{\partial}{\partial t}G = \partial_+\partial_- G. \tag{4.36}$$

The solution of this equation, which satisfies our boundary condition, is well known [7]:

$$G = G_0 = \frac{1}{\pi t}e^{-|x-y|^2/t}. \tag{4.37}$$

So, in the case when $\varrho \neq 1$, we can seek the Green functions in the form (4.34) and (4.35). Moreover, the unknown functions $a_0(x,y)$ and $\hat{a}_0(x,y)$

must satisfy $a_0(x,x) = \hat{a}_0(x,x) = 1$. Looking at our expressions for the $\zeta_Q(0)$ and $\delta\zeta'_Q(0)$, we see that we need to calculate $a_1(x,x)$ and $\hat{a}_1(x,x)$ as

$$\zeta_Q(0) = \frac{1}{\pi}\int \mathrm{d}^2x\, a_1(x,x)(1-\Pi), \tag{4.38}$$

and

$$\delta\zeta'_Q(0) = \frac{2}{\pi}\int \mathrm{d}^2x\, \delta\ln\varrho[a_1(x,x)(1-\Pi) - \hat{a}_1(x,x)]. \tag{4.39}$$

The calculation of $a_1(x,x)$ and $\hat{a}_1(x,x)$ is given in Appendix C, where it is shown that

$$a_1(x,x) = -\hat{a}_1(x,x) = -\partial_+\partial_- \ln\varrho = -\tfrac{1}{4}\partial^2\ln\varrho. \tag{4.40}$$

Thus we see that

$$\zeta_Q(0) = \frac{1}{\pi}\int \mathrm{d}^2x \tfrac{1}{4}\partial^2\ln\varrho - N(k) = k - N(k), \tag{4.41}$$

where $N(k)$ denote the number of zero modes, and

$$\delta\zeta'_Q(0) = -\frac{1}{\pi}\int \mathrm{d}^2x\,\delta\ln\varrho\,\partial^2\ln\varrho + 2\int \mathrm{d}^2x\,\delta\ln\varrho\,\Pi(x,x). \tag{4.42}$$

We call the two terms in (4.42) δI_1 and δI_2 respectively.

We consider, first, δI_2:

$$\delta I_2 = 2\int \mathrm{d}^2x\,\delta\ln\varrho\,\Pi(x,x). \tag{4.43}$$

In what follows, for simplicity, we shall assume that $c = 1$ in our expression for f_0. First of all we have to determine the form of $\Pi(x,x)$. We recall that

$$V_l = \frac{d^2}{\varrho}\frac{\partial f_0}{\partial p_l} = \frac{1}{\varrho}\sum_{m=1}^{2k+1} R_{lm}\, x_+^{m-1}, \tag{4.44}$$

and so

$$\int \mathrm{d}^2x\, V_l^* V_n = \sum_{mr} R^*_{lm}\, M_{mr}\, R_{nr}, \tag{4.45}$$

where

$$M_{mr} = \int \mathrm{d}^2x\,\frac{x_-^{m-1}\,x_+^{r-1}}{\varrho^2}. \tag{4.46}$$

Thus

$$N_{ln} = (R^*\, M\, R^T)_{ln}, \tag{4.47}$$

and so we have

$$\begin{aligned}\Pi(x,y) &= \sum_{l,n} V_l(x)\,(R^* M R^T)^{-1}_{ln}\, V^*_n(y) \\ &= \sum V(x)\,(R^T)^{-1}\, M^{-1}\,(R^*)^{-1}\, V(y).\end{aligned} \tag{4.48}$$

However, as

$$\sum_l V_l\,(R^T)^{-1}_{lm} = \frac{1}{\varrho}\, x_+^{m-1}, \tag{4.49}$$

we see that

$$\Pi(x,x) = \sum_{mr} \frac{1}{\varrho^2}\,(x_+)^{m-1}\,(M^{-1})_{mr}\,(x_-)^{r-1}, \tag{4.50}$$

and so

$$\begin{aligned}\int \mathrm{d}^2x\,\delta\ln\varrho\,\Pi(x,x) &= \sum_{m,r}\int \mathrm{d}^2x\,\frac{\delta\varrho}{\varrho^3}\,(x_+)^{m-1}(M^{-1})_{mr}(x_-)^{r-1} \\ -\tfrac{1}{2}\sum_{m,r}\delta\, M_{rm}(M^{-1})_{mr} &= -\tfrac{1}{2}\,\delta\,\mathrm{Tr}\ln M,\end{aligned} \tag{4.51}$$

thus showing that

$$\delta I_2 = -\delta\ \ln\det M. \tag{4.52}$$

Next we calculate

$$\delta I_1 = -\frac{1}{\pi}\int \mathrm{d}^2x\,\delta\ln\varrho\,\partial^2\ln\varrho, \tag{4.53}$$

which we rewrite as

$$\int \mathrm{d}^2x\,\delta\ln\varrho\,\partial^2\ln\varrho = \delta\int \mathrm{d}^2x\ \ln\varrho\,\partial^2\ln\varrho - \int \mathrm{d}^2x\ \ln\varrho\,\partial^2\delta\ln\varrho. \tag{4.54}$$

We then rewrite the second term as

$$\int \mathrm{d}^2x\ \ln\varrho\,\partial^2\delta\ln\varrho = \int \mathrm{d}^2x\,\partial_\mu[\ln\varrho\,\partial_\mu\delta\ln\varrho] - \int \mathrm{d}^2x\,\partial_\mu\ln\varrho\,\partial_\mu\delta\ln\varrho$$

$$= -\int \mathrm{d}^2x\,\partial_\mu[\partial_\mu\ln\varrho\,\delta\ln\varrho] + \int \mathrm{d}^2x\partial^2\ln\varrho\,\delta\ln\varrho = \int \mathrm{d}^2x\partial^2\ln\varrho\,\delta\ln\varrho,$$

where we have dropped the surface terms, as the variation is assumed to vanish at infinity. So we find that

$$\int \mathrm{d}^2x\,\delta\ln\varrho\,\partial^2\ln\varrho = \tfrac{1}{2}\delta\int \mathrm{d}^2x\ \ln\varrho\,\partial^2\ln\varrho. \tag{4.55}$$

Next we calculate

$$\int \mathrm{d}^2x \, \ln \varrho \, \partial^2 \ln \varrho. \tag{4.56}$$

As

$$\varrho = \prod_{i=1}^{k} |x_+ - a_i|^2 + \prod_{i=1}^{k} |x_+ - b_i|^2, \tag{4.57}$$

we split the expression as follows. We write

$$\ln \varrho = \ln[1 + |f_0|^2] + \ln \prod_{i=1}^{k} |x_+ - b_i|^2 \tag{4.58}$$

and set

$$\begin{aligned} \ln \varrho \, \partial^2 \ln \varrho &= \ln(1 + |f_0|^2) \, \partial^2 \ln \varrho + \ln \varrho \, \partial^2 \ln \prod_{i=1}^{k} |x_+ - b_i|^2 \\ &+ \ln \prod_{i=1}^{k} |x_+ - b_i|^2 \, \partial^2 \ln \varrho - \ln \varrho \, \partial^2 \ln \prod_{i=1}^{k} |x_+ - b_i|^2, \end{aligned} \tag{4.59}$$

and proceed to calculate all three contributions separately (treating the last two terms as one contribution) .

Let us look first at the last contribution. The two terms that form it, together, represent a total divergence:

$$\partial_\mu \left(\ln \prod_{i=1}^{k} |x_+ - b_i|^2 \, \partial_\mu \ln \varrho - \ln \varrho \, \partial_\mu \ln \prod_{i=1}^{k} |x_+ - b_i|^2 \right), \tag{4.60}$$

and so using the divergence theorem we find that

$$\int \mathrm{d}^2x \, \partial_\mu \left(\ln \prod_{i=1}^{k} |x_+ - b_i|^2 \, \partial_\mu \ln \varrho - \ln \varrho \, \partial_\mu \ln \prod_{i=1}^{k} |x_+ - b_i|^2 \right)$$
$$= -2\pi \, 2k \ln 2.$$

Next we consider the second contribution in (4.59). As

$$\ln \left| \prod_{i=1}^{k} (x_+ - b_i) \right|^2 = \sum_{i=1}^{k} \ln |x_+ - b_i|^2,$$

and

$$\partial^2 \ln |x_+ - b_i|^2 = 2\pi\delta(|x_+ - b_i|^2), \tag{4.61}$$

we find that

$$\int \mathrm{d}^2x \, \ln \varrho \, 2\pi \, \delta(|x_+ - b_i|^2) = 2\pi \, \ln[|b_i - a_i|^2], \tag{4.62}$$

and so we obtain

$$\sum_{i=1}^{k} 2\pi \ln |b_i - a_i|^2. \tag{4.63}$$

Finally we look at the first contribution in (4.59). We observe that

$$\partial^2 \ln \varrho = 4 \frac{1}{[1+|f_0|^2]^2} \partial_+ f_0 \partial_- f_0^*, \tag{4.64}$$

which is, in fact, the action density of our k-instanton configuration. So we have

$$\int \mathrm{d}^2x \, \ln[1+|f_0|^2] \frac{\partial_+ f_0 \partial_- f_0^*}{[1+|f_0|^2]^2}. \tag{4.65}$$

We observe that, if we now change our variables of integration from x_+ to $f_0(x_+)$ (and from x_- to $f_0^*(x_-)$), we obtain

$$k \int \mathrm{d}f_0 \, \mathrm{d}f_0^* \frac{\ln[1+|f_0|^2]}{[1+|f_0|^2]^2}, \tag{4.66}$$

where the integer k comes from the fact that our change of variables corresponds to the mapping of the plane onto itself k times. But, as can be easily checked,

$$\int \mathrm{d}f_0 \, \mathrm{d}f_0^* \frac{\ln[1+|f_0|^2]}{[1+|f_0|^2]^2} = \pi, \tag{4.67}$$

and so we see that the first contribution in (4.59) gives an expression proportional to πk.

So (all the time keeping $c = 1$),

$$\delta\zeta_Q'(0) = -\delta \ln \det M - 2\delta \ln \prod_{i=1}^{k} |b_i - a_i|^2 + \text{constant}. \tag{4.68}$$

Hence, combining all the expressions, we have

$$\begin{aligned} \det N \, [\det{}'(-Q/\mu^2)]^{-1} =& (\mu)^{k-N(k)} \exp\Bigg(- \sum_{i,j} \ln |a_i - b_i|^2 \\ &+ \sum_{i<j} \ln |a_i - a_j|^2 + \sum_{i<j} \ln |b_i - b_j|^2 \Bigg), \end{aligned} \tag{4.69}$$

and so finally,

$$Z_k \sim \left(\frac{1}{\pi g^2}\right)^{2k+1} \mu^{2k}\, e^{-2\pi k/g^2} \frac{1}{k!} \int \prod_i \mathrm{d}^2 a_i \mathrm{d}^2 b_i \mathrm{d}^2 c\, \Phi(a_i, b_i, c) e^{-V(a_i,b_i)}, \tag{4.70}$$

where $V(a_i, b_i)$ is the expression above, and where we have inserted $c \neq 1$.

We recognise that Z_k resembles a partition function describing a gas consisting of $2k$ particles (k of them, located at a_i, of positive electric charge and the other k, located at b_j, negatively charged) interacting with each other via the two-dimensional Coulomb potential $V(a_i, b_j)$. The Coulomb potential is entirely induced by the quantum corrections. Notice, that if we scale $\mu = \lambda\mu'$, then

$$\mu^{2k} e^{-2\pi k/g^2} = (\mu')\, e^{2k \ln \lambda}\, e^{-2k\pi/g^2} = (\mu')^{2k}\, e^{-2k\pi/g^2(\lambda)}, \tag{4.71}$$

where

$$\frac{1}{g^2(\lambda)} = \frac{1}{g^2} - \frac{\ln \lambda}{\pi}, \tag{4.72}$$

and we see that $g^2(\lambda)$ is the effective running coupling constant.

Also we see that

$$m^2 = \mu^2\, e^{-2\pi/g^2} \tag{4.73}$$

is a renormalisation group invariant. Observe that

$$m = \int \frac{\mathrm{d}^2 p}{2\pi}\, e^{-p^2/2m}, \tag{4.74}$$

and so we can interpret m^{2k} as

$$\prod_{i=1}^{2k} \int \frac{\mathrm{d}^2 p_i}{2\pi}\, e^{-p_i^2/2m}, \tag{4.75}$$

which is the kinetic energy for a system of $2k$ particles of momentum p_i and mass m. Hence we see that, effectively, we are looking at the statistical mechanics of a neutral gas of electrically charged particles evaluated at the singular temperature $\beta = 1$. Notice that, before the quantum effects are taken into account, a_i and b_j are simply parameters of our instanton solutions. In fact, if we look at one instanton solution (i.e. $k = 1$), the action density is proportional to

$$\frac{|b-a|^2}{[|x_+ - \frac{1}{2}(a+b)|^2 - \frac{1}{4}|b-a|^2]^2}, \tag{4.76}$$

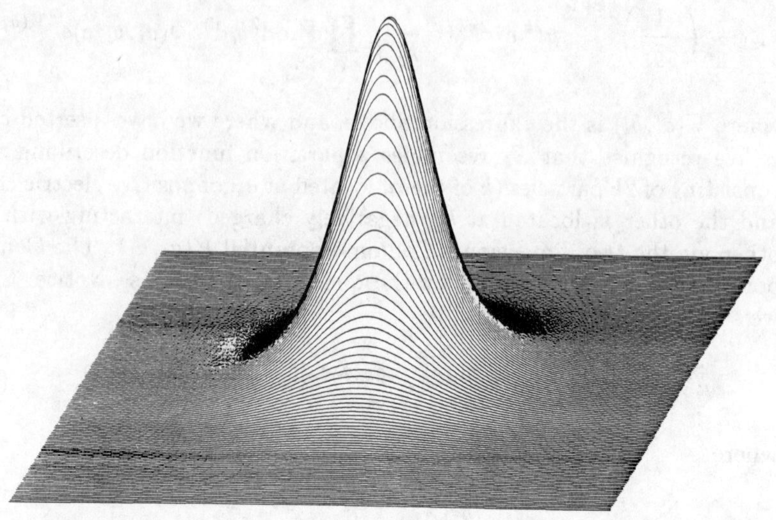

Figure 4.1

the plot of which is shown in figure 4.1.

We observe that the action density has a maximum at $\frac{1}{2}(a+b)$, and $|a-b|$ controls the size of the region in which the action density is significantly different from zero. Thus $\frac{1}{2}(a+b)$ is often referred to as the position of our instanton, and $|a-b|$ as its size, and so we see that although a and b can be described as constituents of instantons, they, by themselves, do not play any direct role. However, when quantum corrections are taken into account these contituents start playing a more important role, as they start interacting with each other. For this reason Fateev *et al* called them 'instanton quarks' [8].

4.2 Noninstanton Case

As we have already said in the previous chapter, noninstanton solutions do not correspond to the minima of the action; instead they correspond to the saddle points of the action and so the fluctuation operator possesses negative modes. Unfortunately very little is known about these modes in general — in fact, we do not even know their number. In the next chapter we will show how the solution of the associated Dirac problem leads to the determination of some of these negative modes; but unfortunately we do not get them all this way. Consequently very little can be done in general. There is one case, however, albeit very special, when we know not only the negative, but in fact, all the modes of the fluctuation operator. Thus we can study this case to get some insight into what to expect in more general cases. Unfortunately, even this case is quite difficult to analyse. So, instead of giving many details, we will just discuss briefly how one can proceed, and where the main problems lie.

The special solution, for which all the fluctuation modes are known, corresponds to an embedding of an $O(3)$ instanton as a (real) solution of the $\mathbb{C}P^2$ model. It is given by

$$z_\alpha^0 = \frac{1}{1+x^2}\begin{pmatrix} x_+ + x_- \\ i(x_+ - x_-) \\ 1 - x^2 \end{pmatrix}, \tag{4.77}$$

where $x^2 = x_+ x_-$, and in which all the parameters have been set at convenient values. This solution represents a topological charge mixture of two instantons and two anti-instantons (all located at the origin), and the total action is 8π. All our instantons and anti-instantons are located at the origin, and so the plot of the action density exhibits one large peak (like the peak in figure 4.1 but four times larger). The plots of more general solutions of this class (i.e. corresponding to two instantons and two anti-instantons) exhibit, in general, four peaks centred at positions of each instanton and anti-instanton. A plot of this type is shown in figure 4.2.

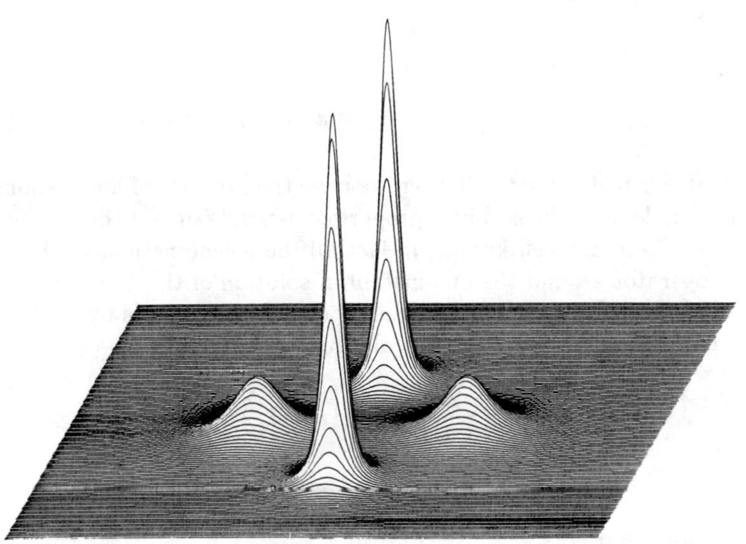

Figure 4.2

Our special solution (4.77) fits into our scheme of all the solutions; it can be obtained by the application of the P_+ operator to

$$f = \begin{pmatrix} 1 - x_+^2 \\ -i(1 + x_+^2) \\ -2x_+ \end{pmatrix}. \tag{4.78}$$

To proceed further we look at the fluctuations around (4.77). As in the previous chapter, we write z_α, a general field in the neighbourhood of z_α^0, as

$$z_\alpha = z_\alpha^0 (1 - |\psi|^2)^{1/2} + \psi_\alpha, \tag{4.79}$$

where ψ_α is a small complex fluctuation:

$$\psi_\alpha = \psi_\alpha^1 + i\psi_\alpha^2, \tag{4.80}$$

fulfilling

$$z_\alpha^0 \cdot \psi_\alpha = 0. \tag{4.81}$$

The action, to second order in ψ_α, turns out to be

$$S = S_0 + V(\psi^1) + V(\psi^2) - 2 \int \mathrm{d}^2x \, L_0 \, (\psi^2)^2, \tag{4.82}$$

where S_0 denotes the action of the z_α^0 configuration, whose Lagrangian density L_0 is given by

$$L_0 = (\partial_\mu z_\alpha^0)^2,$$

and where

$$V(\psi^i) = \int \mathrm{d}^2x \, [-\psi^i \partial^2 \psi^i - L_0 \, (\psi^i)^2]. \tag{4.83}$$

Now, if ψ is real, i.e. $\psi^2 = 0$, then we have the problem of fluctuations of the $O(3)$ model and the stability properties depend only on the properties of $V(\psi^1)$. These are well known; in fact all the eigenfunctions of the fluctuation operator around the one-instanton solution of the $O(3)$ model are explicitly known [9]. One way to think about these modes is by projecting our (compactified) R^2 stereographically onto S^2 by $(x_1, x_2) \to (r_1, r_2, r_3)$, with

$$\begin{aligned} r_\mu &= \frac{2x_\mu}{1 + x^2}, \qquad (\mu = 1, 2), \\ r_3 &= \frac{1 - x^2}{1 + x^2}. \end{aligned} \tag{4.84}$$

Then we find that

$$\partial^2 = -\frac{4}{(1 + x^2)^2} L^2, \tag{4.85}$$

where L is the angular momentum operator given in terms of the r_i variables.

Since the spherical harmonics Y_{pm} are eigenfunctions of L^2, it is easy to see that we can rewrite z^0_α as

$$z^0_\alpha = \left(\frac{4\pi}{3}\right)^{1/2} \sum_{m=-1}^{1} U_{\alpha m} Y_{1m}. \tag{4.86}$$

The explicit form of the matrix elements of the unitary matrix U, which appears in the expression above in order to guarantee the reality of z^0_α, can be determined easily, but in fact, they are not needed in what follows. The Lagrangian density on R^2 of (4.77) is

$$L_0 = \frac{8}{(1+x^2)^2}, \tag{4.87}$$

which, on the sphere S^2, translates into a constant (equal to 2), since the factor $4/(1+x^2)^2$ gets absorbed into the measure on the sphere:

$$\mathrm{d}\Omega = \frac{4}{(1+x^2)^2}\mathrm{d}^2x. \tag{4.88}$$

In consequence, the fluctuation potential takes the form:

$$V(\phi) = \int \mathrm{d}\Omega\,[\phi\, L^2\, \phi - 2\,\phi^2] \tag{4.89}$$

and a convenient set of functions, in terms of which we can expand our fluctuations ψ^i, are given by the spherical tensors [10]:

$$Y^{lp}_{JM} = \sum_{m=-l}^{l} \sum_{i=-p}^{p} C^{JM}_{lmpi}\, Y_{lm}\, \chi_{pi}, \tag{4.90}$$

where C^{JM}_{lmpi} are Clebsch–Gordan coefficients, and χ_{pi} are the Cartesian components of a rank-p spinor which, since p is an integer, is a $(2p+1)$-dimensional vector. The Y^{lp}_{JM}, of course, arise from adding two angular momenta l and p to get an angular momentum J, and so we have the restriction

$$J = |l-p|, \ldots, l+p.$$

The spherical tensors are eigenfunctions of L^2 (as well as J^2). We use them, remembering that in our case $p = 1$.

Next we expand our fluctuation ψ^1 in terms of the spherical tensors:

$$\psi^1_\alpha = \sum_{JMLj} a_{JML}\, U_{\alpha j}\, (Y^{L1}_{JM})_j, \tag{4.91}$$

where U_{ij} is the unitary matrix introduced before (4.86), and where we have used the conventional definition (for integer p) of the covariant component of Y_{JM}^{lp}:

$$(Y_{JM}^{lp})_i = (-1)^i \, C_{l,M+i,p,-i}^{JM} \, Y_{l,M+i}. \tag{4.92}$$

Now, using the reality properties of $(Y_{JM}^{lp})_j$,

$$(Y_{JM}^{lp*})_j = (-1)^{j+M+l+p-J} \, (Y_{J,-M}^{lp})_{-j}, \tag{4.93}$$

we see that the reality property of ψ^1 is guaranteed, if the a_{JML} satisfy

$$a_{JML}^* = a_{J,-M,L}(-1)^{M+L+p+J}. \tag{4.94}$$

Moreover, a straightforward, although lengthy, calculation shows that

$$V(\psi^1) = \sum_{JML} |a_{JML}|^2 \, \epsilon_L, \tag{4.95}$$

where $\epsilon_L = (L-1)(L+2)$. However, not all coefficients a_{JML} can be chosen independently; they have to be chosen in such a way that the constraint (4.81) is satisfied:

$$\psi_\alpha^1 \cdot z_\alpha^0 = \sum_{JMLim} a_{JML} (UY_{JM}^{L1})_i \left(\frac{4\pi}{3}\right)^{1/2} U_{im}^* \, Y_{1m}^* = 0, \tag{4.96}$$

which, using the orthogonality properties of spherical tensors, can be recast as

$$\sum_{JML} a_{JML} \left(\frac{2L+1}{2J+1}\right)^{1/2} C_{L010}^{J0} \, Y_{JM} = 0. \tag{4.97}$$

As Y_{JM} are linearly independent functions, this equation has to be satisfied for each component independently, and we see that the condition of orthogonality imposes the restrictions:

$$\sum_{L=|J-1|}^{J+1} C_{L010}^{J0} \, a_{JML} \, (2L+1)^{1/2} = 0, \tag{4.98}$$

which have to be imposed on a_{JML} for all J and M. Of course, a_{JML} also have to satisfy their reality condition (4.94), which, for the nonzero terms, due to the vanishing of C_{L010}^{J0} for $J+L$ being even, becomes

$$a_{JML}^* = (-1)^M a_{J,-M,L}. \tag{4.99}$$

Looking at our modes again we see that they come in two classes. The first class involves the unconstrained modes a_{JMJ} ($J = 1, 2, \ldots,$ $M =$

$-J, \ldots, J$), with an eigenvalue $\epsilon_L = (J-1)(J+2)$. The second class involves the constrained modes a_{JMJ-1} and a_{JMJ+1}, with eigenvalues $(J-2)(J+1)$ and $J(J+3)$ respectively, and the constraint is $\delta(\alpha a_{JMJ-1} + \beta a_{JMJ+1})$, with

$$\alpha = \sqrt{\frac{2J-1}{2J+1}} C^{J0}_{J-1010} \qquad \text{and} \qquad \beta = \sqrt{\frac{2J+3}{2J+1}} C^{J0}_{J+1010}. \tag{4.100}$$

In evaluating the determinant of the fluctuation operator, we have to integrate $\exp(-\epsilon_L a^2_{JML})$ over the modes a_{JML}. A short calculation shows that (for each J) the integration over the two constrained modes ($L = J \pm 1$) leads to the same result as the integration over the unconstrained mode ($L = J$); i.e. effectively we have a doubling of the unconstrained modes with eigenvalues $\epsilon_J = (J-1)(J+2)$. In particular, we notice the presence of six zero modes corresponding to taking $J = 1$, $M = -1, 0, 1$. A little thought shows that these six modes correspond to the six parameters which characterise the one-instanton solution of the $O(3)$ sigma model (or equivalently of the $\mathbb{C}P^1$ model), namely, complex c, a and b.

Next we look at the fluctuations governed by the imaginary fluctuation ψ^2_α. The appearance of the extra negative term, which, on the sphere, contributes $-4\sum_{JML} |b_{JML}|^2$, shows that we can repeat most of the previous discussion, the only difference being that the eigenvalues are now $\epsilon_L = (L-2)(L+3)$. Again there are two classes of modes: the unconstrained a_{JMJ} with eigenvalue $(J-2)(J+3)$ and the constrained ones, which, as before, again effectively correspond to the unconstrained ones. Looking at the eigenvalues we notice, in particular, the presence of six negative modes corresponding to $J = 1$, $M = -1, 0, 1$ (remember the doubling) and ten zero modes (corresponding to $J = 2$, $M = -2, -1, 0, 1, 2$).

Leaving aside, for the moment, the complications with negative and zero modes, we can now proceed to evaluating the positive modes contribution to the total fluctuation operator. Regularising the resultant determinant by the zeta function technique, the answer, using the spectrum found above, can be related to the function:

$$\zeta(s) = 2\sum_{j=1}^{\infty} [(2j+3)\{j(j+3)\}^{-s} + (2j+5)\{j(j+5)\}^{-s}] \tag{4.101}$$

continued, as usual, to $s = 0$. This result should be compared with the corresponding expression describing the (nonzero mode) vacuum determinant, which is given by

$$\zeta_0(s) = 4\sum_{j=1}^{\infty} (2j+1)\{j(j+1)\}^{-s}. \tag{4.102}$$

We can now use these last two expressions to evaluate the positive mode contribution to the (vacuum-normalised) partition function $\hat{Z}^{(1)}$, and, since the calculation resembles very much the one performed for the instanton case, we skip its details, and only state the result:

$$\hat{Z}^{(1)} = \exp\left(-\frac{8\pi}{g} + 12\ln\frac{\mu_0}{\mu} - 12\ln 2 + 5\ln 5 + 4\ln 3 - 3\right), \quad (4.103)$$

where g is the coupling constant and μ and μ_0 are subtraction points.

To find the complete partition function we would have to insert additional factors into (4.103) coming from zero and negative modes. In our discussion above we found that the complete fluctuation operator had altogether $10 + 6 = 16$ zero modes which can be written out explicitly. We could then evaluate their contribution by the introduction of some set of collective coordinates. What would be the most convenient choice? A little thought shows that this would be the set corresponding to the parameters of a general two-instanton solution of the $\mathbb{C}P^2$ model, as our solution (4.77) can be thought of as having been derived from this solution (for a particular choice of its parameters). However, this transformation is quite difficult to perform, as it involves diagonalisation of 16×16 matrices, etc, and in addition there is still the problem of how to treat the negative modes. Here the most obvious suggestion is to extend the summation in (4.102) to comprise also the negative eigenvalues, a procedure which would correspond to an analytic continuation in functional space [11], but unfortunately it is not clear whether this procedure reproduces the correct result. Clearly some further justification is required here.

We finish this chapter by pointing out one more time that, at present, we have no even partially reliable way of assessing the quantum importance of noninstanton solutions. In the general case we do not know the spectrum of the fluctuation operator, in fact we do not know even the number of its negative modes. In the very special case where all the modes are known some progress can be made, but so far no convenient way has been found for dealing with zero modes of the fluctuation operator nor a reliable way for dealing with negative modes.

References

[1] Berg B and Luscher L 1979 Computations of quantum fluctuations around multi-instanton fields from exact Green's functions: the $\mathbb{C}P^{N-1}$ case *Commun. Math. Phys.* **69** 57

[2] Fateev V, Frolov I and Schwartz A 1979 Quantum fluctuations of instantons in the non-linear σ model *Nucl. Phys.* B **154** 1

[3] 't Hooft G 1976 Computation of the quantum effects due to a four dimensional pseudoparticle *Phys. Rev.* D **14** 3432; Erratum 1978 *Phys. Rev.* D **18** 2199

Ore F 1977 How to compute determinants exactly *Phys. Rev.* D **16** 2577

[4] Corrigan E and Goddard P 1980 Some aspects of instantons *Lecture Notes in Physics* **129** 14
Berg B and Luscher M 1979 Computations of quantum fluctuations around multi-instanton fields from exact Green's functions: the $\mathbb{C}P^{N-1}$ case *Commun. Math. Phys.* **69** 57
Frolov I V and Schwarz A S 1978 Contribution of instantons to the correlation functions of a Heisenberg ferromagnet *JETP Lett.* **28** 249

[5] Hawking S W 1977 Zeta function regularization of path integrals in curved spacetime *Commun. Math. Phys.* **55** 133

[6] Dowker J S and Critchley R 1977 Stress-tensor conformal anomaly for scalar, spinor and vector fields *Phys. Rev.* D **16** 3390

[7] Ramond P 1981 *Field Theory: a Modern Primer* (New York: Benjamin-Cummings)

[8] Fateev V A, Frolov I V and Schwarz A S 1980 Quantum fluctuations of instantons in a two-dimensional nonlinear anisotropic sigma model *Sov. J. Nucl. Phys.* **32** 153

[9] Din A M and Zakrzewski W J 1980 Stability properties of classical solutions to non-linear σ models *Nucl. Phys.* B **168** 173

[10] Varschalovitsch D A, Moskalev A N and Cherconski V K 1975 *Quantum Theory of Angular Momentum* (Moscow: Nauka) (in Russian)

[11] Nahm W 1980 *Semiclassical Non-perturbative Effects in Quantum Field Theories. Bad Honnef Institute, 293*

5 Sigma Models with Fermions

5.1 $\mathbb{C}P^{N-1}$ Model with Fermions

Grassmannian models discussed before can be generalised to include fermions. As there are important differences between spinors which are used to describe fermions depending on whether they are defined in Minkowski or Euclidean space (besides the obvious analytical continuation), this time we have to be precise as to the nature of the space we are going to be working in. As we are interested in the path integral formulation of the theory and its classical solutions we choose to work in the Euclidean regime.

There are many ways of including fermions into grassmannian models, of which the most interesting one is that which renders the theory supersymmetric. In this case the most convenient way of proceeding is through the use of superspace. In our discussion here we will restrict ourselves to considering only the supersymmetric $\mathbb{C}P^{N-1}$ models.

To define them, we start with the introduction of a two-dimensional superspace $(x, y, \theta_1, \theta_2)$, where the anticommuting θ_1 and θ_2 denote two components of a Majorana spinor θ, and so can be thought of as being real. Then we consider [1]

$$\Phi_\alpha(x, y, \theta_1, \theta_2) = z_\alpha(x, y) + i\theta_i \chi^i_\alpha(x, y) + i\theta_1\theta_2 F_\alpha(x, y), \tag{5.1}$$

where χ^i_α is a complex spinor field. Its components anticommute with each other and with θ_i. The hermitian conjugate of the superfield Φ is given by

$$\Phi^\dagger_\alpha = z^\dagger_\alpha + i\theta_i \chi^{i\dagger}_\alpha + i\theta_1\theta_2 F^\dagger_\alpha. \tag{5.2}$$

By analogy with the $\mathbb{C}P^{N-1}$ model we impose on the superfield Φ_α the constraint

$$\Phi^\dagger\Phi = 1, \tag{5.3}$$

which, when expanded in power series in θ, translates into the following conditions on the fields z_α, χ^i_α and F_α:

$$\begin{aligned} z^\dagger_\alpha z_\alpha &= 1 \\ \chi^i_\alpha z^\dagger_\alpha + \chi^{i\dagger}_\alpha z_\alpha &= 0 \\ F^\dagger_\alpha z_\alpha + F_\alpha z^\dagger_\alpha &= i[\chi^{1\dagger}_\alpha \chi^2_\alpha - \chi^{2\dagger}_\alpha \chi^1_\alpha]. \end{aligned} \tag{5.4}$$

We introduce two two-dimensional γ matrices, which we choose to be given by

$$\gamma_1 = \begin{pmatrix} 1 & 0 \\ 0 & -1 \end{pmatrix} \qquad \gamma_2 = \begin{pmatrix} 0 & 1 \\ 1 & 0 \end{pmatrix}, \tag{5.5}$$

and also the two-dimensional analogue of γ_5, which we will still call γ_5:

$$\gamma_5 = \begin{pmatrix} 1 & 0 \\ 0 & -1 \end{pmatrix}\begin{pmatrix} 0 & 1 \\ 1 & 0 \end{pmatrix} = \begin{pmatrix} 0 & 1 \\ -1 & 0 \end{pmatrix}. \tag{5.6}$$

It follows that

$$\theta^T \gamma_5 \theta = (\theta_1, \theta_2)\begin{pmatrix} 0 & 1 \\ -1 & 0 \end{pmatrix}\begin{pmatrix} \theta_1 \\ \theta_2 \end{pmatrix} = 2\theta_1\theta_2. \tag{5.7}$$

Also we introduce the generator of the supertranslations ∂':

$$\partial' = \partial_\theta + i\not{\partial}\theta.$$

Using the explicit form of our gamma matrices (5.5) we see that

$$\partial' = \begin{pmatrix} \partial_{\theta_1} + i(\theta_1\partial_x + \theta_2\partial_y) \\ \partial_{\theta_2} + i(\theta_1\partial_y - \theta_2\partial_x) \end{pmatrix}. \tag{5.8}$$

Next we covariantise ∂' by defining

$$D' = \partial' - (\Phi^\dagger \partial' \Phi). \tag{5.9}$$

Observe that, as in Chapter 2, our operators (∂' and D') are fermionic in nature (i.e. grassmann odd). Hence in all calculations we have to remember to use their anticommuting properties.

To define our model we have to decide what to take for our Lagrangian density in superspace. Guided by the analogy with the $\mathbb{C}P^{N-1}$ model, we take

$$\mathcal{L} = (D'\Phi)^\dagger \gamma_5 D'\Phi = (\partial'\Phi)^\dagger \gamma_5 \partial'\Phi - (\Phi^\dagger \partial' \Phi)^T \gamma_5 (\Phi^\dagger \partial'\Phi). \tag{5.10}$$

Moreover, we can also define the topological charge density:

$$\mathcal{Q} = (D'\Phi)^\dagger D'\Phi. \tag{5.11}$$

Let us observe that when we integrate $\mathcal{L}$ and $\mathcal{Q}$ over $d\theta_1 d\theta_2$ we obtain the conventional Lagrangian and the topological charge densities. To obtain the explicit form of these densities is straightforward, though very tedious. To do this we first calculate the explicit form of $A = (\Phi^\dagger_\alpha \partial' \Phi_\alpha)$, and then observe that

$$\mathcal{L} = (\partial' \Phi)^\dagger \gamma_5 \partial' \Phi - A^\dagger \gamma_5 A. \tag{5.12}$$

To calculate the contribution of each term in the expression above, we use such properties of our gamma matrices as $\gamma_1\gamma_5\gamma_1 = \gamma_5$, $\gamma_2\gamma_5\gamma_2 = -\gamma_5$ and $\gamma_2\gamma_5\gamma_1 = 1$. This allows us to write $(\theta\gamma_\mu)^\dagger \gamma_5 \gamma_\nu = 2\delta_{\mu\nu}\theta_1\theta_2$. Then, after a couple of pages of algebra, we find that

$$\int d\theta_1\, d\theta_2\, (\partial'\Phi)^\dagger \gamma_5 \partial' \Phi = 2[\partial_\mu z^\dagger_\alpha \partial_\mu z_\alpha + F^\dagger_\alpha F_\alpha + i(\chi^i_\alpha)^\dagger \not\partial \chi^i_\alpha]. \tag{5.13}$$

The calculation of the remaining term is even more tedious. However, after a few pages of algebra we find that the total expression for the Lagrangian density is given by

$$\begin{aligned} L = {} & |\partial_\mu z_\alpha|^2 - i\,\psi^\dagger_\alpha \not\partial \psi_\alpha + i(\psi^\dagger_\alpha \gamma_\mu \psi_\alpha)(z^\dagger_\alpha \not\partial z_\alpha) \\ & -(G_\alpha + \rho^T \gamma_5 \psi_\alpha)^\dagger (G_\alpha - \rho^T \gamma_5 \psi_\alpha) + (z^\dagger_\alpha \partial_\mu z_\alpha)(z^\dagger_\alpha \partial_\mu z_\alpha) \\ & + \tfrac{1}{4}[(\psi^\dagger_\alpha \psi_\alpha)^2 + (\psi^\dagger_\alpha \gamma_5 \psi_\alpha)^2 - (\psi^\dagger_\alpha \gamma_\mu \psi_\alpha)^2], \end{aligned} \tag{5.14}$$

where

$$\begin{aligned} \psi_\alpha &= \chi_\alpha - z_\alpha (z^\dagger_\beta \chi_\beta) \\ G_\alpha &= F_\alpha - z_\alpha (z^\dagger_\beta F_\beta) \\ \rho &= \tfrac{1}{2} i (z_\alpha \chi^*_\alpha - z^\dagger_\alpha \chi_\alpha). \end{aligned} \tag{5.15}$$

Next we can eliminate the dummy field G_α, using its equation of motion $G_\alpha = \rho^T \gamma_5 \psi_\alpha$, and we see that ρ disappears at the same time. We are left with

$$\begin{aligned} S = \int d^2x\, [& (D_\mu z_\alpha)^\dagger D_\mu z_\alpha - i(\psi^\dagger_\alpha \not{D} \psi_\alpha) \\ & + \tfrac{1}{4}[(\psi^\dagger_\alpha \psi_\alpha)^2 + (\psi^\dagger_\alpha \gamma_5 \psi_\alpha)^2 - (\psi^\dagger_\alpha \gamma_\mu \psi_\alpha)^2], \end{aligned} \tag{5.16}$$

while the remaining fields are constrained by

$$z^\dagger_\alpha z_\alpha = 1, \qquad z^\dagger_\alpha \psi_\alpha = \psi^\dagger_\alpha z_\alpha = 0. \tag{5.17}$$

Our final action (in two-dimensional space) is invariant under the following gauge transformations:

$$\begin{aligned} z_\alpha &\to z'_\alpha = z_\alpha\, e^{i\Lambda} \\ \psi_\alpha &\to \psi'_\alpha = \psi_\alpha\, e^{i\Lambda}, \end{aligned} \tag{5.18}$$

and under the supersymmetry transformations:

$$\begin{aligned} \delta z_\alpha &= i\epsilon\psi_\alpha \\ \delta\psi &= -\tfrac{1}{2}i\epsilon z_\alpha[\psi^\dagger_\beta\psi_\beta + \tfrac{1}{2}i\gamma_5\epsilon z_\alpha(\psi^\dagger_\beta\gamma_5\psi_\beta)] \\ &\quad +\gamma_\mu\epsilon[D_\mu z_\alpha - \tfrac{1}{2}iz_\alpha(\psi^\dagger_\alpha\gamma_\mu\psi_\alpha)]. \end{aligned} \tag{5.19}$$

In fact our supersymmetric $\mathbb{C}P^{N-1}$ model has a further supersymmetry in such a way that its symmetry group is an extended $O(2)$ supersymmetry. The simplest way to exhibit this symmetry explicitly is by rewriting the model in terms of complex θ_i''s. However, as we will not exploit this symmetry, we will not discuss it here; the interested reader can find a good discussion of this symmetry in the original paper by d'Adda *et al* [1].

5.2 Dirac Background Problem

Now that we have a theory involving bosons and fermions, we can try to calculate some effects due to the coupling of the fermions. However, the fermion fields in our Lagrangian density (5.16) appear in a complicated way, and so all the calculations will be rather involved.

The first thing we will study is the external background field Dirac-like equation. This equation can be thought of as having come from the full equation for the spinor ψ, in which all higher than linear terms in ψ are ignored. Thus the equation is given by

$$(\not{D}\psi)_\alpha - z_\alpha(z^\dagger_\beta\not{D}\psi_\beta) = 0, \tag{5.20}$$

together, of course, with the constraint equations:

$$z^\dagger_\alpha z_\alpha = 1, \qquad z^\dagger_\alpha\psi_\alpha = \psi^\dagger_\alpha z_\alpha = 0. \tag{5.21}$$

The best way to proceed is to resolve ψ_α into the eigenstates of γ_5 (i.e. two-dimensional analogues of helicity states),

$$\psi = \begin{pmatrix}1\\ i\end{pmatrix}\psi_+ + \begin{pmatrix}1\\ -i\end{pmatrix}\psi_-, \tag{5.22}$$

where we have suppressed the $\mathbb{C}P^{N-1}$ index α. We observe that

$$\gamma_5\psi = i\begin{pmatrix}1\\ i\end{pmatrix}\psi_+ - i\begin{pmatrix}1\\ -i\end{pmatrix}\psi_-. \tag{5.23}$$

Thus, using the fact that $D_+ = \frac{1}{2}(D_1 - iD_2)$ and $D_- = \frac{1}{2}(D_1 + iD_2)$, we find that (5.20) separates into a pair of equations for $\psi_\pm$:

$$\begin{aligned} D_-\psi_+ &= \lambda_+\, z \\ D_+\psi_- &= \lambda_-\, z, \end{aligned} \tag{5.24}$$

where $\lambda_\pm$ are some overall functions (i.e. independent of α) of x and y.

We will solve these equations for ψ with the external fields z_α satisfying their own equations of motion ($D_- D_+ z_\alpha = \lambda z_\alpha$) (with $\psi = 0$). Hence the name background field Dirac-like problem. To do this we observe that

$$z_\alpha = \frac{\hat{z}_\alpha}{|\hat{z}|}, \tag{5.25}$$

where $\hat{z}_\alpha = (P_+^k f)_\alpha$ for some k, and that

$$D_+ = |\hat{z}|\partial_+\Big(\frac{1}{|\hat{z}|}\cdot\Big) \qquad D_- = \frac{1}{|\hat{z}|}\partial_-\big(|\hat{z}|\cdot\big). \tag{5.26}$$

Thus we see that writing

$$\psi_- = |\hat{z}|G_-, \qquad \psi_+ = \frac{1}{|\hat{z}|}G_+, \tag{5.27}$$

our equations become

$$\partial_+ G_{-\alpha} = \lambda_+ z_\alpha, \qquad \partial_- G_{+\alpha} = \lambda_- z_\alpha, \tag{5.28}$$

where $G_{\pm\alpha}$ must be orthogonal to z_α (in the $\mathbb{C}P^{N-1}$ space). To solve these equations we recall from the construction of the solutions of the $\mathbb{C}P^{N-1}$ models that, when the projector corresponding to our solution was given by $\mathbb{P} = e_k e_k^\dagger$, then a projector Q defined by $Q = \sum_{i=1}^{k-1} e_i e_i^\dagger$ (i.e. a projector on all the previous solutions) satisfied

$$\partial_+ Q + \mathbb{P}\partial_+\mathbb{P} = 0. \tag{5.29}$$

Taking this into consideration, we see that, if we set

$$G_{-\alpha} = Q_{\alpha\beta} h_\beta, \tag{5.30}$$

and we require that h_α is a function of x_- only, our equations will be satisfied given that $z_\alpha^\dagger G_{-\alpha} = 0$, i.e. by the orthogonality of e_i and

$$\partial_+ G_{-\alpha} = -(\mathbb{P}\partial_+\mathbb{P}h)_\alpha = -z_\alpha\lambda_+, \tag{5.31}$$

where $\lambda_+ = z_\beta^\dagger(\partial_+\mathbb{P})_{\beta\rho}h_\rho$. In the same way we find that, to solve the equation for $G_{+\alpha}$, we can set

$$G_{+\alpha} = (1 - Q - \mathbb{P})_{\alpha\beta}g_\beta, \tag{5.32}$$

where g_β is a function of x_+ only.

What can we say about the arbitrary vectors $h_\alpha = h_\alpha(x_-)$ and $g_\alpha = g_\alpha(x_+)$? Moreover, what happens when our solution z_α^k corresponds to instantons or anti-instantons (i.e. when $e_k \to e_1$ or e_N respectively and so $Q = 0$ or $1 - Q - \mathbb{P} = 0$? The answer lies in the fact that, so far, we have not imposed any restrictions on ψ. If we require that ψ be normalisable on the sphere, i.e. such that

$$\int_\infty^\infty \mathrm{d}^2x \, \frac{|\psi|^2}{1+x^2} < \infty, \tag{5.33}$$

then it can be shown that h_α and g_α have, once again, to be given by polynomials of their respective arguments. We can then count the number of parameters on which ψ_+ and ψ_- depend. Let us call these numbers ξ_+ and ξ_- respectively. Clearly they are both integers. Moreover, a detailed discussion [2] shows that

$$\xi_+(P_+^{k-1}f) = \frac{1}{2\pi}\sum_{i=1}^{k-1}(Q_i - Q_k) = \alpha_{k-1} - \frac{(k-1)}{2\pi}Q_k, \tag{5.34}$$

where, as before, $\alpha_0 = 0$ and where Q_l denotes the topological charge of $(P_+^{l-1}f)$.

In a similar way we can show that

$$\xi_-(P_+^{k-1}f) = \frac{1}{2\pi}\sum_{i=k+1}^{N}(Q_k - Q_i) = \alpha_k + \frac{N-k}{2\pi}Q_k. \tag{5.35}$$

(Recall that $\alpha_N = 0$.) We can now evaluate

$$\xi_+ - \xi_- = \alpha_{k-1} - \alpha_k - \frac{N-1}{2\pi}Q_k = -\frac{N}{2\pi}Q_k, \tag{5.36}$$

as $Q_k = (1/2\pi)(\alpha_k - \alpha_{k-1})$.

This last result is an example of an important index theorem proved by Atiyah *et al* [3]. This theorem relates the number of zero modes of the Dirac operator, i.e. solutions of (5.20), which are normalisable on the sphere, to the topological properties of the background field itself. The theorem states that the difference of the numbers of solutions of each chirality, which in our case means $\xi_+ - \xi_-$, is proportional to the topological charge of the background field, i.e. Q_k.

Observe that, for instantons ($k = 1$),

$$\xi_+ = 0 \quad \text{and} \quad \xi_- = \alpha_1 + (N-1)\alpha_1 = N\alpha_1$$

and so we see that only $\psi_{+\alpha}$ solutions are normalisable on the sphere. For anti-instantons ($k = N$), it is other way round:

$$\xi_- = 0, \quad \xi_+ = \alpha_{N-1} + (N-1)\alpha_{N-1} = N\alpha_{N-1},$$

and only ψ_- solutions are normalisable on the sphere.

Note that, had we considered the case of a minimally coupled ψ field to the bosonic z field (i.e. had we dropped the orthogonality conditions $\bar{\psi}_\alpha z_\alpha = \psi \bar{z}_\alpha = 0$), our equations would have reduced to

$$D_\pm \psi_\mp = 0, \tag{5.37}$$

for which solutions take the simple form:

$$\begin{aligned} \psi_+ &= \frac{1}{|\hat{z}|} g(x_-) \\ \psi_- &= |\hat{z}| h(x_+). \end{aligned} \tag{5.38}$$

The condition of the normalisability on the sphere then gives

$$\begin{aligned} Q > 0 \quad &\Rightarrow \quad \xi_+ = 0 \quad \xi_- = NQ \\ Q < 0 \quad &\Rightarrow \quad \xi_+ = -NQ \quad \xi_- = 0, \end{aligned} \tag{5.39}$$

i.e. we would have satisfied the index theorem in a minimal way. In our case, (5.20), the theorem is satisfied in a nonminimal way; both $\xi_\pm$ are nonzero and their difference satisfies (5.36).

The above discussion generalises to the case of more general grassmannian models; it is relatively straightforward for the generic solutions, but rather more complicated for the embeddings. Let us discuss briefly the case of generic solutions.

As z is now a matrix, so are $\psi_\pm$. Working as before, we write

$$z = \hat{z}(M)^{-1/2}, \tag{5.40}$$

where $\hat{z}^\dagger \hat{z} = M$. Then we find that

$$\begin{aligned} D_-(M)^{-1/2} &= 0 \\ D_+(M)^{1/2} &= 0, \end{aligned} \tag{5.41}$$

and so we can set

$$\psi_\pm = \phi_\pm \cdot (M)^{\mp 1/2} \tag{5.42}$$

(notice that we have here a matrix multiplication). Then, if

$$z = (e_\beta, e_{\beta+1}, \ldots, e_{\beta+M-1}), \tag{5.43}$$

then, defining $\mathbb{P}_\beta$ as a projector on the subspace spanned by $\{e_1, \ldots, e_\beta\}$, we have

$$\mathbb{P} = \mathbb{P}_{\beta+M-1} - \mathbb{P}_{\beta-1}. \tag{5.44}$$

Then, as

$$\partial_+ \mathbb{P}_{\beta-1} = -\mathbb{P}\, \partial_+ \mathbb{P}, \tag{5.45}$$

we can set

$$\begin{aligned} \phi_- &= \mathbb{P}_{\beta-1} h(x_-) \\ \phi_+ &= (1 - \mathbb{P}_{\beta+M-1}) g(x_+), \end{aligned} \tag{5.46}$$

and we see that our equations are satisfied. Again, looking in detail at the number of parameters on which $\psi_\pm$ can depend, we see that the index theorem is satisfied.

5.3 Full Supersymmetric Model

5.3.1 Instantons

In this section we will discuss, in some detail, classical solutions of the full supersymmetric model (i.e. solutions of the equations of motion following from (5.16)) [1].

Let us recall that in the last section we found it convenient to resolve the spinor ψ fields into the eigenstates of γ_5, i.e. the fields $\psi_\pm$. If we do the same for the spinor θ, i.e. we define

$$\begin{aligned} \theta_+ &= \theta_1 + i\theta_2 \\ \theta_- &= \theta_1 - i\theta_2, \end{aligned} \tag{5.47}$$

and also analogously for the spinors χ_α, we find that our superfield Φ_α has the expansion

$$\Phi_\alpha = z_\alpha + 2i(\theta_+ \chi_{+\alpha} + \theta_- \chi_{-\alpha}) - \tfrac{1}{2}\theta_+ \theta_- F_\alpha, \tag{5.48}$$

where we have used

$$\theta_1 \chi_{1\alpha} + \theta_2 \chi_{2\alpha} = 2(\theta_+ \chi_{+\alpha} + \theta_- \chi_{-\alpha}). \tag{5.49}$$

From now on, in this chapter, we will suppress the $\mathbb{C}P^{N-1}$ index α and we will denote the scalar product in the $\mathbb{C}P^{N-1}$ space by a '$\cdot$'.

The constraints on our fields (which follow from $\Phi^\dagger \cdot \Phi = 1$) are now given by

$$z^\dagger \cdot z = 1, \qquad z^\dagger \cdot \chi_+ + z \cdot \chi_-^\dagger = 0 \quad \text{and c.c.}, \tag{5.50}$$

and

$$F^\dagger \cdot z + F \cdot z^\dagger - 2(\chi_-^\dagger \cdot \chi_- - \chi_+^\dagger \cdot \chi_+). \tag{5.51}$$

We can now introduce supercovariant derivatives

$$\check{D}_\pm = \check{\partial}_\pm - (\Phi^\dagger \cdot \check{\partial}_\pm \Phi), \tag{5.52}$$

where

$$\check{\partial}_\pm = -i\partial_{\theta_\pm} + \theta_\pm \partial_\pm \tag{5.53}$$

are the generators of the supersymmetry translations. We notice that we can also write

$$\check{D}_\pm = -iD_{\theta_\pm} + \theta_\pm D_\pm. \tag{5.54}$$

In terms of these quantities, our $\mathcal{L}$ takes the form

$$\mathcal{L} = 2\{|\check{D}_+\Phi|^2 - |\check{D}_-\Phi|^2\}, \tag{5.55}$$

where we have used the notation that $|a|^2 = a^\dagger \cdot a$, and

$$\mathcal{Q} = 2\{|\check{D}_+\Phi|^2 + |\check{D}_-\Phi|^2\}. \tag{5.56}$$

(Notice that in superspace the signs between the two terms in $\mathcal{L}$ and $\mathcal{Q}$ are different as compared to the usual case.) Of course, the usual 'physical' Lagrangian density is given by $L = \int \mathrm{d}\theta_+ \mathrm{d}\theta_- \, \mathcal{L}$.

The equations of motion, corresponding to $\mathcal{L}$, are given by

$$\check{D}_+\check{D}_-\Phi + |\check{D}_-\Phi|^2\,\Phi = 0, \tag{5.57}$$

(or correspondingly with $+ \leftrightarrow -$), and the equations of self-duality in superspace are given by

$$\check{D}_-\,\Phi = 0 \tag{5.58}$$

(for antiself-duality $+ \leftrightarrow -$). To solve these self-dual equations (5.58) we write

$$\Phi = \frac{\omega}{|\omega|}, \tag{5.59}$$

and we find that the equation is satisfied if

$$\omega = \omega(x_+, \theta_+). \tag{5.60}$$

If we now expand this superfield, we find

$$\omega(x_+, \theta_+) = f(x_+) + i\theta_+ g(x_+), \tag{5.61}$$

and we see that this corresponds to the supersymmetric generalisation of the holomorphic curve we encountered in the $\mathbb{C}P^{N-1}$ model. Of course the arbitrary function $g(x_+)$ is a grassmann function of its variable.

We can now substitute our expression for ω into Φ, expand, and obtain explicit expressions for our fields z, $\chi_\pm$ and F. They are given by

$$\begin{aligned} z &= \frac{f}{|f|}, \qquad \chi_- = -\tfrac{1}{4} f \frac{(g^\dagger \cdot f)}{|f|^3} \\ \chi_+ &= \tfrac{1}{2}\frac{g}{|f|} - \tfrac{1}{4} f \frac{(f^\dagger \cdot g)}{|f|^3} \\ F &= g\frac{(g^\dagger \cdot f)}{|f|^3} - f\frac{(g^\dagger \cdot g)}{|f|^3} + \tfrac{3}{2} f \frac{(g^\dagger \cdot f)(g \cdot f^\dagger)}{|f|^5}. \end{aligned} \tag{5.62}$$

By analogy with the conventional $\mathbb{C}P^{N-1}$ model we will refer to this configuration as describing *supersymmetric instantons.* Changing to $\psi_\pm$ given by

$$\psi_\pm = \chi_\pm - z(z^\dagger \cdot \chi_\pm), \tag{5.63}$$

we find

$$\psi_- = 0 \qquad \psi_+ = \frac{1}{2}\frac{1}{|f|}\left(g - f\frac{(f^\dagger \cdot g)}{|f|^2}\right) \tag{5.64}$$

and

$$G = F - z(z^\dagger \cdot F) = -\frac{(g^\dagger \cdot f)}{|f|^3}\left(g - f\frac{(f^\dagger \cdot g)}{|f|^2}\right). \tag{5.65}$$

Let us look next at the Lagrangian density due to these solutions. To do this, first of all, we look at the fermionic part of the Lagrangian density:

$$L_f = -i(\psi^\dagger \cdot \not{D}\psi) + \tfrac{1}{4}\{(\psi^\dagger \cdot \psi)^2 + (\psi^\dagger \cdot \gamma_5\psi)^2 - (\psi^\dagger \cdot \gamma_\mu\psi)^2\}. \tag{5.66}$$

As $\psi^\dagger \cdot \psi = 2(\psi_+^\dagger \cdot \psi_+ + \psi_-^\dagger \cdot \psi_-)$, $\psi^\dagger \cdot \gamma_5\psi = 2i(\psi_+^\dagger \cdot \psi_+ - \psi_-^\dagger \cdot \psi_-)$, etc, we find that we get

$$\begin{aligned} L_f = &-4i\{\psi_+^\dagger \cdot D_+\psi_- + \psi_-^\dagger \cdot D_-\psi_+\} \\ &+ 4\{(\psi_+^\dagger \cdot \psi_+)(\psi_-^\dagger \cdot \psi_-) - (\psi_+^\dagger \cdot \psi_-)(\psi_-^\dagger \cdot \psi_+)\}. \end{aligned} \tag{5.67}$$

As for the supersymmetric instantons $\psi_- = 0$, we see that, in this case, the fermionic part of the action vanishes (the bosonic part is the same as for the instantons in the conventional $\mathbb{C}P^{N-1}$ model). If we look at the topological charge density for supersymmetric instantons (both in superspace and in ordinary space) we find that it is equal to the corresponding Lagrangian density — again by analogy with the result of the conventional $\mathbb{C}P^{N-1}$ model. Moreover, when considered as a contribution in ordinary space, the fermionic contribution vanishes. This result generalises to any Kähler manifold when we consider their supersymmetric extensions; for supersymmetric instanton solutions the fermionic contribution to the topological charge vanishes.

5.3.2 Noninstanton Case

In this subsection we will find the more general (noninstanton) solutions of the equations of motion (5.57) [4]. Our procedure parallels the one we have used to find solutions of the conventional $\mathbb{C}P^{N-1}$ model. In that construction we started with a holomorphic vector f, used it to form a sequence of further holomorphic vectors $\partial_+ f$, $\partial_+^2 f, \cdots \partial_+^k f$, and then Gramm–Schmidt orthonormalised them, obtaining a sequence of vectors which solved the equations of motion. In the supersymmetric case the role of $f(x_+)$ is played by $\omega(x_+, \theta_+)$ and the role of ∂_+ by $\check{\partial}_+$.

However, the most naive approach does not work. To see the problems we run into, let us consider the second vector in our expected sequence. If $\Phi = \omega/|\omega|$ is the first vector, then the next one is derived from the normalisation of

$$\hat{e}_2 = \check{\partial}_+\omega - \omega\frac{(\omega^\dagger \cdot \check{\partial}_+\omega)}{|\omega|^2}, \tag{5.68}$$

and this vector is not normalisable! Let us explore this point in more detail. As

$$\omega = f(x_+) + \theta_+ g(x_+)$$

then

$$\check{\partial}_+\omega = -ig(x_+) + \theta_+\partial_+ f(x_+).$$

Clearly $\check{\partial}_+\omega$ is fermionic (grassmann odd) and so, strictly speaking, is not normalisable (neither is $\hat{e}_2$, which is obtained from $\check{\partial}_+\omega$ by projections!). On the other hand, the next term in the sequence is bosonic as:

$$\check{\partial}_+^2\omega = -i\check{\partial}_+\omega = -i(\partial_+ f + \theta_+\partial_+ g). \tag{5.69}$$

As we are trying to construct a basis we need both sets of vectors, but we need them to be bosonic. To proceed further we take a set of grassmann quantities $\epsilon_\pm^{(k)} = \epsilon_1^{(k)} \pm i\epsilon_2^{(k)}$, and then consider the sequence

$$\omega,\ \epsilon_+\check{\partial}_+\omega,\ \check{\partial}_+^2\omega,\ \epsilon_+^{(2)}\check{\partial}_+^3\omega, \ldots. \tag{5.70}$$

All the vectors are bosonic now, although, strictly speaking, some of them are just grassmann even, i.e. are not given by ordinary functions. In particular, the leading term of $\epsilon_+\check{\partial}_+\omega$ is $\epsilon_+ g$ and is not an ordinary complex vector, but an even nilpotent element of a grassmann algebra.

Next we will make the assumption that we can effectively replace $\epsilon_+ g$ by an ordinary vector and so treat it as such. This assumption corresponds to the extension of the projective space symmetry to the grassmann even numbers. We should stress that this is a *formal* assumption, which in a sense, is no more formal than the superspace formalism itself, and we will show that this assumption leads to correct results. In particular, if we consider the $S\mathbb{C}P^1$ model, we see that we have two classes of solutions (supersymmetric instantons and anti-instantons). The space of solutions is spanned by two vectors, for which we can take ω and $\epsilon_+\check{\partial}_+\omega$. Then we can show that

$$\hat{e}_1 = (\epsilon_+\check{\partial}_+\omega) - \omega\frac{(\omega^\dagger \cdot \epsilon_+\check{\partial}_+\omega)}{|\omega|^2} = \begin{pmatrix} -\omega_2^* \\ \omega_1^* \end{pmatrix} \frac{\epsilon_+\check{\partial}_+\omega_1\,\omega_2 - \epsilon_+\check{\partial}_+\omega_2\,\omega_1}{|\omega|^2}, \tag{5.71}$$

and so we see that $\hat{e}_1$ represents, up to an irrelevant overall (grassmann even) factor, a supersymmetric anti-instanton solution. Had we not put in our ϵ_+ factors, the result would still have been true, but the overall factor would have been grassmann odd. We see that, as in the conventional $\mathbb{C}P^1$ model, we have only supersymmetric instantons and anti-instantons.

In a more general case we proceed as before; i.e. we Gramm–Schmidt orthonormalise the first N-independent vectors in the sequence (5.70), where, of course, $\omega = f + i\theta_+ g$. Then, as before, we can show that

$$e_k = \Phi^k = \frac{\hat{e}_k}{|\hat{e}_k|}, \qquad \text{where} \quad \hat{e}_k = \frac{(\hat{h}^{k-1})^\dagger \cdot \hat{h}^k}{(\hat{h}^{k-1})^\dagger \cdot \hat{h}^{k-1}}, \tag{5.72}$$

where $k = 1, 2, \ldots, N$ and where $\hat{h}^1 = \omega$,

$$\hat{h}^{i+1} = \omega \wedge \epsilon_+^1 \check{\partial}_+ \omega \wedge \ldots \wedge a_i \check{\partial}_+^i \omega, \tag{5.73}$$

and where $a_i = 1$ if i is even and $a_i = \epsilon_+^p$ if i is odd $(i = 2p - 1)$.

These solutions have similar properties to those of the conventional $\mathbb{C}P^{N-1}$ model. In particular

$$\check{D}_+ |\hat{e}_k| = 0, \qquad \check{D}_- \frac{1}{|\hat{e}_k|} = 0, \tag{5.74}$$

and

$$\mathcal{Q} = 2\{|\check{D}_+ \Phi|^2 + |\check{D}_- \Phi|^2\} = 2\{\check{\partial}_-(\Phi^\dagger \cdot \check{\partial}_+ \Phi) + \check{\partial}_+(\Phi^\dagger \cdot \check{\partial}_- \Phi)\}. \tag{5.75}$$

Hence,

$$\mathcal{Q}_k = 2\check{\partial}_+ \check{\partial}_- \ln |\hat{e}_k|^2, \tag{5.76}$$

and so we see that after the integration over θ_1 and θ_2, we find

$$Q_k = \int d\theta_+ d\theta_- \, \mathcal{Q}_k = 2\partial_+ \partial_- \ln |\hat{e}_k|^2 \Big|_{\theta=0}. \tag{5.77}$$

Moreover, as

$$|\check{D}_+ \Phi^k|^2 = \frac{|\hat{e}_{k+1}|^2}{|\hat{e}_k|^2}, \qquad |\check{D}_- \Phi^k|^2 = \frac{|\hat{e}_k|^2}{|\hat{e}_{k-1}|^2}, \tag{5.78}$$

we also find that

$$L_k = 2\partial_+ \partial_- \ln |\hat{e}_k|^2 \Big|_{\theta=0} + 4 \sum_{i=0}^{k-1} \partial_+ \partial_- \ln |\hat{e}_i|^2 \Big|_{\theta=0}. \tag{5.79}$$

Thus we see that, as in the purely bosonic case, the value of the action and of the topological charge is determined by the behaviour of the fields at spatial ∞.

One way to think about the model we have just studied (in which, effectively, all fermionic fields commute) is to consider the following interacting boson–fermion model:

$$\begin{aligned} L = 2(|D_+z|^2 + |D_-z|^2) + 4i(\psi_-^\dagger \cdot D_-\psi_+ + \psi_+^\dagger \cdot D_+\psi_-) \\ -4\{|\psi_+|^2\,|\psi_-|^2 - (\psi_-^\dagger \cdot \psi_+)(\psi_+^\dagger \cdot \psi_-)\}, \end{aligned} \tag{5.80}$$

together with the constraints $|z|^2 = 1$ and $z^\dagger \cdot \psi_\pm = 0$. In this model all the fields are classical (i.e. they commute) but the distinction between the bosonic and fermionic fields is preserved by the way the fields appear in the Lagrangian density. Notice that the fermionic contribution to the Lagrangian density comes in with an opposite sign to the more conventional expression, which appears in the supersymmetric Lagrangian (5.66). This change of the sign of the fermionic part is motivated by convenience, and it guarantees that the contribution of the fermionic fields on shell (i.e. when the equations of motion are satisfied) is positive in parallel with the contribution of the bosonic fields.

The equations of motion for our system (5.80) are highly nonlinear:

$$\begin{aligned} D_+D_-z + |D_-z|^2 z + i\{(\psi_-^\dagger \cdot \psi_+)\,D_-z + (\psi_+^\dagger \cdot \psi_-)\,D_+z\} \\ -i\{z \cdot (D_+\psi_-)^\dagger\,\psi_+ + z \cdot (D_-\psi_+)^\dagger\,\psi_-\} = 0, \end{aligned} \tag{5.81}$$

and

$$\begin{aligned} D_-\psi_+ - (z^\dagger \cdot D_-\psi_+)z + i\{|\psi_+|^2\psi_- - \psi_+(\psi_+^\dagger \cdot \psi_-)\} = 0 \\ D_+\psi_- - (z^\dagger \cdot D_+\psi_-)z + i\{|\psi_-|^2\psi_+ - \psi_-(\psi_+^\dagger \cdot \psi_-)\} = 0. \end{aligned} \tag{5.82}$$

Our solutions, given before, are solutions of these highly nonlinear equations. Notice that, for the instanton solutions, $\psi_- = 0$ and so the equations above reduce to

$$D_+D_-z + |D_-z|^2 z = 0, \tag{5.83}$$

and

$$D_-\psi_+ - (z^\dagger \cdot D_-\psi_+)z = 0, \tag{5.84}$$

i.e. the equations of motion of the $\mathbb{C}P^{N-1}$ model and the background Dirac-like equation studied before. Moreover, we see that the quartic self-coupling term of the ψ fields vanishes, and the Lagrangian density is simply that of the purely bosonic $\mathbb{C}P^{N-1}$ model. In fact, when we evaluate the Lagrangian density for the solutions to our equations (5.81) and (5.82), we find

$$L = 2(|D_+z|^2 + |D_-z|^2) + 4\{|\psi_+|^2|\psi_-|^2 - |\psi_+^\dagger \cdot \psi_-|^2\}, \tag{5.85}$$

which explicitly shows the need for both chirality components ψ_+ and ψ_- for the action to depend on the fermionic fields.

It is very easy to write down the explicit form of our (noninstantonic) solutions. For example the first (simplest) noninstantonic solution is given by

$$z = \frac{g_\perp}{|g_\perp|}, \qquad \psi_+ = -f\frac{|g_\perp|}{|f|^2},$$

and

$$\psi_- = -\frac{i}{|g_\perp|}\left(f'_\perp - g_\perp\frac{(g_\perp^\dagger \cdot f'_\perp)}{|g_\perp|^2}\right),$$

where $f' = \partial_+ f$, and where

$$f'_\perp = f' - f\frac{(f^\dagger \cdot f')}{|f|^2}, \qquad \text{and} \quad g_\perp = g - f\frac{(f^\dagger \cdot g)}{|f|^2}.$$

Notice that we have now dropped the grassmannian ϵ factors since all the functions, including g, are treated as c-number fields. Observe that the solution above represents a case where there is a real coupling between the z and ψ fields, together with the nonvanishing self-coupling of ψ. One simplifying feature occurs, namely that ψ_+ and ψ_- are mutually orthogonal.

It is easy to derive the explicit expressions for the Lagrangian and the topological charge densities for this solution. We find

$$L = 2\partial_+\partial_- \log |g_\perp|^2 + 4\partial_+\partial_- \log |f|^2, \tag{5.86}$$

and

$$Q = 2\partial_+\partial_- \log |g_\perp|^2, \tag{5.87}$$

which (as is easy to see) agree completely with the expressions derived before (5.77), as $|\hat{e}_1|^2_{\theta=0} = |g_\perp|^2$ and $|\hat{e}_0|^2_{\theta=0} = |\omega|^2_{\theta=0} = |f|^2$. In general, it is quite easy to see that

$$\begin{aligned} |\hat{e}_{2i+1}|^2_{\theta=0} &= |\partial_+^i g - \partial_+^i g \downarrow \{f, g, \partial_+ f, \partial_+ g, \cdots, \partial_+^i f\}|^2, \\ |\hat{e}_{2i}|^2_{\theta=0} &= |\partial_+^i f - \partial_+^i f \downarrow \{f, g, \partial_+ f, \partial_+ g, \cdots, \partial_+^{i-1} g\}|^2, \end{aligned} \tag{5.88}$$

which, when inserted into (5.77) and (5.79), will give the explicit expressions for the action density L_k and topological charge density Q_k for any k.

The expressions for the Lagrangian density and for the topological charge density are even more simple if we use the wedge product notation of (5.70) and (5.71). Then

$$\hat{e}_i|_{\theta=0} - \left.\frac{\hat{h}_i^\dagger \hat{h}_{i+1}}{\hat{h}_i^\dagger \hat{h}_i}\right|_{\theta=0}, \tag{5.89}$$

and so we see that $\hat{e}_i|_{\theta=0}$ is given by (5.70) in which all the $\hat{h}_i$ are replaced by $\hat{h}_i|_{\theta=0}$, i.e. by their bosonic projections. Thus, to determine the Lagrangian and topological charge densities for the solutions of the equations of motion, we need only to consider the bosonic components of the model.

We finish this chapter by giving the explicit expressions for the ψ_+, ψ_- and z fields, which solve our equations (5.79) and (5.80). They are given by

$$\begin{aligned} z^{(i)} &= \frac{A}{\sqrt{A^\dagger A}}, \\ \psi_-^{(i)} &= \frac{1}{\sqrt{A^\dagger A}}\left(B - A\frac{(A^\dagger B)}{(A^\dagger A)}\right), \\ \psi_+^{(i)} &= \frac{-1}{\sqrt{A^\dagger A}}\left(C - A\frac{(A^\dagger C)}{(A^\dagger A)}\right), \end{aligned} \tag{5.90}$$

where we have defined

$$\begin{aligned} A &= (F_1 \wedge F_2 \wedge \ldots \wedge F_i)^\dagger (F_1 \wedge F_2 \wedge \ldots \wedge F_{i+1}), \\ B &= (F_1 \wedge F_2 \wedge \ldots \wedge F_i)^\dagger K_{i+1}, \\ C &= K_i^\dagger (F_1 \wedge F_2 \wedge \ldots \wedge F_{i+1}), \end{aligned} \tag{5.91}$$

and where $\hat{\partial}_+^i \omega = F_i + i\theta_+ G_i$, and

$$K_i = \sum_{l=1}^{i} \bigwedge_{j=1,\ldots,i} F'_j, \quad F'_{j=l} = G_l \quad F'_{j\neq l} = F_j, \tag{5.92}$$

i.e. K_i is the fermionic component of the superfield

$$\omega \wedge \hat{\partial}_+ \omega \wedge \ldots \wedge \hat{\partial}_+^i \omega. \tag{5.93}$$

It is easy to check that the fields (5.90) satisfy our equations (5.79) and (5.80).

References

[1] D'Adda A, Di Vecchia P and Luscher M 1979 Confinement and chiral symmetry breaking in $\mathbb{C}P^{N-1}$ models with quarks *Nucl. Phys.* B **152** 125

[2] Din A M and Zakrzewski W J 1981 Fermion solutions in the supersymmetric $\mathbb{C}P^{N-1}$ model *Phys. Lett.* **101B** 166
Fujii K, Koikawa T and Sasaki R 1984 Classical solutions for the supersymmetric Grassmannian sigma models in two dimensions *Prog. Theor. Phys.* **71** 388

[3] Atiyah M F, Drinfeld V G, Hitchin N G and Manin Yu I 1978 Construction of instantons *Phys. Lett.* **65A** 185

[4] Din A M, Lukierski J and Zakrzewski W J 1982 General classical solutions of a supersymmetric non-linear coupled boson–fermion model in two dimensions *Nucl. Phys.* B **194** 157

6 1/N Limits

6.1 $1/N$ Expansion in Quantum Mechanics

In this section we follow closely some work of Witten [1], who was one of the first to realise the importance of seeking new and unorthodox parameters which could serve to define new perturbation expansions in field theories. He considered also such problems in quantum mechanics and so he discussed its $1/N$ expansion. The quantum mechanical case is slightly different from the field theory ones, but, as the philosophy is the same, we start this chapter with a discussion of Witten's $1/N$ expansion in quantum mechanics [1]. To motivate this expansion let us consider the hydrogen atom. Its energy levels are given by the eigenvalues of the Hamiltonian describing the system, i.e.

$$H = \frac{p^2}{2m} - \frac{e^2}{r}. \tag{6.1}$$

As is well known, this problem can be solved exactly; however, we shall ignore this fact and seek an approximate solution. Then, as e^2 is small, it would seem obvious to treat the problem perturbatively, i.e. treat e^2/r as a perturbation. However, when we say that e^2 is small we have to specify the scale, and it is easy to show that, on the energy scale available, e^2 is not small; in fact, it serves to define the overall scale of energy eigenvalues, and so it makes little sense to base a perturbation expansion on e^2. To see this, observe that, if we rescale $r \to \lambda r$, $p \to p/\lambda$, with $\lambda = 1/me^2$ then

$$H = me^4 \left(\frac{p^2}{2} - \frac{1}{r}\right)$$

and so we see that as e^2 appears only in me^4 it can be absorbed into the rescaling of time.

However, as Witten pointed out, the theory does possess an additional parameter, a hidden one, namely the dimension of space. Of course the actual dimension of space is 3, but we may consider it as a parameter, called N, and then study possible $1/N$ expansions. Unfortunately 3 is not very large, but there may be quantities, which are insensitive to, or not very sensitive to, the exact value of N; for them the results obtained in the $1/N$ perturbation expansion may give very good approximations to the exact values. What is interesting is that the results are not bad even for the ground state energies, where, at first sight, there is little chance of success.

Let us look at the ground state energy of the hydrogen atom. The quantum mechanical ground state wavefunction ψ is required to satisfy the radial Schrödinger equation in N dimensions:

$$\left[-\frac{1}{2m}\left(\frac{\mathrm{d}^2}{\mathrm{d}r^2}+\frac{N}{r}\frac{\mathrm{d}}{\mathrm{d}r}\right)-\frac{e^2}{r}\right]\psi = E\psi, \tag{6.2}$$

where we have used the fact that we have an s-state and that, to leading order, $N-1$ should be replaced by N. Observe that, although we study the problem in N dimensions, we have kept the usual form of the potential term (i.e. $V = e^2/r$).

Next, to eliminate the first-order derivatives, we change our variables to

$$\psi = r^{-N/2}\tilde{\psi}$$

and find

$$H \to H' = -\frac{1}{2m}\frac{\mathrm{d}^2}{\mathrm{d}r^2} + \frac{N^2}{8mr^2} - \frac{e^2}{r}. \tag{6.3}$$

We rescale the radial coordinate r by setting $r = N^2R$ and find that the Hamiltonian changes to

$$H' \to H'' = \frac{1}{N^2}\left(-\frac{1}{2mN^2}\frac{\mathrm{d}^2}{\mathrm{d}R^2} + \frac{1}{8mR^2} - \frac{e^2}{R}\right). \tag{6.4}$$

Hence, apart from the overall $1/N^2$ factor, the only N-dependence appears in the effective mass ($2mN^2$). So we see that the motion resembles that of a particle of mass $m_{\mathrm{eff}} = mN^2$ moving in an effective potential

$$V_{\mathrm{eff}} = \frac{1}{8mR^2} - \frac{e^2}{R}. \tag{6.5}$$

For large values of m_{eff} the particle is at rest at the minimum of the potential. This minimum is given by

$$\frac{\partial V}{\partial R} = 0 \qquad \Rightarrow \qquad R = \frac{1}{4me^2}, \tag{6.6}$$

and so the value of the energy is given by

$$H_{min} = -\tfrac{2}{9}\,me^4, \tag{6.7}$$

where we have used $N = 3$ in setting the overall scale. As is well known, the actual number is $-\frac{1}{2}me^4$, hence the result is encouraging. (In fact, had we not replaced $N-1$ by N in (6.2), we would have obtained, in this special case, an exact answer.) This method can be modified to allow us to calculate energies of other states or to apply it to other quantum mechanical problems [2]. Of course, we are not advocating the use of this expansion in quantum mechanics, where we have many other, more useful expansions; we are only trying to point out that a $1/N$ expansion is possible and that it gives reasonable answers. We want to apply it in field theories, where we do not have so many other possibilities.

6.2 1/N Expansion in $\lambda\phi^4$ Theory

In his Erice lectures Coleman [3] shows very clearly how to use the $1/N$ expansion to calculate various quantities in a $\lambda\phi^4$ theory. As in the next section we are going to apply this expansion to the $\mathbb{C}P^{N-1}$ model, we repeat some of his results here.

To define a $\lambda\phi^4$ model we consider a set of N fields ϕ^a, where $a = 1,\ldots,N$ and for a Lagrangian density we take (the Euclidean case):

$$L = \tfrac{1}{2}\partial_\mu\phi^a\partial_\mu\phi^a + \tfrac{1}{2}\mu_0^2\phi^a\phi^a + \tfrac{1}{8}\lambda_0(\phi^a\phi^a)^2. \tag{6.8}$$

This time we cannot find the Green functions exactly, so we have to rely on some expansion scheme and we will consider a $1/N$ expansion. Observe that this time we keep the dimension of spacetime fixed; our expansion will be in $1/N$, where N is the number of our fields ϕ^a. To do this, let us look, first of all, at some diagrams which arise in a more usual perturbation theory, namely the one based on the expansion in powers of λ_0. Looking at the four-point function, we find the following lowest order diagrams:

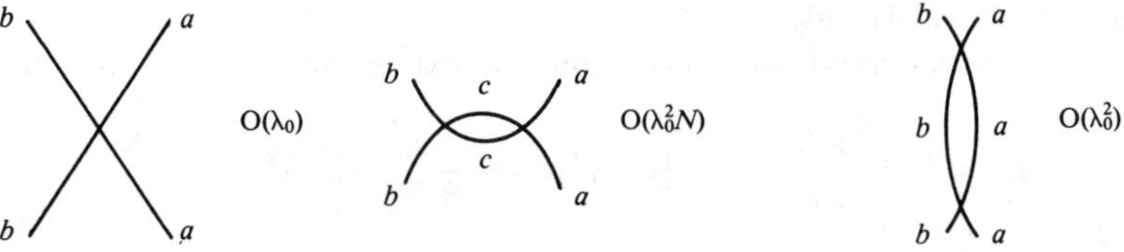

We see that, of the two diagrams of order λ_0^2, the diagram in the middle is of the order of $\lambda_0^2 N$, as the line c in it can represent any of the fields ϕ^a. Thus we see that the contribution of the diagram on the right is suppressed by a factor $1/N$ relative to the contribution of the diagram in the middle, and they both contain a further factor λ_0 relative to the diagram on the left. If we now introduce a new coupling constant g_0, defined by $g_0 = \lambda_0 N$, and consider the limit $\lambda_0 \to 0, N \to \infty$, such that g_0 is kept fixed, we see that the diagram on the left and the one in the middle are of the first order in $1/N$, while the diagram on the right is of the second order and so its contribution can be neglected. The $1/N$ expansion relies on this fact; observe that the lowest order contribution corresponds to an ∞ number of diagrams of the conventional perturbation theory. There are no terms which correspond to N^α, where α is a positive power and the leading perturbative terms are of order $1/N$. The lower order terms (i.e. of order $1/N^\beta$, where $\beta > 1$) can then be ignored or their contribution can be calculated perturbatively.

To study the model further, it is convenient to introduce an auxiliary field σ by modifying the Lagrangian

$$L \to L' = L + \frac{1}{2}\frac{N}{g_0}\left(\sigma - \frac{i}{2}\frac{g_0}{N}\phi^a\phi^a\right)^2. \tag{6.9}$$

Notice that the addition of this new term has no effect on the dynamics, as

$$\begin{aligned}&\int \mathrm{d}[\sigma]\, \exp\left[-\int \mathrm{d}^d x\, \frac{1}{2}\frac{N}{g_0}\left(\sigma - \frac{i}{2}\frac{g_0}{N}\phi^a\phi^a\right)^2\right] \\ &\qquad = \int \mathrm{d}[\tilde{\sigma}]\, \exp\left(-\int \mathrm{d}^d x\, \frac{1}{2}\frac{N}{g_0}\tilde{\sigma}^2\right) = K,\end{aligned} \tag{6.10}$$

where K is some number which depends on N, g_0 but which is independent of ϕ^a. Hence the effect of introducing the σ field can be incorporated into the change of our functional measure. Moreover, observe that the Euler–Lagrange equations for the σ field give

$$\sigma = \frac{i}{2}\frac{g_0}{N}\phi^a\phi^a, \tag{6.11}$$

and as they do not involve the time derivative, this equation is just an equation of constraint.

On the other hand, we can expand the expressions in (6.9) and find

$$\begin{aligned}L' &= L + \frac{1}{2}\frac{N}{g_0}\sigma^2 + \tfrac{1}{2}i\phi^a\phi^a\sigma^2 - \frac{1}{8}\frac{g_0}{N}(\phi^a\phi^a)^2 \\ &= \tfrac{1}{2}\partial_\mu\phi^a\partial_\mu\phi^a + \tfrac{1}{2}\mu_0^2\phi^a\phi^a + \frac{1}{2}\frac{N}{g_0}\sigma^2 + \tfrac{1}{2}i\phi^a\phi^a\sigma.\end{aligned} \tag{6.12}$$

We notice that the term $(\phi^a\phi^a)^2$ has been eliminated. Although our introduction of the σ field has not modified the theory it has changed its

Feynman rules — we have no ϕ_0^4 vertex; instead we have a $\phi^a\phi^a\sigma$ vertex and a trivial g_0/N propagator for the σ field. Moreover, all the $1/N$ factors are associated with the σ field, all the closed ϕ^a loops give a factor N and index a is conserved. Hence our previous diagrams now look as follows:

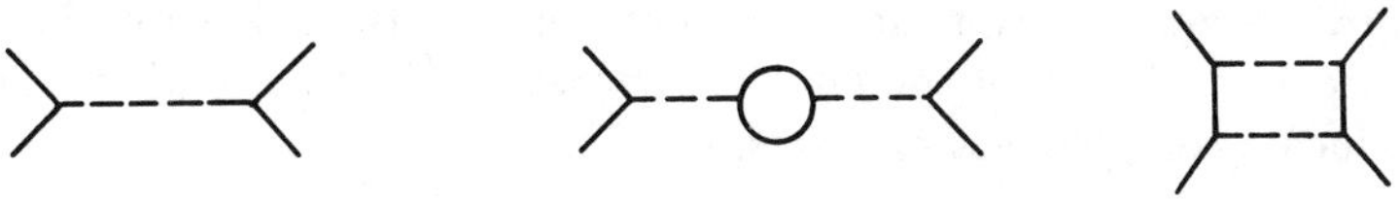

where --- denotes the σ propagator.

As there are no ϕ^4 terms in our modified Lagrangian, in principle, we can now perform the integration over the ϕ^a fields obtaining

$$\begin{aligned}\int \mathrm{d}[\phi^a] \exp\{\tfrac{1}{2}\partial_\mu\phi^a\partial_\mu\phi^a + \tfrac{1}{2}\mu_0^2\phi^a\phi^a + \tfrac{1}{2}i\phi^a\phi^a\sigma\} \\ \sim \det\{(-\partial_\mu\partial_\mu + \mu_0^2 + \sigma i)\delta(x-y)\}^{-N/2}.\end{aligned} \tag{6.13}$$

The 'closed form' of the obtained determinant is an illusion. To do anything with it we need to perform an expansion into Feynman diagrams.

If we now look at σ fields and recall that

$$(\det A)^{-N/2} = \exp\{-\tfrac{1}{2}N \operatorname{Tr}\ln A\}, \tag{6.14}$$

we observe that we have an effective, nonlocal, action for the σ field given by

$$S_{\text{eff}} = \int \mathrm{d}^d x \left(\frac{1}{2}\frac{N}{g_0}\sigma^2 + \tfrac{1}{2}N \operatorname{Tr}\ln[(-\partial_\mu\partial_\mu + \mu_0^2 + \sigma i)\delta(x-y)]\right). \tag{6.15}$$

Notice that, due to the explicit N dependence of both terms, we have

$$S_{\text{eff}}(\sigma, N) = N\, S_{\text{eff}}(\sigma, 1). \tag{6.16}$$

What do we actually mean by

$$\operatorname{Tr}\ln[(-\partial_\mu\partial_\mu + \mu_0^2 + \sigma i)\delta(x-y)]?$$

To have a better understanding of this expression we, first of all, rewrite it as

$$\operatorname{Tr}\ln[(-\partial_\mu\partial_\mu + \mu_0^2)\delta(x-y)] + \int \mathrm{d}^d x \ln[1 + i\sigma(x)(-\partial_\mu\partial_\mu + \mu_0^2)^{-1}]. \tag{6.17}$$

Next we realise that the first term in (6.17) is irrelevant, as it does not depend on σ and so it contributes only an overall constant to any functional integral over σ. The second term in (6.17) can be expanded using

$$\ln(1 + a) = a - \tfrac{1}{2}a^2 + \tfrac{1}{3}a^3 - \ldots . \tag{6.18}$$

This expansion, however, has to be performed with some care, due to the appearance in (6.17) of the nonlocal operator $(-\partial_\mu\partial_\mu + \mu_0^2)^{-1}$. Thus we introduce a complete set of states:

$$\int \mathrm{d}^d x\ |x\rangle\langle x| = 1. \tag{6.19}$$

Then we observe that

$$\begin{aligned} \langle x|\,\sigma\,|y\rangle &= \sigma(x)\delta(x - y) \\ \langle x|\,(-\partial_\mu\partial_\mu + \mu_0^2)^{-1}\,|y\rangle &= \Delta(x, y), \end{aligned} \tag{6.20}$$

where $\Delta(x, y)$ is the well known propagator of the scalar field. Thus the second term in (6.17) gives us

$$\begin{aligned} &\langle x|\ln[1 + \sigma i(-\partial_\mu\partial_\mu + \mu_0^2)^{-1}]\,|x\rangle \\ &= i\sigma(x)\Delta(x, x) + \tfrac{1}{2}\int \mathrm{d}^d y\,\sigma(x)\Delta(x, y)\sigma(y)\Delta(y, x) \\ &- \tfrac{1}{3}i\int \mathrm{d}^d y\,\mathrm{d}^d z\,\sigma(x)\Delta(x,y)\sigma(y)\Delta(y, z)\sigma(z)\Delta(z, x) + \ldots . \end{aligned} \tag{6.21}$$

Hence we see that we have a nonlocal interaction of the σ field, with the action given by

$$\begin{aligned} S_{\text{eff}} = \tfrac{1}{2}N\int \mathrm{d}^d x \Bigg(&\frac{1}{g_0}\sigma(x)\sigma(x) + i\sigma(x)\Delta(x, x) \\ &+ \tfrac{1}{2}\int \mathrm{d}^d y\sigma(x)\Delta(x, y)\sigma(y)\Delta(y, x) \\ &- \tfrac{1}{3}i\int \mathrm{d}^d y\,\mathrm{d}^d z\,\sigma(x)\Delta(x, y)\sigma(y)\Delta(y, z)\sigma(z)\Delta(z, x) + \ldots \Bigg), \end{aligned} \tag{6.22}$$

or, diagramatically,

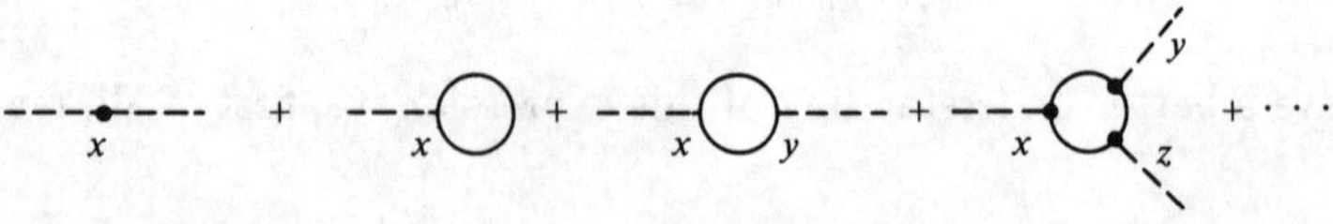

We see that we can interpret these terms as corresponding to local $(\frac{1}{2}Ng_0\,\sigma(x)\sigma(x))$ and nonlocal $(\frac{1}{4}N\,\sigma(x)\Delta(x, y)\sigma(y)\Delta(y, x))$ contributions to the propagator of the σ field and nonlocal vertices involving an arbitrary number of σ fields. As the σ field propagator is, in fact, the inverse of the expression given above, we notice that each propagator contributes a factor $1/N$ and each vertex contributes a factor N. Hence a general diagram with E external σ lines with their propagators, I internal σ lines, V vertices and L loop integrations has a coefficient N^{V-I-E} which, as $L = I - V + 1$, is given by N^{-E-L+1}. (We have included propagators for the external lines as we want to connect these lines to our ϕ^a fields.) Thus the smallest power of $1/N$ is obtained from graphs with no loops and a minimum number of external σ fields.

However, we still have a problem due to the term linear in σ. We eliminate this term by a shift of the σ field

$$\sigma \to \breve{\sigma} = \sigma - \sigma_0, \tag{6.23}$$

where σ_0 is so chosen that $\sigma = \sigma_0$ is a stationary point of S_{eff} and so $\breve{\sigma}$ satisfies

$$\left.\frac{\delta S_{\text{eff}}}{\delta\sigma}\right|_{\breve{\sigma}} = 0. \tag{6.24}$$

Then there are no linear terms, when the effective Lagrangian density is expressed in terms of $\breve{\sigma}$. Of course, we can go back to the original Lagrangian (6.12) and reexpress it in terms of $\breve{\sigma}$. We find

$$L = \tfrac{1}{2}\partial_\mu\phi^a\partial_\mu\phi^a + \mu_1^2\phi^a\phi^a + \frac{1}{2}\frac{N}{g_0}\breve{\sigma}^2 - \tfrac{1}{2}i\breve{\sigma}\phi^a\phi^a + \frac{N}{g_0}\breve{\sigma}\sigma_0 + \text{constant}. \tag{6.25}$$

Here $\mu_1^2 = \mu_0^2 + \sigma_0$ and we have an additional linear vertex, which we use to fix σ_0 at such a value that the total linear vertex vanishes.

Now we can calculate whatever we want; in particular we can calculate the ϕ–ϕ scattering to leading order in $1/N$. Only tree diagrams contribute and we have

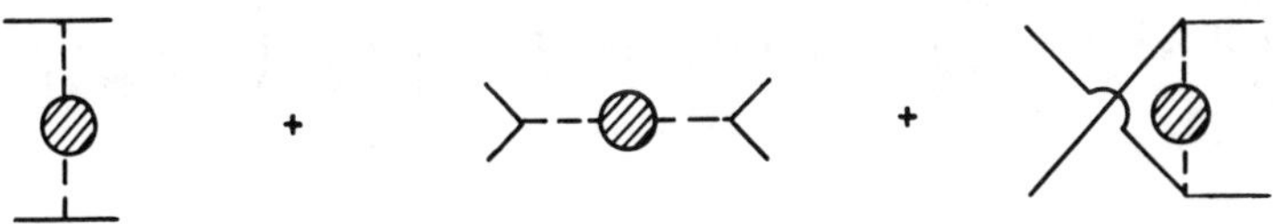

where --◍-- denotes the complete two-point function, which is given by the inverse of the previous expression, i.e. by

$$D^{-1}(p) = -N\left(\frac{1}{g_0} + \int\frac{\mathrm{d}^d k}{(2\pi)^d}\frac{1}{k^2+\mu_1^2}\frac{1}{(p+k)^2+\mu_1^2}\right). \tag{6.26}$$

Of course we can calculate other quantities in the same way.

6.3 $1/N$ Expansion of the $\mathbb{C}P^{N-1}$ Model

In this section we will discuss the $1/N$ limit expansion procedure for the $\mathbb{C}P^{N-1}$ model. This was done in the original paper by D'Adda *et al* [4] and we closely follow their work. We begin by altering the normalisation of the z_α fields; we choose them to satisfy

$$z_\alpha^\dagger \cdot z_\alpha = \frac{N}{2f}, \tag{6.27}$$

where f is the coupling constant. This normalisation brings out the fact that, as in the example discussed in the previous section, the number of fields grows with N.

In contradistinction to the previous example where we considered various Green functions, this time we try to be more general and so we choose to consider a general generating functional $Z[J, J^*, K_\mu]$ defined by

$$\begin{aligned} Z[J, J^\dagger, K_\mu] = \int \mathrm{D}[z]\mathrm{D}[z^\dagger] \prod_x \delta\left(z^\dagger \cdot z - \frac{N}{2f}\right) \\ \times \exp\left(-S + \int \mathrm{d}^2x (J^\dagger z + z^\dagger J + K_\mu A_\mu)\right), \end{aligned} \tag{6.28}$$

where $J_\alpha, J_\alpha^\dagger$ and K_μ are sources of $z_\alpha, z_\alpha^\dagger$ and of the composite $A_\mu = (2f/N)i(z^\dagger \cdot \partial_\mu z)$ field. With the present choice of the variables the action is given by

$$S = \int \mathrm{d}^2x \left(\partial_\mu z_\alpha^\dagger \cdot \partial_\mu z_\alpha + \frac{2f}{N}(z_\alpha^\dagger \cdot \partial_\mu z_\alpha)(z_\beta^\dagger \cdot \partial_\mu z_\beta)\right). \tag{6.29}$$

To proceed further we have to take into account our constraint $z^\dagger z = N/2f$ represented by our functional δ function, and also to eliminate terms proportional to z^4 (i.e. to introduce a σ-like field). We do this by introducing two new fields $\alpha(x)$ and $\lambda_\mu(x)$. First we put

$$\prod_x \left(\delta(z^\dagger \cdot z) - \frac{N}{2f}\right) = \int \mathrm{D}[\alpha] \exp\left\{\int \mathrm{d}^2x \left[\frac{i}{\sqrt{N}}\alpha\left(z^\dagger \cdot z - \frac{N}{2f}\right)\right]\right\}, \tag{6.30}$$

and

$$\begin{aligned} &\exp\left(\frac{f}{N}\int \mathrm{d}^2x[-2(z^\dagger \cdot \partial_\mu z)(z^\dagger \cdot \partial_\mu z) + 2iK_\mu(z^\dagger \cdot \partial_\mu z)]\right) \\ &= \int \mathrm{D}[\lambda_\mu] \exp\left[\int \mathrm{d}^2x \left(-\frac{1}{2f}\lambda_\mu\lambda_\mu \right.\right. \\ &\left.\left. + \frac{1}{\sqrt{N}}\lambda_\mu(K_\mu + 2iz^\dagger \cdot \partial_\mu z) - \frac{f}{2N}K_\mu K_\mu\right)\right]. \end{aligned} \tag{6.31}$$

A few words of explanation are needed at this stage. The first expression (6.30) is really a functional analogue of the Fourier transform, which we discussed in the first chapter. Also the factor $1/\sqrt{N}$, which appears in it, is really put in 'by hand' (we discard the irrelevant overall factor associated with the change of variables which allows us to include this factor in the exponential in (6.31).) In fact there exists a simple rule governing what factors have to be 'put in by hand'. It is: put $\sqrt{N}$ for all linear terms, 1 for quadratic, $1/\sqrt{N}$ for cubic, etc; all the overall factors get absorbed into the integration measure. Of course, had we made a wrong guess as to what factor to 'put in by hand', we would have obtained an expression which would either vanish or diverge in the $N \to \infty$ limit. The expression for (6.31) follows from the introduction of an overall factor represented by a gaussian functional integral, in which the integration variable is given by

$$\lambda_\mu - K_\mu \frac{f}{\sqrt{N}} - \frac{2if}{\sqrt{N}}(z^\dagger \cdot \partial_\mu z).$$

Moreover, we observe that, using $z^\dagger \cdot z = N/2f$, we can replace $\lambda_\mu \lambda_\mu / 2f$ by $\lambda_\mu \lambda_\mu (z^\dagger z)/N$, and also that we can insert an additional factor $-m^2 z^\dagger z$, which, due to the constraint, corresponds to an irrelevant overall factor. However, as we will see later, the value of m, which is arbitrary at this stage, will be fixed by some consistency considerations. With these modifications we have

$$Z[J, J^\dagger, K_\mu] = \int \mathrm{D}[z]\,\mathrm{D}[z^\dagger]\,\mathrm{D}[\alpha]\,\mathrm{D}[\lambda_\mu] \exp\Bigg[-\int \mathrm{d}^2x \Bigg((z^\dagger \cdot \partial_\mu z) \tag{6.32}$$

$$+ \frac{i\sqrt{N}}{2f}\alpha + (J^\dagger \cdot z) + (z^\dagger \cdot J) + \frac{1}{\sqrt{N}} K_\mu \lambda_\mu - \frac{f}{2N} K_\mu K_\mu \Bigg)\Bigg],$$

where

$$\Delta = -D_\mu D_\mu + M^2 - \frac{i\alpha}{\sqrt{N}} \qquad \text{and} \qquad D_\mu = \partial_\mu + \frac{i}{\sqrt{N}}\lambda_\mu. \tag{6.33}$$

Our new Lagrangian still posesses gauge invariance; it is now represented by the invariance with respect to transformations

$$z_\alpha \to z'_\alpha = e^{i\Lambda} z_\alpha, \quad \alpha \to \alpha' = \alpha, \quad \lambda_\mu \to \lambda'_\mu = \lambda_\mu - \sqrt{N}\partial_\mu \Lambda. \tag{6.34}$$

As our new Lagrangian is quadratic in the z_α fields, we can integrate them out and obtain

$$Z[J, J^\dagger, K_\mu] = \int \mathrm{D}[\alpha]\mathrm{D}[\lambda_\mu] \exp\Bigg[-S_{\text{eff}}$$

$$+ \int \mathrm{d}^2x \left(J^\dagger \cdot \Delta^{-1} \cdot J + \frac{1}{\sqrt{N}} K_\mu \lambda_\mu - \frac{f}{2N} K_\mu K_\mu \right)\Bigg], \tag{6.35}$$

where

$$S_{\text{eff}} = N\,\text{Tr}\ln\Delta + \frac{i\sqrt{N}}{2f}\int \mathrm{d}^2x\alpha(x), \tag{6.36}$$

and where we have changed $z \to z' = z + \Delta^{-1}J$, and similarly for $z^\dagger$. In the derivation of this result we have dropped some expressions which represented total divergences; this corresponds to assuming that the radial components of our fields A_μ and λ_μ vanish at ∞.

We are now ready to perform the $1/N$ expansion of our generating functional. Looking at (6.35) we observe that we can write

$$S = \sum_{\nu=1}^{\infty} N^{1-\nu/2}\,S^{(\nu)} + N\ \text{constant}. \tag{6.37}$$

The constant term is independent of both α and λ_μ and so can be ignored as an overall factor; we are left with having to determine all the other terms. We expand

$$\begin{aligned}\text{Tr}\ln&\left\{\left[-\left(\partial_\mu + \frac{i}{\sqrt{N}}\lambda_\mu\right)\left(\partial_\mu + \frac{i}{\sqrt{N}}\lambda_\mu\right) + m^2 - \frac{i}{\sqrt{N}}\alpha\right]\delta(x-y)\right\}\\ &= \text{Tr}\ln\left\{(-\partial_\mu\partial_\mu + m^2)\delta(x-y)\left[1 - \left(\frac{2i}{\sqrt{N}}\lambda_\mu\partial_\mu + \frac{i}{\sqrt{N}}(\partial_\mu\lambda_\mu)\right.\right.\right.\\ &\quad\left.\left.\left. + \frac{i}{\sqrt{N}}\alpha - \frac{1}{N}\lambda_\mu\lambda_\mu\right)(m^2 - \partial_\mu\partial_\mu)^{-1}\right]\right\}.\end{aligned} \tag{6.38}$$

Once again we use $\ln AB = \ln A + \ln B$, and we disregard the first term ($\text{Tr}\ln[(-\partial_\mu\partial_\mu + m^2)\delta(x-y)]$) (as it gives an irrelevant overall constant factor) and expand all the other terms in the power series in $1/N$.

The lowest terms in the expansion are given by

$$-\frac{i}{\sqrt{N}}\,\text{Tr}\left[\alpha(m^2 - \partial_\mu\partial_\mu)^{-1}\,\delta(x-y)\right], \tag{6.39}$$

and

$$\frac{2i}{\sqrt{N}}\,\text{Tr}\left[(2\lambda_\mu\partial_\mu + \partial_\mu\lambda_\mu)(m^2 - \partial_\mu\partial_\mu)^{-1}\,\delta(x-y)\right]. \tag{6.40}$$

Despite the appearance to the contrary, the terms involving λ_μ do not contribute to these terms. To see this we use the complete set of states (6.19) (except that this time in momentum space), and rewrite the second term (apart from the $2i/\sqrt{N}$ factor and only up to an overall $1/(2\pi)^2$) as

$$\begin{aligned}\int \mathrm{d}^2p \int \mathrm{d}^2x \mathrm{d}^2y \Bigg(&\langle x|\, 2\lambda_\mu \,|y\rangle \,\langle y|p\rangle p_\mu \frac{1}{p^2+m^2} \langle p|x\rangle \\ &+ \langle x|\, \partial_\mu\lambda_\mu \,|y\rangle\, \langle y|p\rangle \frac{1}{p^2+m^2} \langle p|x\rangle \Bigg) \\ = \int \mathrm{d}^2x \int \mathrm{d}^2p \left(2\lambda_\mu(x) \frac{p_\mu}{p^2+m^2} + \partial_\mu\lambda_\mu \frac{1}{p^2+m^2} \right).\end{aligned} \tag{6.41}$$

The first integral vanishes as

$$\int \mathrm{d}^2p\, p_\mu \frac{1}{p^2+m^2} = 0 \tag{6.42}$$

(due to the antisymmetry under $p_\mu \to -p_\mu$), and the second vanishes as

$$\int \mathrm{d}^2x\, \partial_\mu\lambda_\mu(x) = \lambda_\mu(x) n_\mu \Big|_r = 0, \tag{6.43}$$

where the subscript r denotes the radial contribution at spacial ∞, which vanishes.

Thus we see that the leading terms of our expansion are given by

$$\begin{aligned} S^{(1)} &= \frac{i}{2f} \int \mathrm{d}^2x\, \alpha(x) - i\, \mathrm{Tr}\left[\alpha(m^2 - \partial_\mu\partial_\mu)^{-1}\delta(x-y)\right] \\ &= \breve{\alpha}(0) \left(\frac{1}{2f} - \int \frac{\mathrm{d}^2q}{(2\pi)^2} \frac{1}{m^2+q^2} \right), \end{aligned} \tag{6.44}$$

where $\breve{\alpha}(p) = \int \mathrm{d}^2x\, e^{-ipx} \alpha(x)$.

The last integral is clearly divergent and so it needs some regularisation. If we introduce a 'Pauli–Villars' cut-off Λ, we have

$$\int^\Lambda \frac{\mathrm{d}^2q}{(2\pi)^2} \frac{1}{m^2+q^2} = \int_0^\Lambda \frac{q\,\mathrm{d}q}{2\pi} \frac{1}{m^2+q^2} = \frac{1}{4\pi} \ln \frac{m^2+\Lambda^2}{m^2}. \tag{6.45}$$

Hence we choose $\breve{\Lambda}^2 = m^2 + \Lambda^2$ and obtain

$$\frac{1}{4\pi} \ln \frac{\Lambda^2}{m^2}$$

(dropping the ˘). So, if we let the bare coupling constant f depend on the cut-off Λ as

$$\frac{2\pi}{f} = \ln \frac{\Lambda^2}{\mu^2} + \frac{2\pi}{f_R(\mu)}, \tag{6.46}$$

where μ is the renormalisation point and $f_R(\mu)$ is the renormalised coupling constant, we obtain

$$S^{(1)} = i\breve{\alpha}(0)\left(\frac{1}{2f_R(\mu)} + \frac{1}{4\pi}\ln\frac{m^2}{\mu^2}\right). \tag{6.47}$$

We see that, as $N \to \infty$, $\sqrt{N}S^{(1)}$ changes vary rapidly, showing that unless $S^{(1)}$ vanishes our $1/N$ expansion does not make sense. However, we can choose the previously arbitrary parameter m to be such that $S^{(1)}$ vanishes. Thus we impose the so-called *saddle-point condition*:

$$m^2 = \mu^2 \exp\left(-\frac{2\pi}{f_R(\mu)}\right). \tag{6.48}$$

With this choice $S^{(1)}$ vanishes. Before we go any further let us make a few comments about our conditions (6.46) and (6.48).

First of all, observe that (6.46) implies that as $\Lambda \to \infty$, $f \to 0$. The fact that the bare coupling constant goes to zero implies that our theory is *asymptotically free*. To see this more clearly we rewrite (6.46) as

$$f_R(\mu) = 2\pi\left(\ln\frac{\mu^2}{\Lambda^2} + \frac{2\pi}{f}\right)^{-1/2},$$

and recall [5] that, in the general setting,

$$\beta(\lambda) = \lim_{\epsilon\to 0}\mu\frac{\partial\lambda}{\partial\mu}, \tag{6.49}$$

where λ is a coupling constant. In our case $\epsilon \to 0$ corresponds to $\Lambda \to \infty$ and $\lambda \sim f_R(\mu)$. Then

$$\mu^2\frac{\partial f_R}{\partial\mu^2} = \frac{\partial f}{\partial\ln\mu^2} = -\frac{1}{2\pi}f_R^2, \tag{6.50}$$

and so we see that our beta function is given by

$$\beta(\lambda) = -\frac{1}{\pi}\lambda^2, \tag{6.51}$$

thus exhibiting the behaviour (for $\lambda \to 0$) corresponding to asymptotic freedom (see the figure below).

Moreover, the condition (6.48) describes the effective generation of mass. Initially the parameter m was inserted 'by hand' and nothing depended on its value; now we find that we can satisfy our 'saddle-point condition' only if this parameter takes the specific value given by (6.48). In addition, its value is proportional to μ^2 — and we have an example of the generation

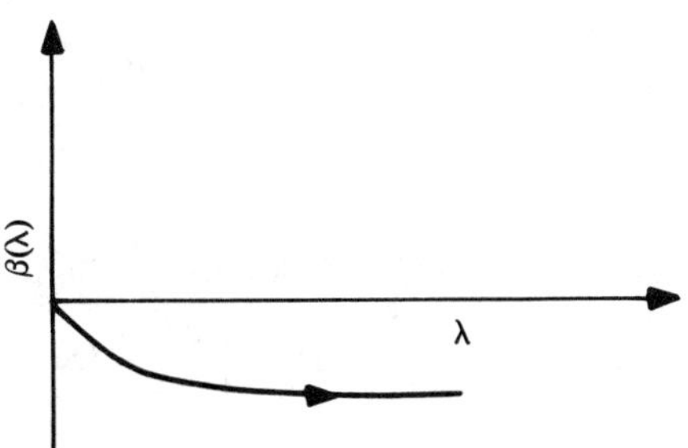

of mass by quantum effects. The classical model is conformally invariant and so has no scale. The quantum corrections break this invariance and introduce a dimensionful parameter μ, allowing the complete model to exhibit mass-like terms, whose scale is set by μ.

Next we look at terms which are $\sim O(1)$ in $1/N$. They are given by $S^{(2)}$:

$$\begin{aligned} S^{(2)} &= \operatorname{Tr}\{\lambda_\mu\lambda_\mu(m^2-\partial_\mu\partial_\mu)^{-1}\delta(x-y)\} \\ &+\tfrac{1}{2}\operatorname{Tr}\{(i\alpha+i\partial_\mu\lambda_\mu+2i\lambda_\mu\partial_\mu)(m^2-\partial_\mu\partial_\mu)^{-1} \\ &(1+\alpha+i\partial_\mu\lambda_\mu+2i\lambda_\mu\partial_\mu)(m^2-\partial_\mu\partial_\mu)^{-1}\delta(x-y)\}. \end{aligned} \tag{6.52}$$

Consider first the terms in $S^{(2)}$ which involve only the α fields. We call them $S^{(2)}_{(\alpha,\alpha)}$. They are given by:

$$\begin{aligned} S^{(2)}_{(\alpha,\alpha)} &= -\tfrac{1}{2}\int \mathrm{d}^2x\mathrm{d}^2y\mathrm{d}^2z\ \langle x|\,\alpha\,|y\rangle\,\Delta(y,z)\alpha(z)\Delta(z,x) \\ &= -\tfrac{1}{2}\int \mathrm{d}^2x\mathrm{d}^2y\,\alpha(x)\alpha(y)\,F^\alpha(x,y), \end{aligned} \tag{6.53}$$

where

$$\begin{aligned} F^\alpha(x,y) &= \Delta(x,y)\Delta(y,x) \\ &= \int\frac{\mathrm{d}^2q}{(2\pi)^2}\,e^{iq(x-y)}\int\frac{\mathrm{d}^2p}{(2\pi)^2}\,e^{ip(y-x)}\frac{1}{q^2+m^2}\frac{1}{p^2+m^2} \\ &= \int\frac{\mathrm{d}^2q'}{(2\pi)^2}\,e^{iq'(x-y)}\int\frac{\mathrm{d}^2p}{(2\pi)^2}\frac{1}{p^2+m^2}\frac{1}{(p+q')^2+m^2}. \end{aligned} \tag{6.54}$$

Thus we see that $F^\alpha(x,y)$ is a Fourier transform of

$$\tilde{F}^\alpha(q') = \int\left(\frac{\mathrm{d}^2p}{(2\pi)^2}\right)\frac{1}{p^2+m^2}\frac{1}{(p+q')^2+m^2}. \tag{6.55}$$

In a similar way we analyse the $(\lambda_\mu \lambda_\mu)$ case. From the first term in (6.52) we have

$$\int \mathrm{d}^2x \mathrm{d}^2y \mathrm{d}^2z \, \langle x| \lambda_\mu |y\rangle \langle y| \lambda_\mu |z\rangle \Delta(z,x) = \int \mathrm{d}^2x \lambda_\mu(x)\lambda_\mu(x)\Delta(x,x). \tag{6.56}$$

The second term gives

$$\begin{aligned} &-\tfrac{1}{2}\,\mathrm{Tr}\,\{[2\lambda_\mu\partial_\mu + (\partial_\mu\lambda_\mu)](m^2 - \partial_\mu\partial_\mu)^{-1}[2\lambda_\mu\partial_\mu + (\partial_\mu\lambda_\mu)] \\ &(m^2 - \partial_\mu\partial_\mu)^{-1}\delta(x-y)\} = -\tfrac{1}{2}\int \mathrm{d}^2x\mathrm{d}^2y \frac{\mathrm{d}^2q\mathrm{d}^2p}{(2\pi)^4} \langle x| 2\lambda_\mu\partial_\mu + (\partial_\mu\lambda_\mu) |p\rangle \\ &\frac{1}{p^2+m^2}\langle p|y\rangle \langle y| 2\lambda_\mu\partial_\mu + (\partial_\mu\lambda_\mu) |q\rangle \langle q|x\rangle \frac{1}{q^2+m^2} \\ &= -\tfrac{1}{2}\int \mathrm{d}^2x\mathrm{d}^2y \frac{\mathrm{d}^2p\mathrm{d}^2q}{(2\pi)^4} \langle x|p\rangle(2\lambda_\mu(x)ip^\mu + \partial_\mu\lambda_\mu(x))\frac{1}{p^2+m^2}\langle p|y\rangle \\ &\langle y|q\rangle[2\lambda_\mu(y)iq_\mu + (\partial_\mu\lambda_\mu(y)]\frac{1}{q^2+m^2}\langle q|x\rangle. \end{aligned} \tag{6.57}$$

However, the integration by parts gives us

$$\int \mathrm{d}^2x\langle x|p\rangle\partial_\mu\lambda_\mu(x)\langle q|x\rangle = -i(p_\mu - q_\mu)\lambda_\mu(x)\langle x|p\rangle\langle q|x\rangle,$$

where we have assumed that $\lambda_\mu \to 0$ as $|x| \to \infty$. So we obtain

$$\begin{aligned} &-\tfrac{1}{2}\int \mathrm{d}^2x\mathrm{d}^2y \frac{\mathrm{d}^2p\mathrm{d}^2q}{(2\pi)^4}\langle x|p\rangle(p_\mu+q_\mu)(p_\nu+q_\nu)\frac{1}{p^2+m^2}\frac{1}{q^2+m^2}\langle p|y\rangle\langle q|x\rangle \\ &\times \langle y|q\rangle\lambda_\mu(x)\lambda_\nu(y) = \tfrac{1}{2}\int \mathrm{d}^2x\mathrm{d}^2y\lambda_\mu(x)\lambda_\nu(y) \\ &\times \int \frac{\mathrm{d}^2p'}{(2\pi)^2}e^{ip'(x-y)}\int \frac{\mathrm{d}^2q'}{(2\pi)^2}\frac{4q'_\mu q'_\nu}{[(q'-\frac{1}{2}p')^2+m^2][(q'+\frac{1}{2}p')^2+m^2]}, \end{aligned} \tag{6.58}$$

where $p' = p - q$ and $q' = q + \frac{1}{2}p'$.

Collecting all the terms together for the $(\lambda_\mu, \lambda_\mu)$ contribution we have

$$S^{(2)}_{\lambda,\lambda} = \tfrac{1}{2}\int \mathrm{d}^2x \int \mathrm{d}^2y\, \lambda_\mu(x)\lambda_\nu(y)\, \check{F}^{(\lambda)}_{\mu\nu}(x-y), \tag{6.59}$$

where

$$\check{F}^{(\lambda)}_{\mu\nu}(x,y) = \int \frac{\mathrm{d}^2q}{(2\pi)^2}\, e^{ip(x-y)}\, \check{F}^{(\lambda)}_{\mu\nu}(p), \tag{6.60}$$

with

$$\begin{aligned} \check{F}^{(\lambda)}_{\mu\nu}(p) = 2\delta_{\mu\nu}&\int \frac{\mathrm{d}^2q}{(2\pi)^2}\frac{1}{q^2+m^2} \\ &-\int \frac{\mathrm{d}^2q'}{(2\pi)^2}\frac{4q'_\mu q'_\nu}{[(q'-\frac{1}{2}p)^2+m^2][(q'+\frac{1}{2}p)^2+m^2]}. \end{aligned} \tag{6.61}$$

In fact we can calculate explicitly both $\check{F}^{(\alpha)}(p)$ and $\check{F}^{(\lambda)}_{\mu\nu}(p)$. This calculation is completely routine and we find

$$\check{F}^{(\alpha)}(p) = \frac{1}{2\pi}\frac{1}{\sqrt{p^2(p^2+4m^2)}}\ln\frac{\sqrt{p^2+4m^2}+\sqrt{p^2}}{\sqrt{p^2+4m^2}-\sqrt{p^2}} \tag{6.62}$$

and

$$\check{F}^{(\lambda)}_{\mu\nu}(p) = \left(\delta_{\mu\nu}-\frac{p_\mu p_\nu}{p^2}\right)\left(-\frac{1}{\pi}+(p^2+4m^2)\check{F}^{(\alpha)}(p)\right). \tag{6.63}$$

We give details of this calculation in Appendix D.

Finally we look at the mixed (α,λ) terms. They are given by

$$2\,\mathrm{Tr}\,\{[2\lambda_\mu\partial_\mu+(\partial_\mu\lambda_\mu)](m^2-\partial_\mu\partial_\mu)^{-1}\,\alpha\,(m^2-\partial_\mu\partial_\mu)^{-1}\delta(x-y)\}, \tag{6.64}$$

and so correspond to

$$2\int \mathrm{d}^2x\mathrm{d}^2y\frac{\mathrm{d}^2p\mathrm{d}^2q}{(2\pi)^4}\left(\alpha(x)e^{ipx}\frac{1}{p^2+m^2}e^{-ipy}\right.$$
$$\left.\times[2\lambda_\mu(y)\partial_\mu(y)\,+\,\partial_\mu(y)\lambda_\mu(y)]e^{iqy}\frac{1}{q^2+m^2}e^{-iqx}\right). \tag{6.65}$$

Integrating by parts, as before, we obtain

$$2\int \mathrm{d}^2x\int \mathrm{d}^2y\,\alpha(x)\,\lambda_\mu(y)\,\Gamma_\mu(x-y), \tag{6.66}$$

where

$$\Gamma_\mu(x) = \int\frac{\mathrm{d}^2\mathrm{d}}{(2\pi)^2}\,e^{ipx}\,\check{\Gamma}_\mu(p), \tag{6.67}$$

and

$$\check{\Gamma}_\mu(p)\int\frac{\mathrm{d}^2q'}{(2\pi)^2}\frac{q'_\mu}{[(q'+\frac{1}{2}p)^2+m^2][(q'-\frac{1}{2}p)^2+m^2]} = 0 \tag{6.68}$$

by symmetrical integration. Hence all the mixed terms vanish.

We can also calculate the further contributions, S^ν for $\nu\geq 3$, to the effective action. They are all finite and their calculation does not cause any serious problems.

We can now calculate any Green function of z_α, $z^\dagger_\alpha$ and A_μ. We obtain

$$\langle z_{\alpha_1}(x_1),\ldots,A_{\mu_l}(x_l)\rangle$$
$$=\frac{1}{Z(0,0,0)}\int \mathrm{D}[\alpha]\mathrm{D}[\lambda_\mu]\,e^{-S^{(2)}}\exp\left(-\sum_{\nu=3}^{\infty}N^{-\nu/2+1}S^{(\nu)}\right)$$
$$\times\frac{\delta}{\delta J_{\alpha_1}(x_1)}\cdots\frac{\delta}{\delta K_{\mu_l}(x_l)}\exp\left(\int \mathrm{d}^2x\,J^\dagger\Delta^{-1}J+\frac{1}{\sqrt{N}}K_\mu\lambda_\mu\right), \tag{6.69}$$

where we have dropped the $fK_\mu K_\mu/2N$ term as $f \to 0$ as $\Lambda \to \infty$. Observe that Δ^{-1} is still the complicated expression given before, as Δ is given by (6.33). Next we expand all the terms in the power series in $1/\sqrt{N}$ and perform the gaussian integrals over α and λ_μ (at this stage we can fix the gauge so that $\partial_\mu\lambda_\mu = 0$). The outcome of the calculation can be expressed in terms of Feynman integrals built from the z, α and λ_μ lines and vertices. They are:

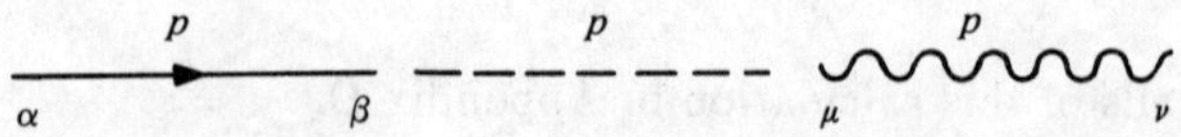

$\delta_{\alpha\beta}G(p)$ — z propagator; $D^\alpha(p)$ — α propagator; $(\delta_{\mu\nu} - \frac{p_\mu p_\nu}{p^2})D^\lambda(p)$ — λ_μ propagator

and the vertices are given by

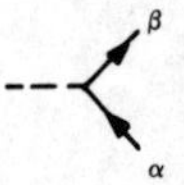

$$\frac{i}{\sqrt{N}}\delta_{\alpha\beta} \qquad -\frac{1}{\sqrt{N}}(p_\mu + p'_\mu)\delta_{\alpha\beta}$$

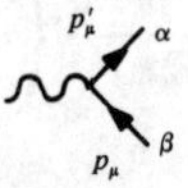

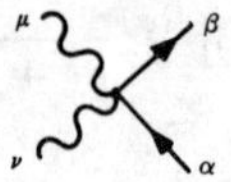

$$-\frac{1}{N}\delta_{\mu\nu}\delta_{\alpha\beta},$$

(these vertices come from the consideration of Δ) plus further vertices (more suppressed in $1/\sqrt{N}$), which come from $S^{(3)}$, etc.

The simplest way to see how to derive these Feynman rules would correspond to the reintroduction of the z fields through

$$\exp\left(-\int \mathrm{d}^2x(J^\dagger\Delta^{-1}J)\right) \sim \int \mathrm{D}[z^\dagger]\mathrm{D}[z]\exp\left(-\int \mathrm{d}^2x(z^\dagger\Delta z + J^\dagger z + Jz^\dagger\right),$$

with appropriate factors. Notice that Green functions involving the A_μ field correspond to Feynman diagrams with external λ_μ lines, as effectively $A_\mu = \lambda_\mu/\sqrt{N}$. Observe that the subdiagrams shown below will never appear in our expansion (their effects have already been included in the definition of propagators, etc).

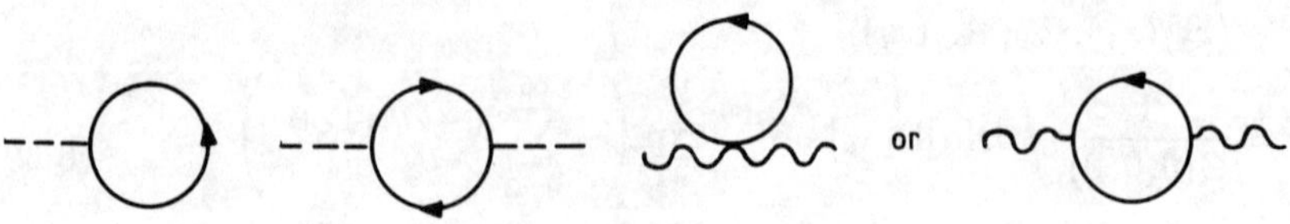

or

We can summarise what we have found by saying that in the $1/N$ limit the model describes interacting massive charged particles z_α. The mass of these particles is due to quantum effects associated with the fact that the lowest order diagrams in the $1/N$ expansion correspond to an infinite number of conventional diagrams. It is this partial summation of conventional perturbation theory that leads to the generation of mass for the z_α particles. Moreover, this mass depends nonperturbatively on the renormalised coupling constant:

$$m^2 = \mu^2 \exp\left(\frac{-2\pi}{f_R(\mu)}\right). \tag{6.70}$$

The z_α particles interact with each other by interchanging α and λ_μ quanta.

etc

Are there α and λ_μ particles; i.e. do the interactions involving the α and λ_μ exchange correspond to an exchanges of particles?

To answer this question we have to check whether their propagators have particle poles (notice that the α propagator is a gauge invariant quantity). Thus we have to look at $\check{F}^{(\alpha)}(p)$ and $\check{F}^{(\lambda)}_{\mu\nu}(p)$ and check whether they can vanish for positive p^2. But as

$$\check{F}^{(\alpha)}(p) = \frac{1}{2\pi}\frac{1}{\sqrt{p^2(p^2+4m^2)}} \ln \frac{\sqrt{p^2+4m^2}+\sqrt{p^2}}{\sqrt{p^2+4m^2}-\sqrt{p^2}}, \tag{6.71}$$

we see that $\check{F}^{(\alpha)}(p) > 0$ for $p^2 \geq 0$. Moreover, as

$$\begin{aligned}\lim_{p\to 0} \check{F}^{\alpha}(p) = \lim_{p\to 0} & \frac{1}{4\pi m\sqrt{p}}\Big(\ln(\sqrt{4m^2+p^2}+\sqrt{p^2}) \\ & - \ln(\sqrt{4m^2+p^2}-\sqrt{p^2})\Big) \\ & = \lim_{p\to 0}\frac{1}{4\pi m|p|}\left(\sqrt{\frac{p^2}{m}}+\cdots\right) = \frac{1}{4m^2\pi},\end{aligned} \tag{6.72}$$

we see that

$$\lim_{p\to 0} \left(-\frac{1}{\pi} + (p^2+4m^2)\check{F}^{(\alpha)}(p)\right) = 0. \tag{6.73}$$

Thus the λ_μ propagator is singular at $p = 0$. Calculating it in more detail we find that

$$\check{F}^{(\lambda)}_{\mu\nu} = \left(\delta_{\mu\nu} - \frac{p_\mu p_\nu}{p^2}\right)\left(\frac{12\pi m^2}{p^2}\right)^{-1}, \tag{6.74}$$

which shows that the propagator for the λ_μ field (remember, the propagator is given by the inverse of $\check{F}^{(\lambda)}_{\mu\nu}$) has a pole at $p^2 = 0$, which in turn generates a long-range force between the z particles.

Notice that the appearance of this singularity can be thought of as corresponding to the $F_{\mu\nu}F^{\mu\nu}$ term generated quantum mechanically, as such a term would give $D^{(A_\mu)}(p) \sim 1/p^2$ as $p^2 \to 0$. Thus, quantum mechanically, the composite A_μ field starts propagating; this mechanism has been suggested to operate in some, more realistic, supergravity models [6].

We can now go further and look at $1/N$ corrections to our result. This was done by Abdalla *et al* [7], who proved that the pole in $D^{(\lambda)}$ persists to all orders in $1/N$ and that the infrared divergences caused by internal λ_μ-lines in higher order diagrams cancel in Green functions of gauge-invariant quantities such as $z^\dagger_\alpha z_\beta$. To analyse whether this implies the existence of a λ_μ particle, we look a little more closely at the confinement in two dimensions.

For this discussion let us use as a model [3] the '$\mathbb{C}P^{N-1}$-electrodynamics', i.e. a model based on

$$L = \tfrac{1}{2}F_{01}^2 + e(z^\dagger_\alpha \overleftrightarrow{\partial}_\mu z_\alpha)A_\mu + \ldots, \tag{6.75}$$

where $F_{01} = \partial_0 A_1 - \partial_1 A_0$. We choose the axial gauge $A_1 = 0$ and

$$L = \tfrac{1}{2}(\partial_1 A_0)^2 + e(z^\dagger_\alpha \overleftrightarrow{\partial}_\mu z_\alpha)A_0 + \ldots. \tag{6.76}$$

As no time derivative of A_0 appears in L, we see that A_0 is not a dynamical field but a constrained variable which should be eliminated from the theory before we recast it in a canonical form, ready for quantisation. A_0 is determined by

$$\partial_1{}^2 A_0 = e\,(z^\dagger_\alpha \partial_0 z_\alpha) = e j_0. \tag{6.77}$$

The general solution of this equation is given by

$$A_0(x_0, x_1) = \tfrac{1}{2}e \int \mathrm{d}y_1 |x_1 - y_1|\, j_0(x_0, y_1) + Bx_1 + C. \tag{6.78}$$

However, C can be set to zero by gauge symmetry, and $B \neq 0$ corresponds to the existence of a constant background electric field (i.e. the existence of classical charges at ∞). Substituting this expression into the equation above, we see that we obtain

$$L = L_0 + \tfrac{1}{2}e^2 \int \mathrm{d}y_1 |x_1 - y_1|\, j_0(x_0, y_1)\, j_0(x_0, x_1), \tag{6.79}$$

where L_0 stands for the part involving only the z field.

We see that we have a *linear potential* between the charges. This becomes even more clear if we replace $(z^\dagger_\alpha \overleftrightarrow{\partial}_\mu z_\alpha)$ by $\psi^\dagger\gamma_\mu\psi$. Only gauge-invariant propagators (like our α propagator) provide information about the existence or non-existence of particle poles. A_μ propagator is gauge variant and so the discussion is more complicated. However, in two dimensions there are no transverse degrees of freedom and so no particle poles for A_μ. Even though there are no particle poles, the effects of the A_μ (λ_μ) fields are non-negligible; their exchange by the z fields forces them to form bound states.

There is one more field in the $\mathbb{C}P^{N-1}$ model which still should be mentioned, and this is the $\epsilon_{\mu\nu}\partial_\nu\lambda_\mu$. This field is gauge invariant but it can be shown [4] that its two-point function is analytic in momentum space for $\mathrm{Re}(p^2) > -4m^2$ and so consequently falls off exponentially in position space.

6.4 $\mathbb{C}P^{N-1}$ Fields with Fermions

When fermions are included (in a supersymmetric way or not), the $1/N$ discussion becomes much more complicated. D'Adda *et al* have analysed it in their paper [8]; here, for completeness, we repeat their discussion in a very abbreviated form.

Consider an action given by

$$S = \int \mathrm{d}^2x \Bigg((D_\mu z)^\dagger D_\mu z + \psi^\dagger(\not{D} - M)\psi \, \frac{1}{2N}(g_s + e^2 f)(\psi^\dagger\gamma_\mu\psi)^2 - \frac{g_v}{2N}[(\psi^\dagger\tau^i\psi)^2 + (\psi^\dagger\gamma_5\tau^i\psi)] \Bigg), \tag{6.80}$$

where

$$\begin{aligned} D_\mu z_\alpha &= \partial_\mu z_\alpha - \frac{f}{N}(z^\dagger \overleftrightarrow{\partial}_\mu z)\, z_\alpha \\ D_\mu \psi^a_\alpha &= \partial_\mu \psi^a_\alpha - e\frac{f}{N}(z^\dagger \overleftrightarrow{\partial}_\mu z)\psi^a_\alpha, \end{aligned} \tag{6.81}$$

and where τ^i ($\tau^0 = 1$) are the $n \times n$ generators of a 'flavour' symmetry of our ψ fields (they act on the 'upper' index of ψ^a_α fields), and, as before, the fields are constrained by

$$z^\dagger z = \frac{N}{2f}, \qquad z^\dagger\psi = \psi^\dagger z = 0. \tag{6.82}$$

(The supersymmetric case corresponds to $e^2 = 1$, $M = 1$, $g_s = 0$ and $g_v = f$.) We can now repeat all the arguments of the previous section, and find that the effective action is given by

$$
\begin{aligned}
S = & N \operatorname{Tr} \ln\{(\Delta_B + c^\dagger \Delta_F^{-1} c/N)\delta(x-y)\} \\
& - N \operatorname{Tr} \ln[\Delta_F \delta(x-y)] + \int \mathrm{d}^2 x \left(\frac{i\sqrt{N}}{2f}\alpha + \frac{1}{2g_v}(\phi^i\phi^i + \phi_5^i\phi_5^i) \right),
\end{aligned} \tag{6.83}
$$

where α, c and $c^\dagger$ fields enforce the constraints $z^\dagger z = N/2f$ and $z^\dagger\psi = 0$ and $\psi^\dagger z = 0$, and where the ϕ^i, ϕ_5^i and λ_μ fields were introduced to eliminate $(\psi^\dagger\tau^i\psi)$, $(\psi^\dagger\gamma_5\tau^i\psi)$ and $(z^\dagger \overleftrightarrow{\partial}_\mu z) - (\psi^\dagger\gamma_\mu\psi)$ respectively. Moreover, Δ_B and Δ_F are given by

$$
\begin{aligned}
\Delta_B &= -D_\mu D_\mu + m^2 - \frac{i}{\sqrt{N}}\alpha \\
\Delta_F &= \not{D} - M - \frac{1}{\sqrt{N}}(\phi^i + \gamma_5\phi_5^i)\tau^i,
\end{aligned} \tag{6.84}
$$

where $D_\mu = \partial_\mu + (i/\sqrt{N})\lambda_\mu$, when applied to the boson fields and $D_\mu = \partial_\mu + (ie/\sqrt{N})\lambda_\mu$ when applied to the fermion fields. When we now look at the lowest order action ($\sim \sqrt{N}S^{(1)}$) and try to impose our 'saddle-point condition' $S^{(1)} = 0$, we find that this is more difficult. The condition would appear to be

$$
\breve{\alpha}(0)\left(\frac{1}{2f} - \int \frac{\mathrm{d}^2 q}{(2\pi)^2}\frac{1}{q^2+m^2}\right) - \breve{\phi}^0(0)\left(\int \frac{\mathrm{d}^2 q}{(2\pi)^2}\frac{1}{q^2+m^2}\right). \tag{6.85}
$$

We find that to proceed further we need to impose the conditions:

$$
\begin{aligned}
\langle\phi^0\rangle &= \sqrt{nN}M_1 \qquad M_1 > 0 \\
\langle\phi^i\rangle &= 0 \quad i \neq 0 \quad \langle\phi_5^i\rangle = 0,
\end{aligned} \tag{6.86}
$$

which correspond to the spontaneous generation of the fermion mass. Now we can change our variables to

$$
\varphi^0 = \phi^0 - \sqrt{nN}M_1, \quad \varphi^i = \phi^i\,(i \neq 0), \quad \text{and} \quad \varphi_5^i = \phi_5^i, \tag{6.87}
$$

and find that now

$$
\begin{aligned}
S^{(1)} &= i\breve{\alpha}(0)\left(\frac{1}{2f} - \int \frac{\mathrm{d}^2 q}{(2\pi)^2}\frac{1}{q^2+m}\right) \\
&+ \breve{\varphi}^0(0)\left(\frac{M_1}{2g_v} - M_t\int \frac{\mathrm{d}^2 q}{(2\pi)^2}\frac{1}{q^2+M_t}\right),
\end{aligned} \tag{6.88}
$$

where $M_t = M + M_1$. Now we can satisfy the condition of the vanishing of $S^{(1)}$; we introduce a cut-off Λ, and require that the bare coupling constants f and g_v vary with Λ as

$$\frac{2\pi}{f} = \ln\frac{\Lambda^2}{m^2}, \qquad \frac{2\pi}{g_v} = \frac{M_t}{M_1}\ln\frac{\Lambda^2}{M_t^2}. \tag{6.89}$$

Notice that in the supersymmetric case we have $M = 0$, $f = g_v$, $n = 1$ and so $M_1 = m$, i.e. both the ψ and the z fields acquire the same mass. However, this is not the end of the story. It can be shown [8] that we also need to vary M with Λ as

$$M(\Lambda) = \epsilon M \frac{2\pi}{\ln(\Lambda^2/M^2)}.$$

Then as $\Lambda \to \infty$, the total fermionic mass coincides with M_1.

We should now consider two cases corresponding to whether ϵ vanishes or not. If $\epsilon \neq 0$ then $M \neq 0$, and we can show that the exchange of the λ_μ gives rise to a linear Coulomb potential (as in the purely bosonic $\mathbb{C}P^{N-1}$ model). This potential confines both the ψ and the z fields. If, however, $\epsilon \neq 0$ (i.e. $M = 0$ and we have, for instance, the supersymmetric case), the pole in $D^{(\lambda)}$ disappears and the z and ψ fields are no longer confined. Such a mechanism corresponds to a screening effect. Its existence, in physically more relevant theories, would have profound implications on whether we can hope to base our theory on classically composite models.

References

[1] Witten E 1980 The $1/N$ expansion in atomic physics *Recent Developments in Gauge Theories* ed G 't Hooft *et al* (New York: Plenum) p 403

[2] Jevicki A and Papanicolaou N 1980 Classical dynamics in the large N limit *Nucl. Phys.* B **171** 360
Jevicki A and Levine H 1981 Large N classical equations and their quantum significance *Ann. Phys., NY* **136** 113

[3] Coleman S 1980 *Lectures given at the Erice Summer School of Subnuclear Physics, 1979* ed Zichichi (New York: Academic)

[4] D'Adda A, Di Vecchia P and Luscher M 1978 A $1/N$ expandable series of non-linear σ models with instantons *Nucl. Phys.* B **146** 63

[5] Politzer H D 1979 Asymptotic freedom: an approach to strong interactions *Phys. Rep.* **14** 129

[6] Ellis J, Gaillard M K and Zumino B 1980 A grand unified theory obtained from broken supergravity *Phys. Lett.* **94B** 343
Davis A C *et al* 1983 Composite gauge bosons in sigma models and supergravity *Phys. Lett.* **125B** 151

[7] Abdalla E, Abdalla M C B and Gomes M 1982 Anomaly cancellation in the supersymmetric $\mathbb{C}P^{N-1}$ model *Phys. Rev.* D **25** 452; 1983 Nonlocal charge

of the $\mathbb{C}P^{N-1}$ model and its supersymmetric extension to all orders *Phys. Rev.* D **27** 825

[8] D'Adda A, Di Vecchia P and Luscher M 1979 Confinement and chiral symmetry breaking in $\mathbb{C}P^{N-1}$ models with quarks *Nucl. Phys.* B **152** 125

7 Chiral Models

7.1 Introduction

So far we have discussed models based on coset spaces. The reason for this is that by possessing topological properties they resemble four-dimensional nonabelian gauge theories. However, other two-dimensional models are also interesting; and among them particularily interesting are models where the target space is a group manifold. The simplest one of them corresponds to the $U(N)$ σ model; the so-called $U(N)$ chiral model.

This model can be defined as follows; take $g \in U(N)$, where $U(N)$ denotes the group of $N \times N$ unitary matrices, and consider the Lagrangian density given by

$$\mathrm{Tr}\,\{\partial_\mu Q^\dagger\, \partial_\mu Q\}. \tag{7.1}$$

As $Q \in U(N)$, we see that $Q^\dagger = Q^{-1}$ and so that Q satisfies the constraint $Q^\dagger Q = QQ^\dagger = 1$. Notice that, if we further restrict $Q = Q^\dagger$, then Q would correspond to a grassmannian field, as we would then be able to put $Q = A(1 - 2P)$, where A is a constant matrix and P is a projector of some rank. This time however we do not impose this hermiticity condition and so our field is more general. Notice however, that for $N = 2$ our model becomes equivalent to the $O(4)$ σ model, which we mentioned briefly in chapter 3, and which we will discuss in more detail in chapter 9. To see this we observe that we can parametrise the fields of the $U(2)$ model as

$$g = \phi_0 + i \sum_k \phi_k \tau_k, \tag{7.2}$$

where τ_k $(k = 1, \ldots, 3)$ are the familiar Pauli matrices.

Returning to the general case we observe that the chiral model, (7.1), is invariant under

$$Q \rightarrow Q' = h^{-1} Q k, \tag{7.3}$$

where both h and k are constant $U(N)$ matrices. Hence, the real invariance of the model is $U(N) \times U(N)$, and it is this invariance which is responsible for the adjective 'chiral' in the name of these models.

The Euler–Lagrange equations for this model (classical equations of motion) are given by

$$\partial_\mu(Q^{-1}\partial_\mu Q) = 0, \tag{7.4}$$

of course together with the constraint $Q^{-1} = Q^\dagger$.

This model and its generalisations have recently been the subject of several papers [1]. Among them particularly interesting is a paper by Uhlenbeck [2], in which she showed that there is a relation between solutions to the classical equations of motion of the chiral model and those of all grassmannians obtained from it by the imposition of the hermiticity condition $Q = Q^\dagger$. Moreover, her approach allows the explicit determination of the solutions to the equations of motion. We will discuss this construction and some properties of these solutions in the next sections.

7.2 Uhlenbeck's Equations and their Simple Solutions

It has been known for some time that all the solutions of grassmannian models are also solutions of the chiral model (as the grassmannian subspace is totally geodesic in $U(N)$) [3], but not much has been known about other solutions. Recently, however, Uhlenbeck made great progress by proving a very interesting factorisation theorem [2]. Namely, she managed to prove a theorem showing that all the classical solutions of the chiral model are of the form

$$Q = K\prod_{i=1}^{k}(1 - 2R_i), \tag{7.5}$$

where k is some number (called by her the *uniton* number), K is a constant matrix and R_i are projectors which satisfy some first-order differential equations.

This theorem provides a convenient method of generating new solutions from the old ones, which she calls the *addition of a uniton*. Namely, one writes

$$Q = Q_0(1 - 2R), \tag{7.6}$$

and then, as she showed, Q satisfies the equation of motion if Q_0 does, and if R satisfies the following two equations

$$\begin{aligned} RA_-(1 - R) &= 0 \\ (1 - R)[\partial_- R + A_- R] &= 0, \end{aligned} \tag{7.7}$$

where

$$A_\pm = \tfrac{1}{2} Q_0^\dagger \partial_\pm Q_0. \tag{7.8}$$

If $Q_0 = K$, the equations reduce to $\partial_- RR = 0$, i.e. the equation for the instantons of the grassmannian models. For $Q_0 \neq K$ we have more general solutions, which include noninstanton solutions of grassmannian models and also nongrassmannian solutions.

Moreover, Uhlenbeck proved that all the solutions of the $U(N)$ model can be obtained from the constant solutions by adding to them k unitons, where k is an integer smaller than $N-1$. As a direct consequence, the most general solutions of the $U(2)$ are the one-uniton solutions, or, in other words, the instanton solutions of the $\mathbb{C}P^1$ model. As this model is equivalent to the $O(4)$ model, we see that all the finite action solutions of the $O(4)$ model correspond to the embeddings of the $\mathbb{C}P^1$, i.e. $O(3)$ model. This agrees with the old result of Borchers and Garber [4].

To construct all the solutions of the $U(N)$ model, one can start from a constant solution and add to it one uniton. Then, to this one-uniton solution, one can add a second uniton, then a third one, and so on. It is important to note that, strictly speaking, the number of unitons is not well defined. By this we mean, that if we add a uniton to, say, a two uniton solution the resulting expression may be equivalent to a one-uniton solution. For this reason, Uhlenbeck defined the minimal uniton number, i.e. the minimal number of unitons that are required to construct a given solution. In the rest of this chapter, by a uniton we will always mean the minimal uniton.

In addition, there is a further consequence of the Uhlenbeck construction which is easy to derive [5]: namely one can show that for all k uniton solutions

$$A_- = \sum_{i=1}^{k} \partial_- R_i. \tag{7.9}$$

We will prove this result further on in this chapter in the section in which we will discuss properties of our solutions and in particular their stability.

The main aspect of the Uhlenbeck construction is that one has to solve a first-order nonlinear partial differentrial equation coupled with an nonlinear algebraic equation. This last equation admits two obvious solutions, namely $RA_- = 0$ and $A_-(1-R) = 0$. Following Wood [6], we will call them respectively basic and antibasic unitons.

Moreover, in his paper [6] Wood has managed to show that any uniton factor R can be decomposed into a product of basic unitons as follows:

$$(1-2R) = (1-2R_1)(1-2R_2)\ldots(1-2R_k) \tag{7.10}$$

for some $k \leq N$, where R_1 is a basic uniton for Q, and R_i are basic unitons for the solution $Q_i = Q(1-2R_1)\ldots(1-2R_{i-1})$. We see thus that there are three types of unitons: the basic, the anti-basic and the remaining ones which we shall call mixed. Moreover, each of those unitons can be of any rank.

In this and the next sections we will reproduce the general construction of solutions of the chiral model for $N=3$ and 4 recently given in [7]. (The $N=2$ case is trivial as its only solutions are the embeddings of the $\mathbb{C}P^1$ instantons.)

As will become clearer later on, the specific form of each uniton describing a particular solution depends, through the construction of Uhlenbeck, on the form of all the previous unitons in the solution. Hence, to describe a particular uniton we need to develop a notation which takes this interdependence into account. Thus we will denote each k uniton solution by a Q with two sets of indices. Those in the superscript will give the rank of each uniton (reading from left to right), and those in subscript will describe the type of each uniton, with the letters a,b and m standing for, respectively, antibasic, basic and mixed unitons. As all one uniton solutions are of both basic and antibasic type at the same time, we will leave their type blank. Thus, for example $Q^{(2,3,4)}_{(\ ,a,m)}$ corresponds to a three-uniton solution, for which the first, second and third unitons are respectively of rank 2, 3 and 4, and of the type: undefined, anti-basic and mixed. To refer to a particular uniton in the solution we will use the same notation, the only difference being that the properties of the projector corresponding to the uniton itself will be written outside the parenthesis. Thus, the third uniton of our previous example will be written as $R^{(2,3)4}_{(\ ,a)m}$. When all this information is not relevant, we will denote a k uniton by R_k.

When one tries to add a basic or an anti-basic uniton to a given solution, the Uhlenbeck equations (7.7) simplify. For the basic uniton they become

$$\begin{aligned} RA_- &= 0 \\ (1-R-P(I))\,\partial_- R &= 0, \end{aligned} \tag{7.11}$$

where I denotes the subspace spanned by the image of A_-, and $P(I)$ is the projector on this subspace. Similarly, for the anti-basic unitons $R=1-S$,

$$\begin{aligned} A_- S &= 0 \\ (1-S-P(H))\,\partial_+ S &= 0, \end{aligned} \tag{7.12}$$

where H denotes the subspace spanned by the image of $(A_-)^\dagger$, and $P(H)$ corresponds to the projector on this subspace. To see this we multiply the second Uhlenbeck equation (7.7) by $P(I)$ from the left, and put the result,

$$P(I)\,\partial_- R = -A_-\,R, \tag{7.13}$$

into the remaining Uhlenbeck equation.

To prove the complete equivalence between the equations (7.11) and the Uhlenbeck equations (7.7) we have to take the equation satisfied by A^1_+ [2]

$$\partial_- A^1_+ + [A^1_-\,, A^1_+] = 0, \qquad A_- = -(A_+)^\dagger, \tag{7.14}$$

and multiply it from the left by R. A simple algebraic manipulation then shows that this equation then implies the equation (7.13), which establishes the equivalence between the two sets of equations. The equivalence between the Uhlenbeck equations (7.7) and (7.12) can be proved in a very similar way. As an immediate consequence of the equation (7.11), we see that $(1 - P(I))$ and $P(H)$ are automatically projectors corresponding to, respectively, basic and anti-basic unitons, and that we can construct more general solutions of both types by considering projectors of smaller rank.

As we have seen before, the one-uniton solutions are of the form

$$Q = K(1 - 2R_1), \tag{7.15}$$

where R_1 satisfies

$$(1 - R_1)\,\partial_- R_1 = 0. \tag{7.16}$$

These solutions are the instanton solutions of grassmannian models discussed in chapter 3. The most general solutions of this class for R_1 are given by

$$R_1 = F\,(F^\dagger F)^{-1} F^\dagger, \tag{7.17}$$

where F is a holomorphic matrix (i.e. whose entries are functions of only $x + iy$) of a maximal rank. Observe that, when F is of rank 1 (i.e. is a vector), R_1 is a self-dual solution of the $\mathbb{C}P^{N-1}$ model.

In the next sections we will derive and discuss more general solutions of the $U(3)$ and $U(4)$ models. To give explicit forms of these solutions we will need convenient vectors of basis (in the three-dimensional and four-dimensional complex space respectively), and, as will become clear from the discussion, different choices of basis will be more convenient in different cases. However, most of these choices are based on, and related to, the choice made in the construction of the solutions to the $\mathbb{C}P^{N-1}$ models which we discussed in chapter 3.

In chapter 3, in addition to an analytic vector f, we introduced also

$$P_+ f = \partial_+ f - f\,\frac{(f^\dagger \partial_+ f)}{|f|^2}, \tag{7.18}$$

and further vectors $P_+^k f$ defined by induction:

$$P_+^k f = P_+(P_+^{k-1} f). \tag{7.19}$$

We also proved various properties of these vectors and in particular:

$$\partial_- P_+^k f = -P_+^{k-1} f\,\frac{|P_+^k f|^2}{|P_+^{k-1} f|^2}, \qquad \partial_+ \frac{P_+^k f}{|P_+^k f|^2} = \frac{P_+^{k+1} f}{|P_+^k f|^2}. \tag{7.20}$$

We will use these results below.

In the construction of solutions to the $U(3)$ and $U(4)$ models, we will use both the projectors and the vectors onto which they project. Thus, to exhibit this property, we will introduce the following notation: for any vector V the $P(V)$ defined by

$$P(V) = \frac{V V^\dagger}{|V|^2} \tag{7.21}$$

will denote the corresponding projector onto this vector. Also, as they will appear often in our construction, we introduce

$$P_0 = P(f), \quad \ldots, \quad P_k = P(P_+^k f). \tag{7.22}$$

7.3 Solutions of the $U(3)$ Model

7.3.1 One-uniton Solutions

As from Uhlenbeck's theorem we know that for $U(N)$ the largest uniton number is less than N, we see that all the solutions of the $U(3)$ model correspond to either one or two unitons. In the previous section we showed that one-uniton solutions are given by

$$Q^{(1)}_{(\;)} = (1 - 2P_0) \tag{7.23}$$

and

$$Q^{(2)}_{(\;)} = (1 - 2(P_0 + P_1)). \tag{7.24}$$

Here we have set the arbitrary constant matrix K in (7.15) equal to 1. Moreover, we will assume that $P_+^2 f$ does not vanish. This effectively eliminates all the $U(2)$ embeddings from our construction. Clearly they have to be added to complete the set of all the solutions of the $U(3)$ model. We see that, to complete the set of nonembedding solutions, all we have to do now is to add a further general uniton to our one-uniton solutions.

7.3.2 Two-uniton Solutions

We start by trying to add a uniton to $Q^{(2)}_{(\;)}$, for which

$$A_- = \frac{P_+ f \, P_+^2 f^\dagger}{|P_+ f|^2}. \tag{7.25}$$

As this matrix is of rank one, looking at (7.7) we see that the only uniton we can add must be either of basic or anti-basic type.

So let us first try to add an anti-basic uniton R_2. The first equation of (7.12) tells us that $S_2 = 1 - R_2$ must be orthogonal to $P_+^2 f$, and so the second equation reads

$$(1 - S_2 - P_2)\,\partial_+ S_2 = 0. \tag{7.26}$$

However, using the fact that P_2 corresponds to an anti-instanton solution, we observe that (7.26) implies that

$$(1 - S_2 - P_2)\,\partial_+(S_2 + P_2) = 0, \tag{7.27}$$

which in turn shows that

$$\begin{aligned} Q_2 &= (1 - 2(P_0 + P_1)(1 - 2R_2) \\ &= -(1 - 2(P_2 + S_2)) \end{aligned} \tag{7.28}$$

corresponds to a one-uniton solution. So adding an anti-basic uniton to $Q^{(2)}_{(\ )}$ does not give us any new solutions. A similar computation shows that the addition of a basic uniton to the solution $Q^{(1)}_{(\ )}$ also leads to a solution which corresponds to one uniton.

So we see that we are left with having to consider only two further cases. The first corresponds to adding a basic uniton to $Q^{(2)}_{(\ )}$. However, if R_2 is of rank 2, then it is equal to $P_0 + P_2$ which automatically satisfies (7.11). The resulting solution is then given by $Q^{(2,2)}_{(\ ,a)} = (1 - 2P_0)$, which again is a one-uniton solution. Finally, we try to construct $R^{(2)1}_{(\ )b}$. As $R^{(2)1}_{(\ )b}$ must be orthogonal to $P_+ f$, we can introduce a vector V_2 which lies in the image of this projector. This vector is given by

$$V_2 = \alpha f + \beta P_+^2 f. \tag{7.29}$$

Then defining W_2, a vector orthogonal to V_2 and to $P_+ f$, by

$$W_2 = \beta^* \frac{f}{|f|^2} - \alpha^* \frac{P_+^2 f}{|P_+^2 f|}, \tag{7.30}$$

we see that the second equation of (7.11) becomes equivalent to

$$W_2^\dagger\,\partial_- V_2 = 0. \tag{7.31}$$

Using the relations between the $P_+^k f$ and their derivatives, it is easy to prove that α and β must satisfy

$$\partial_- \left(\frac{\alpha}{\beta}\right) = 0, \tag{7.32}$$

which means that up to an irrelevant common factor α and β must be holomorphic.

A similar computation to the one we have just performed shows that $Q^{(1,2)}_{(\ ,a)}$ is actually a one-uniton solution, and that $Q^{(1,1)}_{(\ ,a)}$ is equivalent to $Q^{(2,1)}_{(\ ,b)}$. In conclusion, we see that the set of all the solutions of the $U(3)$ model is, up to a multiplication from the left by an arbitrary constant matrix, given by

$$\begin{aligned} Q^{(1)}_{(\ )} &= (1 - 2\,P_0) \\ Q^{(2)}_{(\ )} &= (1 - 2\,(P_0 + P_1)) \\ Q^{(2,1)}_{(\ ,b)} &= (1 - 2\,(P_0 + P_1))(1 - 2\,P(V_2)), \end{aligned} \tag{7.33}$$

where V_2 is given by (7.29) with α and β holomorphic (i.e. functions of $x + iy$).

7.4 Solutions of the $U(4)$ Model

Let us observe, first of all, that all the fields of the $U(3)$ model can be embedded into the $U(4)$ model, and so the $U(3)$ solutions are automatically also solutions of the $U(4)$ model. In what follows we will recover some of them as special cases of the derived expressions, or we will have to make some assumptions which will exclude them from our construction. In this case they will have to be added extra to the genuine $U(4)$ solutions.

From Uhlenbeck' s theorem we know that all the solutions of the $U(4)$ model can be expressed as either one-, two- or three-uniton configurations. Thus, first of all, we will describe the one-uniton solutions which, being instanton solutions, were discussed in chapter 3. Then we will add to them a second uniton and, because of the nonuniqueness of the uniton number, we will have to check whether the resultant configurations are really different from the one-uniton ones. We will show that most of the constructed configurations are in fact either one-uniton solutions or are equivalent to each other. Thus in the end we will be left with only four different types of solutions. To these, we will have to add a third uniton, and we will find that there are only two classes of genuine three-uniton solutions. As in the previous sections we will set the arbitrary constant matrix K in (7.15) equal to 1.

7.4.1 *One-uniton Solutions*

The first class of one-uniton solutions R^2_2 is given by (7.17), where F is a rank two matrix for which $(1 - P(F))\,\partial_+ F$ is of maximal rank. The other class corresponds to solutions which are given by

$$R^1_1 = P_0 \qquad R^2_1 = P_0 + P_1 \qquad R^3_1 = P_0 + P_1 + P_2. \tag{7.34}$$

Here we have changed our notation. From now on the first entry in the subscript (which was left blank before) will denote the rank of the A_- matrix corresponding to the first uniton factor. We choose our vectors in such a way that the vector $P_+^3 f$ does not vanish. This effectively eliminates all $U(3)$ and $U(2)$ embeddings from our construction. Clearly they have to be added to complete the set of solutions.

From Uhlenbeck's theorem [2] we know that, to obtain further solutions, we must now add a uniton to our one-uniton configurations.

7.4.2 Two-uniton Solutions

We start with the one-uniton solution (R_2^2) from the first class, and observe that in this case

$$A_- = F\,(F^\dagger F)^{-1}((1 - P(F))\,\partial_+ F)^\dagger \tag{7.35}$$

is a rank two matrix. Then, by an argument similar to the one we used in the $U(3)$ case, it is easy to show that any attempt to add a basic uniton to this solution produces a configuration which is equivalent to a one-uniton solution. By Wood's theorem [6], this implies that all two-uniton solutions which can be constructed for this class of configurations will be either one-uniton solutions or two-uniton solutions which can be constructed by the addition of one uniton to a more general one-uniton configuration, which we will discuss next.

Next we consider the other class of solutions, and we find that all the one-uniton solutions in this class fall into three categories with the A_- matrix in each case being given respectively by

$$A_-^1 = \frac{f\,P_+ f^\dagger}{|f|^2} \qquad A_-^2 = \frac{P_+ f\,P_+^2 f^\dagger}{|P_+ f|^2} \qquad A_-^3 = \frac{P_+^2 f\,P_+^3 f^\dagger}{|P_+^2 f|^2}. \tag{7.36}$$

As all these matrices are of rank one, we observe that the only unitons we can add to them are either basic or anti-basic. So the different types of solutions we should look at are

$$\begin{matrix} Q_{(1,b)}^{(1,i)} & Q_{(1,b)}^{(2,i)} & Q_{(1,b)}^{(3,i)} \\ Q_{(1,a)}^{(1,i)} & Q_{(1,a)}^{(2,i)} & Q_{(1,a)}^{(3,i)} \end{matrix} \tag{7.37}$$

where i can be 1, 2 or 3.

Repeating our previous arguments, we can prove that $Q_{(1,b)}^{(1,i)}$ and $Q_{(1,a)}^{(3,i)}$ are equivalent to one-uniton solutions. Moreover, when $i = 3$ all these solutions are either constant or one-uniton solutions. To see this, take for example $Q_{(1,b)}^{(2,3)}$: then $R_{(1)b}^{(2)3} = P_0+P_2+P_3$ and $Q_{(1,b)}^{(2,3)} = (1-(P_1+P_2+P_3)) = -(1-2P_0)$, which is a one-uniton solution.

If we want to construct $Q^{(1,i)}_{(1,a)}$ where $i = 1$ or 2, we must solve

$$(1 - S_2 - P_1)\,\partial_+ S_2 = 0, \tag{7.38}$$

where $S_2 = (1 - R^{(1)i}_{(1)a})$. In this case we can rewrite $Q^{(1,i)}_{(1,a)}$ as

$$Q^{(1,i)}_{(1,a)} = (1 - 2(P_0 + P_1))(1 - 2R_2), \tag{7.39}$$

where $R_2 = (1 - P_1 - S_2)$ satifies $R_2\,P_1 = 0$ and $(1 - P_1 - R_2)\partial_- R_2 = 0$. This implies that $Q^{(1,i)}_{(1,a)}$ is equivalent to $Q^{(2,j)}_{(1,b)}$ for some j related to i. Similarly it is easy to show that $Q^{(2,i)}_{(1,a)}$ is equivalent to $Q^{(3,j)}_{(1,b)}$. This means that we are now left with having to consider only two classes of solutions: $Q^{(2,i)}_{(1,b)}$ and $Q^{(2,i)}_{(1,a)}$.

To proceed further we recall a fact known from the construction of solutions of the $\mathbb{C}P^{N-1}$ model discussed in chapter 3, namely that

$$g = \frac{P_+^3 f}{|P_+^3 f|} \tag{7.40}$$

is an antiholomorphic vector, and that, in the same way as we have constructed our basis starting from a holomorphic vector f, we can use g to construct this basis using $P_-^k g$, $k = 0, \ldots, 3$. Moreover, up to an overall factor, we find that $f \sim P_-^3 g$, $P_+ f \sim P_-^2 g$, $P_+^2 f \sim P_- g$, $P_+^3 f \sim g$. In other words, the solution $Q^{(2)}_{(1)}$ can be determined by starting from either a holomorphic or an antiholomorphic vector. Also, as $(A_-)^\dagger = -A_+$, we see that, if we take a solution $Q^{(2,i)}_{(1,b)}$ and exchange x_+ and x_-, we will obtain a solution of the type $Q^{(2,i)}_{(1,a)}$. This means that the only solutions that we have to consider are $Q^{(2,i)}_{(1,b)}$ for $i = 1$ or 2.

We start by constructing $Q^{(2,1)}_{(1,b)}$. If V_1 is the vector on which $R^{(2),1}_{(1),b}$ projects, we can write, in full generality, $V_1 = \alpha f + \beta v/|v|^2$, where v is orthogonal to f and $P_+ f$. If we now introduce w, a vector which is orthogonal to both f, $P_+ f$ and v, we observe that the equation (7.11) implies that $w^\dagger\,\partial_- V_1 = 0$. From this equation we conclude that $v^\dagger\,\partial_+ w = 0$. However, by construction, $(1 - P(w) - P(v))\,\partial_+ w = 0$, which in turn implies that w must be antiholomorphic. Moreover, as we have seen in (7.40), the two vectors g and $\partial_- g$ form an antiholomorphic basis of the space spanned by $P_+^2 f$ and $P_+^3 f$. So w must be of the form

$$w = a(x_-)\,g + b(x_-)\,\partial_- g \tag{7.41}$$

for some a and b. Thus we see that we can then write

$$v = (1 - P(w))\partial_- h, \tag{7.42}$$

where h is a holomorphic vector spanned by g and $\partial_- g$ not parallel to w. Next we define $W_1 = \beta^* f/|f|^2 - \alpha^* v$, which is orthogonal to V_1, $P_+ f$ and to w, and we find that the equation (7.11) shows that $W_1^\dagger \partial_- V = 0$. Using the fact that v and $\partial_+ v$ are orthogonal to each other, this demonstrates that, up to a common factor, α and β must be holomorphic.

To construct $Q^{(2,2)}_{(1,b)}$, we first define a vector W_2 by $P(W_2) = 1 - P_1 - R^{(2)2}_{(1)b}$. Then the equation (7.11) becomes $R^{(2)2}_{(1)b}\,\partial_+ W_2 = 0$. An argument similar to the one we have just used shows that $R^{(2)2}_{(1)b} = P(v) + P(V_2)$, where $V_2 = \alpha f + \beta w/|w|^2$, with α and β being holomorphic functions, and where w and v are defined by (7.41) and (7.42) respectively.

Thus we see that we have proved that for the $U(4)$ model there are only four types of two-uniton solutions. Two of them are given by

$$\begin{aligned} Q^{(2,1)}_{(1,b)} &= (1 - 2(P_0 + P_1))(1 - 2\,P(V_1)) \\ Q^{(2,2)}_{(1,b)} &= (1 - 2(P_0 + P_1))(1 - 2\,(P(v) + P(V_2)), \end{aligned} \tag{7.43}$$

and the remaining two correspond to interchanging x_+ and x_- in the expressions above.

7.4.3 Three-uniton Solutions

The possible three-uniton solutions would correspond to $Q^{(2,1,i)}_{(1,b,x)}$ and $Q^{(2,2,i)}_{(1,b,x)}$, where x stands for a, b or m and i takes the value 1 or 2. To construct these solutions we have first to compute A_- and its image. As

$$A_{-(1,b)}^{(2,1)} = \partial_- R_1^2 + \partial_- R^{(2)1}_{(1)b} = \frac{P_+ f\, P_+^2 f^\dagger}{|P_+ f|^2} + \partial_- \frac{V_1 V_1^\dagger}{|V_1|^2}, \tag{7.44}$$

and, using the fact that $V_1^\dagger \partial_- V_1 = 0$ and that $W_1^\dagger \partial_+ W_1 = 0$, it is a matter of elementary algebra to show that both W_1 and w are orthogonal to the image of $A_{-(1,b)}^{(2,1)}$, and that $A_{-(1,b)}^{(2,1)}$ is a rank two matrix with its image spanned by $P_+ f$ and by V_1. To construct $R^{(2,1)1}_{(1,b)b}$ we introduce V_3, the vector on which $R^{(2,1)1}_{(1,b)b}$ projects, and which is given by

$$V_3 = \alpha_3 \frac{W_1}{|W_1|^2} + \beta_3 \frac{w}{|w|^2}.$$

Next we introduce W_3, a vector orthogonal to V_3, $P_+ f$ and to V_1 and given by $W_3 = \beta_3^* W_1 - \alpha_3^* w$. We observe that equation (7.11) reduces to $W_3^\dagger \partial_- V_3 = 0$. Moreover, V_2 and V_3 are mutually orthogonal and orthogonal to $P_+ f$, and by construction $W_3^\dagger\, \partial_- V_2 = 0$. Thus we see that

$$(1 - P(V_2) - P(V_3) - P_1)\partial_-(P(V_2) + P(V_3)) = 0. \tag{7.45}$$

Consequently $Q^{(2,1,1)}_{(1,b,b)} = (1 - 2(P_0 + P_1))(1 - 2(P(V_2) + P(V_3))$ is a two-uniton solution. In addition it is also straightforward to show that $Q^{(2,1,2)}_{(1,b,b)}$ in fact describes a one-uniton configuration. Thus, using Wood's theorem, we can conclude that the addition of a uniton to $Q^{(2,1)}_{(1,b)}$ gives no new configurations.

The addition of a uniton to $Q^{(2,2)}_{(1,b)}$ is slightly more complicated. First we compute

$$\begin{aligned} A_{-(1,b)}^{(2,2)} &= \partial_- R_1^2 + \partial_- P(v) + \partial_- P(V_2)) \\ &= \frac{f\,P_+ f^\dagger}{|f|^2} - \partial_-\left(\frac{W_2\,W_2^\dagger}{|W_2|^2}\right). \end{aligned} \tag{7.46}$$

Then, defining

$$\begin{aligned} P_- w &= (1 - P(w))\partial_- w \\ t &= (1 - P(w) - P(P_- w))\partial_-^2 g, \end{aligned} \tag{7.47}$$

it is easy to check that $t^\dagger\,\partial_+ t = 0$ and $t^\dagger\,\partial_- W_2 = 0$. In addition, it is easy to prove that $t^\dagger A_{-(1,b)}^{(2,2)} = W_2^\dagger A_{-(1,b)}^{(2,2)} = 0$, and so that the image of $A_{-(1,b)}^{(2,2)}$ is spanned by V_2 and $P_- w$.

Now we are ready to start considering a more general third uniton that we would like to add to this solution. To do this we must, first of all, solve the first Uhlenbeck equation (7.7):

$$R_3\,A_-^2\,(1 - R_3) = 0. \tag{7.48}$$

We note that we can rewrite

$$A_-^2 = X_1\,Y_1^\dagger + X_1\,a\,Y_2^\dagger + X_2\,b\,Y_2^\dagger, \tag{7.49}$$

where we have introduced

$$\begin{aligned} X_1 &= \frac{V_2}{|V_2|^2} & X_2 &= \frac{P_- w}{|P_- w|^2} \\ Y_1 &= P_+ f & Y_2 &= \frac{W_2}{|W_2|^2} \\ a &= -V_2^\dagger\,\partial_- W_2 & b &= -P_- w^\dagger\,\partial_- W_2, \end{aligned} \tag{7.50}$$

which satisfy the conditions that $X_1^\dagger X_2 = X_1^\dagger Y_1 = X_1^\dagger Y_2 = X_2^\dagger Y_2 = Y_1^\dagger Y_2 = 0$ and that $Y_1 = c_1\,X_2 + c_2\,t$ for some c_1 and c_2.

To proceed further we first consider the case when $R_3 = P(V_4)$ is of rank 1. In this case we can rewrite V_4 as

$$V_4 = \sigma_1 X_1 + \sigma_2\,X_2 + \sigma_3\,t + \sigma_4\,Y_2. \tag{7.51}$$

Multiplying $V_4^\dagger A_-^2(1 - R_3) = 0$ on the right by $W_3^1 = \sigma_3^* X_1/|X_1|^2 - \sigma_1^* t/|t|^2$ shows that σ_1 must vanish. This in turn implies directly that V_4 must be orthogonal to X_2. Or in other words, the only solution of (7.48) for R of rank 1 is given by $R_3 A_-^2 = 0$. In a similar way, it is easy to prove that if R_3 is of rank 3 then the only solution to (7.48) is given by $A_-^2(1 - R_3) = 0$. Moreover, if R_3 is of rank 2, similar manipulations to the ones we have just performed show that, if $Y_1^\dagger X_2 \neq 0$,

$$\begin{aligned} R &= P(Y_2) + P(a t + b X_2) \\ 1 - R &= P(X_1) + P\Big(b^* \frac{t}{|t|^2} - a^* \frac{X_2}{|X_2|}\Big) \end{aligned} \tag{7.52}$$

for some a and b. On the other hand, if $Y_1^\dagger X_2 = 0$, $w = P_+^3 f/|P_+^3 f^2|$ and the image and the kernel of A_- coincide. Observe that in this case our two-uniton solution becomes a grassmannian configuration. Moreover, in this case the solution of the first Uhlenbeck equation gives us

$$\begin{aligned} R &= P(V_6) + P(V_7) \\ (1 - R) &= P(W_6) + P(W_7), \end{aligned} \tag{7.53}$$

where

$$\begin{aligned} x &= a_1 V_2 - a_2 P_+^2 f & y &= b_2^* \frac{P_+ f}{|P_+ f|^2} - b_1^* W_2 \\ V_6 &= b_1 P_+ f + b_2 \frac{W_2}{|W_2|^2} & W_6 &= a_2^* \frac{V_2}{|V_2|^2} + a_1^* \frac{P_+^2 f}{|P_+^2 f|^2} \\ V_7 &= \sigma_1 x + \sigma_2 \frac{y}{|y|^2} & W_7 &= \sigma_2^* \frac{x}{|x|^2} - \sigma_1^* y, \end{aligned} \tag{7.54}$$

and where

$$x^\dagger \partial_- y = 0. \tag{7.55}$$

We see that we have now solved the first Uhlenbeck equation in full generality and we are left with having to solve only her second equation. Thus in what follows we will concentrate on the solutions of that equation.

First we construct the solution $Q_{(1,b,b)}^{(2,2,1)}$. We proceed as before, and so we take V_4 and W_4 given by

$$\begin{aligned} V_4 &= \alpha_4 \frac{W_2}{|W_2|^2} + \beta_4 \frac{t}{|t|^2} \\ W_4 &= \beta_4^* W_2 - \alpha_4^* t. \end{aligned} \tag{7.56}$$

The equation (7.11) reduces to $W_4 \partial_- V_4 = 0$. Moreover, V_4 and V_2 are orthogonal to each other and to v and, by construction, $W_4^\dagger \partial_- V_2 = 0$.

Then we observe that $Q^{(2,2,1)}_{(1,b,b)} = -(1-2P(w))(1-2(P(V_2)+P(V_4))$ again corresponds to a two-uniton solution.

In a similar way we construct $Q^{(2,2,3)}_{(1,b,a)}$. Again, we define V_5 and W_5 such that $W_5^\dagger V_5 = 0$. They are given by

$$\begin{aligned} W_5 &= \alpha_5 \frac{V_2}{|V_2|^2} + \beta_5 v \\ V_5 &= \beta_5 V_2 - \alpha_5 \frac{v}{|v|^2}. \end{aligned} \tag{7.57}$$

The equation (7.12) reduces now to $V_5^\dagger \partial_+ W_5 = 0$. As $(1 - P_1 - P(V_5))\partial_- V_5 = 0$ and $P_1 V_5 = 0$, we see that $Q^{(2,2,3)}_{(1,b,a)} = (1-2(P_0+P_1))(1-2P(V_5))$ describes a two-uniton configuration.

The construction of $Q^{(2,2,2)}_{(1,b,m)}$ is rather more complicated. If $Y_1^\dagger X_2 \neq 0$, we define

$$\begin{aligned} x_3 &= \alpha_6 P_- w + \beta_6 t \\ y_3 &= \beta_6^* \frac{P_- w}{|P_- w|^2} - \alpha_6^* \frac{t}{|t|^2}. \end{aligned} \tag{7.58}$$

We can now take $R^{(2,2)2}_{(1,b)m}$ in the form

$$R^{(2,2)2}_{(1,b)m} = P(W_2) + P(x_3).$$

Using the fact that $w^\dagger \partial_- x_3 = t^\dagger \partial_- W_2 = 0$, it is easy to show that in this case the second Uhlenbeck equation (7.7) reduces to $y_3^\dagger \partial_- x_3 = 0$. However, by construction $(1-(P(x_3)+P(y_3)+P(w)))\partial_+(P(y_3)+P(w)) = 0$, and so we can conclude that $(1-(P(y_3)+P(w)))\partial_+(P(y_3)+P(w)) = 0$, and we see that $Q^{(2,2,2)}_{(1,b,m)} = -(1-2(P_0+P(x_3))) = (1-2(P(y_3)+P(w)))$ describes a one-uniton solution.

When $Y_1^\dagger X_2 = 0$, the second Uhlenbeck equation reduces to

$$W_6^\dagger \partial_- x = 0 \qquad W_6^\dagger \partial_- V_6 = 0 \qquad W_7^\dagger \partial_- V_7 = 0. \tag{7.59}$$

In this case we can choose, without any loss of generality, a_1 to be holomorphic, and we see that $W_6^\dagger \partial_- x = 0$ implies that

$$a_2 = a_1(\Gamma(x_+) - \partial_+ \ln(|P_+^2 f|^2\, |P_+ f|^2\, |f|^2)\, \beta_2), \tag{7.60}$$

where Γ is an arbitrary function of its argument.

Choosing b_1 also to be holomorphic, we observe that $x^\dagger \partial_- y = W_6^\dagger \partial_- V_6 = 0$ implies that

$$b_2 = b_1 \left[\Gamma(x_+) - \beta_2\, \partial_+ \ln\left(|P_+^3 f|^2 |P_+^2 f|^2 |P_+ f|^2 \frac{\beta_2}{\alpha_2} \right) \right]. \tag{7.61}$$

This last equation then tells us that

$$\partial_- \left(\frac{\sigma_2}{\sigma_1}\right) = -a_2\, b_2 \frac{|P_+^2 f|^2}{|P_+ f|^2}, \tag{7.62}$$

where σ_1 and σ_2 are defined by (7.54). Unfortunately this equation is not easy to solve. Nevertheless, in the special case when a_1 and b_1 vanish while a_2 and b_2 are holomorphic, it is a matter of algebra to show that

$$\left(\frac{\sigma_2}{\sigma_1}\right) = -a_2 b_2\, \partial_+ \ln\left(|P_+^2 f|^2 |P_+ f|^2 |f|^2\right) + \sigma_3(x_+), \tag{7.63}$$

where $\sigma_3(x_+)$ is an arbitrary holomorphic function. When a_1 and b_1 do not vanish, the equation (7.62) is difficult to solve explicitly in the general case. Nevertheless, it can be solved in one particular case. If we take, for example,

$$f = \begin{pmatrix} 1 \\ \sqrt{3}x_+ \\ \sqrt{3}x_+^2 \\ x_+^3 \end{pmatrix}, \tag{7.64}$$

then

$$\begin{array}{ll} |f|^2 = (x^2 + 1)^3 & |P_+ f|^2 = 3(x^2 + 1) \\ |P_+^2 f|^2 = 12(x^2 + 1)^{-1} & |P_+^3 f|^2 = 36(x^2 + 1)^{-3}, \end{array} \tag{7.65}$$

where x^2 stands for $x_+ x_-$. In this case it is straightforward to show that

$$\begin{aligned} \left(\frac{\sigma_2}{\sigma_1}\right) = -4\Bigg[&\Gamma\left(\Gamma + \beta_2\, \partial_+ \ln\frac{\alpha_2}{\beta_2}\right)\frac{\ln(1 + x^2)}{x_+} \\ &+ \frac{3\beta_2^2}{x_+^2}\, \partial_+ \ln(\frac{\alpha_2}{\beta_2})\left(\ln(1 + x^2) - \frac{x^2}{(1 + x^2)}\right) \\ &- \frac{9\,\beta_2^2}{x_+^3}\left(\frac{x^2}{(1 + x^2)} - \ln(1 + x^2) + \frac{1}{2}\frac{x^4}{(1 + x^2)^2}\right)\Bigg]. \end{aligned} \tag{7.66}$$

We have thus completed the construction of all the solutions of the $U(3)$ and $U(4)$ model. In both cases the maximal value of the minimal uniton number given by Uhlenbeck is reached. Moreover, we have found that, in both the $U(4)$ and the $U(3)$ cases, all the solutions correspond to the addition of a uniton to some grassmannian solution.

7.5 More General Solutions

We can, in fact generalise the discussion given above and, considering the $U(N)$ case, derive some more general solutions. First of all, it is easy to determine one-uniton solutions. To do this we generalise the notation of the previous section, and for V which is a matrix, we denote by $P(V)$ the projector onto the space it spans. When V is of maximal rank, this projector is given by

$$P(V) = V(V^{\dagger}V)^{-1}V^{\dagger}. \tag{7.67}$$

We can now proceed to determine one-uniton solutions.

7.5.1 One-uniton Solutions

As we showed in the previous section, the one-uniton solutions are of the form

$$Q = K\,(1 - 2\,R_1), \tag{7.68}$$

where R_1 satisfies

$$(1 - R_1)\,\partial_- R_1 = 0. \tag{7.69}$$

The most general solutions for R_1 of this class, as in the previous section, are given by

$$R_1 = P(F), \tag{7.70}$$

where F is a holomorphic matrix of maximal rank.

It is important to note that a given one-uniton solution is not characterised by only one holomorphic matrix F. In fact one can always take holomorphic linear combinations of F without altering the solution. From now on we will say that a projector R is holomorphic or anti-holomorphic if it satisfies $\partial_- RR = 0$ or $\partial_+ RR = 0$ respectively. Thus it is clear that holomorphic bases play an important role in our construction. For this reason, we will say that a set of rectangular matrices

$$V_1,\, V_2,\, \dots V_k, \tag{7.71}$$

is an *orthogonal (anti-)holomorphic basis sequence* if all the V_i are orthogonal to each other,

$$\sum_{i=1}^{k} P(V_i) = 1, \tag{7.72}$$

$\sum_{i=1}^{j} P(V_i)$ is (anti-)holomorphic, and if

$$V_i^{\dagger}\partial_{\pm} V_j = 0 \tag{7.73}$$

for all i, j such that $|i - j| \geq 2$. As a consequence of these properties, we can choose to normalise each V_i in such a way that

$$V_i = \left(1 - \sum_{j=1}^{i-1} P(V_j)\right) F_i, \tag{7.74}$$

where F_i is some (anti-)holomorphic matrix. With this normalisation, we have

$$V_i^\dagger \partial_- V_i = 0 \tag{7.75}$$

valid for all i. As this normalisation plays an important role in the construction, we call it the *natural (anti-)holomorphic normalisation.* When R is a holomorphic projector, then $1 - R$ is an anti-holomorphic one, so that when read from right to left, the orthogonal holomorphic sequence becomes an anti-holomorphic one. Moreover, when V is normalised in the natural holomorphic way, then $V(V^\dagger V)^{-1}$ has the natural anti-holomorphic normalisation. To add a uniton to the one-uniton solution, one needs the following result, which is quite easy to prove: when F is a holomorphic matrix of maximal rank and $R = P(F)$ corresponds to a one-uniton solution, then

$$((1 - P(I))F,\, I,\, H,\, G_R) \tag{7.76}$$

is an orthogonal holomorphic basis sequence, where G_R is the orthogonal complement of the vectors (F, H).

To prove this we observe that, in general, the matrix $(F\,,\, \partial_+ F)$ is not of maximal rank, but we can always split $F = (F_1,\, F_2)$ into two parts such that $(1 - P(F))\partial_+ F_1$ is of maximal rank. As a consequence we can write

$$\partial_+ F_2 = F_1 A + F_2 B + \partial_+ F_1 C \tag{7.77}$$

for some holomorphic matrices A, B and C. So, rather than taking F to construct R, we can use $\tilde{F} = (F_1,\, G_L)$, where $G_L = F_2 - F_1 C$, and so $(1 - P(F))\partial_+ G_L = 0$. Now, using the fact that

$$A_- = \partial_- P(\tilde{F}) = \tilde{F}(\tilde{F}^\dagger \tilde{F})^{-1}((1 - P(\tilde{F}))\partial_+ F_1)^\dagger, \tag{7.78}$$

we see that $G_L^\dagger A_- = 0$, and that the rank of A_- is equal to the rank of $(1 - P(\tilde{F}))\partial_+ F_1$. As a consequence, we see that I is spanned by $(1 - P(G_L))\tilde{F}$, and we can write the sequence given above as $(G_L, (1 - P(G_L))\tilde{F}, (1 - P(\tilde{F}))\partial_+ F_1, G_R)$, which demonstrates the validity of our result.

7.5.2 *Adding a Basic Uniton*

Having determined the most general form of the one-uniton solution we can now construct a two-uniton solution by adding a basic uniton to the one-uniton configuration. Thus generalising the discussion given in chapter 3 we construct an orthogonal holomorphic basis sequence

$$(Y_1, Y_2, Y_3, \ldots, Y_i, \ldots, Y_{2r}, Y_{2r+1}), \tag{7.79}$$

where the sets Y_1 and Y_{2r+1} may be empty. It is now easy to see that

$$Q = (1 - 2R), \tag{7.80}$$

where

$$R = \sum_{i=1}^{r} P(Y_{2i}) \tag{7.81}$$

is a grassmannian solution of the $U(N)$ model (here we have used the same construction as in [8]). Observe that, when $r = 1$ and Y_1 is empty, we recover the instanton solutions described before.

We can now generalise the discussion of the previous sections and so split each set Y_i into various subsets

$$Y_i = (V_i, U_i, W_i, I_i), \tag{7.82}$$

chosen in such a way that

$$M_i^\dagger \partial_\pm N_j = 0 \quad i \neq j, \tag{7.83}$$

where M_i and N_j stand any set from V_i, U_i or W_i. Then, as the detailed calculation shows, we can construct a second uniton factor by choosing our new projector R_1 to be built from projectors onto various subsets V_i, U_j and W_k. Of course the coefficients of the various matrices which arise in the construction of this projector have to appropriately chosen as they have to satisfy various constraints in order for this construction to work. As the construction is quite technical we do not reproduce it here. The interested reader can consult [9] for more details. Instead, we finish this section by presenting an explicit example.

7.5.3 An Explicit Example

When we try to construct solutions of the grassmannian models it is convenient to use holomorphic bases. Let us choose a holomorphic matrix F of maximal rank. Then, as in chapter 3, we can define

$$P_+F = \partial_+F - F(F^\dagger F)^{-1}F^\dagger\partial_+F, \tag{7.84}$$

and use induction to define further matrices P_+^kF:

$$P_+^kF = P_+(P_+^{k-1}F). \tag{7.85}$$

We recall from chapter 3 that all the P_+^kF which correspond to different k's are orthogonal to each other, and that they satisfy the following properties which are essential in our construction:

$$\begin{aligned} \partial_-P_+^kF &= -P_+^{k-1}F\alpha_{k-1}^{-1}\alpha_k \\ \partial_+(P_+^kF\alpha_k^{-1}) &= P_+^{k-1}F\alpha_k^{-1}, \end{aligned} \tag{7.86}$$

where

$$\alpha_k = P_+^kF^\dagger P_+^kF. \tag{7.87}$$

As a consequence of the second equation, we see that the $P_+^i F$ have the natural holomorphic normalisation, and that

$$(F,\, P_+F,\, P_+^2F,\, \ldots,\, P_+^{2r}F) \tag{7.88}$$

is an orthogonal holomorphic basis sequence. Defining projectors onto each of these matrices,

$$P_0 = P(F), \quad \ldots \quad P_k = P(P_+^k F), \tag{7.89}$$

we find that the sum of any of these projectors forms a solution of the type (7.80), in which the sum of consecutive projectors corresponds to a single projector $P(Y_i)$ of our previous construction. To make this clearer, let us consider, for example

$$R = P_2 + P_7 + P_8 + P_9. \tag{7.90}$$

This solution corresponds to the choice $Y_1 = (F,\, P_+F)$, $Y_2 = P_+^2 F$, $Y_3 = (P_+^3F,\, P_+^4F,\, P_+^5F,\, P_+^6F)$, $Y_4 = (P_+^7F,\, P_+^8F,\, P_+^9F)$, $Y_5 = (P_+^{10}F, \ldots, P_+^{2r}F)$. Using relations (7.86) it is now easy to show that $I_1 = P_+F$, $I_2 = P_+^2F$, $I_3 = P_+^6F$, $I_4 = P_+^9F$. An example of a basic uniton that we can add to this solution is given by

$$R = P_3 + P_{10} + P(V_M), \tag{7.91}$$

where

$$V_M = Fa + U_3 b + P_+^7 F c, \tag{7.92}$$

and where a, b and c are holomorphic matrices of maximal rank and U_3 is given by $U_3 = (1 - P_0 - P_1 - P_2 - P_3)(\partial_+^4 F,\, \partial_+^5 F)$. It is interesting to note that, when we take the limit in which two of the three coefficients a, b and c tend to zero, our solution goes over to a grassmannian solution of the type (7.80). Thus we see that the non-grassmannian $U(N)$ solutions we have computed have the surprising property that they interpolate between different grassmannian solutions.

7.6 Properties of the Derived Solutions

Before we look at various properties of the solutions presented let us observe that although we have studied solutions of the $U(N)$ chiral model, our procedure has also given us solutions of the $SU(N)$ chiral model. In the latter case we have to choose appropriately the arbitrary constant matrix K in (7.5).

7.6.1 *Action*

To calculate the value of the action for our solutions, let us observe that, if G and G_0 are both solutions of the equations of motion, and related by

$$G = G_0(1-2R), \tag{7.93}$$

where the grassmannian projector R satisfies the Uhlenbeck equations, then the values of the action for both solutions are related by

$$S(Q) = S(Q_0) + \tilde{Q}(R), \tag{7.94}$$

where $\tilde{Q}(R)$ is the topological charge of the grassmannian projector R. To see this we observe that

$$\begin{aligned}\operatorname{Tr}\partial_+ Q^\dagger \partial_- Q =& 4\operatorname{Tr}\partial_+ R\partial_- R + \operatorname{Tr}\partial_+ Q_0^\dagger \partial_- Q_0 \\ &+ 4\operatorname{Tr} A_+\partial_- R(1-2R) - 4\operatorname{Tr} A_-(1-2R)\partial_+ R.\end{aligned} \tag{7.95}$$

However from the Uhlenbeck equations we find

$$\operatorname{Tr} A_-\partial_+ RR = \operatorname{Tr} RA_- R\partial_+ R = 0, \tag{7.96}$$

and

$$\operatorname{Tr} A_-\partial_+ R = \operatorname{Tr} A_- R\partial_+ R = -\operatorname{Tr}(1-R)\partial_- R\partial_+ R. \tag{7.97}$$

Thus we see that

$$\begin{aligned}\operatorname{Tr}\partial_+ Q^\dagger \partial_- Q + \operatorname{Tr}\partial_- Q^\dagger \partial_+ Q =& \operatorname{Tr}\partial_+ Q_0^\dagger \partial_- Q_0 \\ &+ \operatorname{Tr}\partial_- Q_0^\dagger \partial_+ Q_0 + 8\{\operatorname{Tr}(\partial_+ R\partial_- R - 2(1-R)\partial_- R\partial_+ R)\}.\end{aligned} \tag{7.98}$$

But the last expression can be rewritten as

$$\operatorname{Tr}\partial_+ R\partial_- R - 2\operatorname{Tr}(1-R)\partial_- R\partial_+ R = \operatorname{Tr}(\partial_+ RR\partial_- R - \partial_- RR\partial_+ R), \tag{7.99}$$

which is the topological charge density of the projector R.

Thus for the general solution corresponding to l unitons

$$Q = (1-2R_1)(1-2R_2)\ldots(1-2R_l) \tag{7.100}$$

the value of the action is given by

$$S = \sum_{i=1}^{l} \tilde{Q}_i, \tag{7.101}$$

where $\tilde{Q}_i$ is the topological charge corresponding to the projector R_i, and where we have used the fact that for the one-uniton solution the value of the action and of the topological charge are the same, since this uniton corresponds to an instantonic solution of some grassmannian model. So, to compute the action of a l-uniton solution, all we have to do is to add the topological charges of all the projectors corresponding to each uniton. As a direct consequence, we see that the action of each solution is quantised.

In the construction of our solutions, we have found that the projector defining each uniton factor can be expressed as a sum of some elementary projectors. Thus we see that the computation would simplify if we could express the topological charge of a uniton projector as the sum of the topological charges of each constituent projector. To check whether this is possible, we first of all observe that, if $R = S + T$, where S and T are two orthogonal projectors chosen in such a way that $S\partial_- T = 0$ (note that $-$ can be replaced by $+$ by permuting T and S), then the topological charge of R is given by the sum of the topological charges of T and S:

$$\tilde{Q}(R) = \tilde{Q}(S) + \tilde{Q}(T). \tag{7.102}$$

On the other hand, if $P(V)$ is a projector for which V is a maximal rank matrix with the natural holomorphic normalisation, i.e. given by $V^\dagger \partial_- V = 0$, then

$$\tilde{Q}(P(V)) = \int \mathrm{d}x^2 \,\mathrm{Tr}\left[|V|^{-2}(|P_+V|^2 - |P_-V|^2)\right], \tag{7.103}$$

where, as before,

$$P_\pm V = \partial_\pm V - V|V|^{-2}V^\dagger \partial_\pm V. \tag{7.104}$$

Moreover, using the fact that, for any invertible matrix A,

$$\partial_\mu \,\mathrm{Tr}(\ln(A)) = \mathrm{Tr}(\partial_\mu A\, A^{-1}), \tag{7.105}$$

it is easy to show that

$$\tilde{Q}(P(V)) = \int \mathrm{d}x^2 \partial_+\partial_- \ln\det|V|^2 = 2\pi q, \tag{7.106}$$

where q is the leading power of $\det|V|$ at infinity. From the above results it is now straightforward to show that

$$\begin{aligned}
\tilde{Q}(R_0) &= \int \mathrm{d}x^2 \left(\sum_{i=1}^{s} \partial_+\partial_- \ln\det|Y_{2i}|^2\right) \\
\tilde{Q}(R_1) &= \int \mathrm{d}x^2 \left(\sum_{i=1}^{2s+1} \partial_+\partial_- \ln\det|V_i|^2 + \partial_+\partial_- \ln\det|U|^2\right).
\end{aligned} \tag{7.107}$$

From this result we see that the action is given by an integer multiple of 2π. From now on we will always refer to this integer as the value of the action (remembering that the actual value has to be multiplied by 2π).

Next we use an argument similar to the one used in the grassmannian case discussed in chapter 3, and find that the condition of the finiteness of the action requires all the elements of the matrices F_i's and all the a_i's to be polynomials in x_+.

7.6.2 *Stability*

Next we discuss the stability properties of our solutions. In doing so we generalise the discussion of chapter 3, where we studied the stability properties of classical solutions of grassmannian models. However, before looking at the general case, let us recall that the simplest $U(N)$ chiral model, namely, the $U(2)$ model, is equivalent to the $O(4)$ model and as all the solutions of the $O(4)$ models are the embeddings of the $O(3)$ model solutions, all nontrivial solutions are unstable. To see this it suffices to take the fluctuation in the direction orthogonal to the $O(3)$ manifold and observe that this fluctuation reduces the action and so is a negative mode of the fluctuation operator [10]. This suggests that it is unlikely that any nontrivial solutions of the $U(N)$ models will be stable.

To look at the general case we follow Rawnsley [11], and consider the field Q_t defined by

$$Q_t = Q_0(1-2R)^t = Q_0 \exp(i\pi Rt). \tag{7.108}$$

This field interpolates between the two solutions (corresponding to $t=0$ and $t=1$), and so for $\tilde{Q}>0$ the change of t from 1 to 0 decreases the action, thus demonstrating the instability of Q_1. Moreover, the calculation of the action for Q_t gives

$$S(Q_t) = S(Q_0) + \tilde{Q}(R)\frac{(1-\cos(\pi t))}{2}, \tag{7.109}$$

which shows that the fluctuation in the direction of R is a negative mode. Of course, one can now repeat this procedure, factorise Q_0 and, if the topological charge of the last projector is positive, find a negative mode of this solution. This suggests that there are two negative modes of fluctuations (associated with both directions of descent). Are these modes independent, and can we exhibit a negative mode for a solution in which the last factor corresponds to a grassmannian projector of a negative topological charge? If, instead of using the successive reduction method, we reduce all uniton factors simultaneously, we will establish the general instability of all the solutions. On the other hand, if we can establish the commutativity of the reduction of different uniton factors, this will demonstrate the existence of independent modes.

Let us consider first the general instability due to the simultaneous reduction of all the uniton factors. Thus we take a general solution given by

$$Q^l = \prod_{i=1}^{l}(1 - 2\mathbb{P}_i), \tag{7.110}$$

and consider

$$Q^l_t = \prod_{i=1}^{l}(1 - 2\mathbb{P}_i)^t = \prod_{i=1}^{l} K_i, \tag{7.111}$$

where

$$K_i = 1 + \mathbb{P}_i[\exp(i\pi t) - 1] = 1 - \mathbb{P}_i\phi, \tag{7.112}$$

where $\phi = 1 - \exp(i\pi t)$.

Next we show that

$$\partial_- Q^{l\dagger}_t Q^l_t = \partial_- Q^{l\dagger}_1 Q^l_1 \frac{1 - \exp(-i\pi t)}{2}. \tag{7.113}$$

This we prove by induction. We observe that

$$\begin{aligned} \partial_- Q^{1\dagger}_t Q^1_t &= -\partial_- \mathbb{P}_1 \phi^*(1 - \mathbb{P}_1\phi) \\ = -\partial_- \mathbb{P}_1 \phi^* &= \partial_- Q^{1\dagger}_1 Q^1_1 \frac{\phi^*}{2}. \end{aligned} \tag{7.114}$$

Then we look at $\partial_- Q^{l\dagger}_t Q^l_t$, and using the inductive step, we find that

$$\begin{aligned} \partial_- Q^{l\dagger}_t Q^l_t &= (1 - \mathbb{P}_l\phi^*)\,\partial_- Q^{l-1\dagger}_t Q^{l-1}_t (1 - \mathbb{P}_l\phi) \\ -\partial_- \mathbb{P}_l\phi^*(1 - \mathbb{P}_l\phi) &= (1 - \mathbb{P}_l\phi^*)\partial_- Q^{l-1\dagger}_1 Q^{l-1}_1 \frac{\phi^*}{2}(1 - \mathbb{P}_l\phi) \\ -\partial_- \mathbb{P}_l\phi^*(1 - \mathbb{P}_l\phi) &= (\partial_- Q^{l-1\dagger}_1 Q^{l-1}_1 - 2\partial_- \mathbb{P}_l)\frac{\phi^*}{2} \\ &+ \mathbb{P}_l \partial_- Q^{l-1\dagger}_1 Q^{l-1}_1 \mathbb{P}_l \frac{\phi^*|\phi|^2}{2} - \mathbb{P}_l \partial_- Q^{l-1\dagger}_1 Q^{l-1}_1 \frac{\phi^{*2}}{2} \\ &- \partial_- Q^{l-1\dagger}_1 Q^{l-1}_1 \mathbb{P}_l \frac{|\phi|^2}{2} + \partial_- \mathbb{P}_l \mathbb{P}_l |\phi|^2. \end{aligned} \tag{7.115}$$

Next we consider the sum of the last four terms in the expression above. We observe, that as $|\phi|^2 = \phi + \phi^*$, we can rewrite them as

$$\begin{aligned} &|\phi|^2 \left(\mathbb{P}_l \frac{\partial_- Q^{l-1\dagger}_1 Q^{l-1}_1}{2} \mathbb{P}_l - \frac{\partial_- Q^{l-1\dagger}_1 Q^{l-1}_1}{2} \mathbb{P}_l + \partial_- \mathbb{P}_l \mathbb{P}_l \right) \\ &+ \phi^{*2} \left(\mathbb{P}_l \frac{\partial_- Q^{l-1\dagger}_1 Q^{l-1}_1}{2} \mathbb{P}_l - \mathbb{P}_l \frac{\partial_- Q^{l-1\dagger}_1 Q^{l-1}_1}{2} \right). \end{aligned} \tag{7.116}$$

However, as is very easy to check, the coefficients of ϕ^{*2} and $|\phi|^2$ vanish by the equations of Uhlenbeck as $A_-^l = -\frac{1}{2}\partial_- Q_1^{l-1\dagger} Q_1^{l-1}$. Thus we see that

$$\partial_- Q_t^{l\dagger} Q_t^l = \left(\partial_- Q_1^{l-1\dagger} Q_1^{l-1} - 2\partial_- \mathbb{P}_l\right) \frac{\phi^*}{2}, \tag{7.117}$$

which establishes our result, and also it gives us a byproduct:

$$A_-^l = A_-^{l-1} + \partial_- \mathbb{P}_l, \tag{7.118}$$

which, again by induction, shows that

$$A_-^l = \sum_{i=1}^{l} \partial_- \mathbb{P}_i. \tag{7.119}$$

Now we can look at the Lagrangian density $L(Q_t^n)$. As $Q_t^n = Q_t^{n-1} K_n$ we find

$$\begin{aligned} L(Q_t^n) &= \mathrm{Tr}\, \partial_+ \left[K_n^\dagger Q_t^{n-1\dagger}\right] \partial_- \left[K_n Q_t^{n-1}\right] + \partial_- \left[K_n^\dagger Q_t^{n-1\dagger}\right] \partial_+ \left[K_n Q_t^{n-1}\right] \\ &= L(Q_t^{n-1}) + \mathrm{Tr}\, \{\partial_+ K_n^\dagger \partial_- K_n + \partial_- K_n^\dagger \partial_+ K_n\} \\ &\quad + \mathrm{Tr}\, \{Q_t^{n-1\dagger} \partial_- Q_t^{n-1} K \partial_+ K^\dagger + \partial_+ Q_t^{n-1\dagger} Q_t^{n-1} \partial_- K K^\dagger \\ &\quad + Q_t^{n-1\dagger} \partial_+ Q_t^{n-1} K \partial_- K^\dagger + \partial_- Q_t^{n-1\dagger} Q_t^{n-1} \partial_+ K K^\dagger\}. \end{aligned} \tag{7.120}$$

However, using our results, we have

$$\mathrm{Tr}\, (\partial_+ K_n^\dagger \partial_- K_n + \partial_- K_n^\dagger \partial_+ K_n) = 2\, \mathrm{Tr}\, \partial_+ \mathbb{P}_n \partial_- \mathbb{P}_n |\phi|^2, \tag{7.121}$$

and the last four terms are given by

$$\begin{aligned} &\mathrm{Tr} \left\{ Q_1^{n-1\dagger} \partial_- Q_1^{n-1} K_n \partial_+ K_n^\dagger \frac{\phi^*}{2} + \partial_+ Q_1^{n-1\dagger} Q_1^{n-1} \partial_- K_n K_n^\dagger \frac{\phi}{2} \right. \\ &\left. + Q_1^{n-1\dagger} \partial_+ Q_1^{n-1} K_n \partial_- K_n^\dagger \frac{\phi}{2} + \partial_- Q_1^{n-1\dagger} Q_1^{n-1} \partial_+ K_n K_n^\dagger \frac{\phi^*}{2} \right\}. \end{aligned} \tag{7.122}$$

We can now use the property that $K_n \partial_+ K_n^\dagger = -\partial_+ \mathbb{P}_n \phi^* + \mathbb{P}_n \partial_+ \mathbb{P}_n |\phi|^2$, and so show that

$$\begin{aligned} L(Q_t^n) = L(Q_t^{n-1}) &+ 2\, \mathrm{Tr}\, \partial_+ \mathbb{P}_n \partial_- \mathbb{P}_n |\phi|^2 - \phi^{*2}\, \mathrm{Tr}\, (A_-^{n-1} \partial_+ \mathbb{P}_n) \\ &+ \phi^* |\phi|^2\, \mathrm{Tr}\, (A_-^{n-1} \mathbb{P}_n \partial_+ \mathbb{P}_n) + \phi^2\, \mathrm{Tr}\, (A_+^{n-1} \partial_- \mathbb{P}_n) \\ &- \phi |\phi|^2\, \mathrm{Tr}\, (A_+^{n-1} \partial_- \mathbb{P}_n \mathbb{P}_n) - |\phi|^2\, \mathrm{Tr}\, (A_+^{n-1} \partial_- \mathbb{P}_n) \\ &+ \phi |\phi|^2\, \mathrm{Tr}\, (A_+^{n-1} \mathbb{P}_n \partial_- \mathbb{P}_n) + |\phi|^2\, \mathrm{Tr}\, (A_-^{n-1} \partial_+ \mathbb{P}_n) \\ &- \phi^* |\phi|^2\, \mathrm{Tr}\, (A_-^{n-1} \partial_+ \mathbb{P}_n \mathbb{P}_n). \end{aligned} \tag{7.123}$$

Using the Uhlenbeck equations we find that

$$\mathrm{Tr}\,(A_-^{n-1}\partial_+\mathbb{P}_n\mathbb{P}_n) = \mathrm{Tr}\,(A_+^{n-1}\mathbb{P}_n\partial_-\mathbb{P}_n) = 0 \tag{7.124}$$

and

$$\begin{aligned}\mathrm{Tr}\,(A_-^{n-1}\partial_+\mathbb{P}_n) &= \mathrm{Tr}\,(A_-^{n-1}\mathbb{P}_n\partial_+\mathbb{P}_n) = -\,\mathrm{Tr}\,(A_+^{n-1}\partial_-\mathbb{P}_n\mathbb{P}_n)\\ &= -\,\mathrm{Tr}\,(A_+^{n-1}\partial_-\mathbb{P}_n) = -\,\mathrm{Tr}\,(\partial_+\mathbb{P}_n\mathbb{P}\partial_-\mathbb{P}_n),\end{aligned} \tag{7.125}$$

and so that

$$\begin{aligned}L(Q_t^n) - L(Q_t^{n-1}) &= 2\,\mathrm{Tr}\,\partial_+\mathbb{P}_n\partial_-\mathbb{P}_n|\phi|^2 - 4\,\mathrm{Tr}\,\partial_+\mathbb{P}_n\mathbb{P}_n\partial_-\mathbb{P}_n|\phi|^2\\ &= 4(\mathrm{Tr}\,\partial_+\mathbb{P}_n\partial_-\mathbb{P}_n\mathbb{P}_n - \mathrm{Tr}\,\partial_+\mathbb{P}_n\mathbb{P}_n\partial_-\mathbb{P}_n)(1-\cos(\pi t)).\end{aligned} \tag{7.126}$$

Hence, by induction,

$$L(Q_t^n) = 4(1-\cos(\pi t))\sum_{i=1}^{n}\mathrm{Tr}\,(\partial_+\mathbb{P}_i\partial_-\mathbb{P}_i\mathbb{P}_i - \partial_+\mathbb{P}_i\mathbb{P}_i\partial_-\mathbb{P}_i). \tag{7.127}$$

As the action of our solution is

$$S(Q^l) = \tfrac{1}{4}\int d^2x\, L(Q_1^l) = \left(2\sum_{i=1}^{l}\tilde{Q}_i\right) > 0, \tag{7.128}$$

where $\tilde{Q}_i$ are the topological charges of the constituent projectors, we see that our choice of Q_t^l demonstrates the instability of Q^l. Hence there is always at least one negative mode of the fluctuation.

To study the independence of the negative modes we should consider

$$Q_{t_1\ldots t_l}^l = \prod_{i=1}^{l}(1 - 2\mathbb{P}_i)^{t_i}, \tag{7.129}$$

where t_i are independent. The general discussion is very complicated, so we will restrict ourselves to the case of $l = 2$ and consider

$$Q_{ts} = (1-2\mathbb{P})^t(1-2R)^s. \tag{7.130}$$

Repeating the previous procedure, we find, after a lengthy calculation, that

$$\begin{aligned}S(Q_{ts}) &= \mathrm{Tr}\,(\partial_+\mathbb{P}\partial_-\mathbb{P})(1-\cos(\pi t))\\ &\quad + \mathrm{Tr}\,(\partial_+R\partial_-R)(1-\cos(\pi s)) + \mathrm{Tr}\left[\partial_-\mathbb{P}\partial_+\mathbb{P}R - \partial_+\mathbb{P}\partial_-\mathbb{P}R\right]\\ &\quad \times\left[(1-\cos(\pi t))(1-\cos(\pi s)) + \sin(\pi t)\sin(\pi s)\right],\end{aligned} \tag{7.131}$$

which can also be rewritten as

$$L(Q_{ts}) = \mathrm{Tr}\,(\partial_+\mathbb{P}\partial_-\mathbb{P})(1-\cos(\pi t)) + \mathrm{Tr}\,(\partial_+ RR\partial_- R)(1-\cos(\pi s)) + \mathrm{Tr}\,(\partial_+ R\partial_- RR)\Big[(1-\cos(\pi s))\cos(\pi t) - sin(\pi t)\sin(\pi s)\Big]. \tag{7.132}$$

Looking at the instability locally around the solution we set $t = 1-\epsilon/\pi$, $s = 1-\mu/\pi$, where ϵ and μ are infinitesimal parameters, and we find

$$S(Q_{ts}) = S(Q_{11}) - \int \mathrm{d}^2x \left[\frac{\epsilon^2}{2}\mathrm{Tr}\,\partial_+\mathbb{P}\partial_-\mathbb{P} + \frac{\mu^2}{2}\mathrm{Tr}\,\partial_+ RR\partial_- R + \left(-\epsilon^2 - \frac{\mu^2}{2} + \epsilon\mu\right)\mathrm{Tr}\,\partial_+ R\partial_- RR\right], \tag{7.133}$$

where we have kept only the quadratic terms in the infinitesimal variables μ and ϵ. Using the definitions of the topological charge densities we rewrite the fluctuation as

$$S(Q_{ts}) - S(Q_{11}) = -\left(\frac{\epsilon^2}{2}\tilde{Q}(\mathbb{P}) + \frac{\mu^2}{2}\tilde{Q}(R) + (\epsilon\mu - \epsilon^2)\int \mathrm{d}^2x\,\mathrm{Tr}\,\partial_+ R\partial_- RR\right). \tag{7.134}$$

It is clear that in general the above expression can be of any sign, showing that in general the ϵ and the μ modes are *not* independent.

Let us look then at the simplest nongrassmannian solution, which, as we showed in the previous sections, is given by

$$g = (1 - 2\mathbb{P})(1 - 2R), \tag{7.135}$$

where

$$\mathbb{P} = \frac{ff^\dagger}{|f|^2}, \qquad R = 1 - \frac{hh^\dagger}{|h|^2}, \tag{7.136}$$

and where

$$h = \frac{\alpha(x_-)f}{|f|^2} + \frac{\beta(x_-)P_+^{N-1}f}{|P_+^{N-1}f|^2}. \tag{7.137}$$

In particular, if we consider the $N = 3$ case, we can take

$$f = \begin{Bmatrix} x_+^2 \\ \sqrt{2}x_+ \\ 1 \end{Bmatrix}, \tag{7.138}$$

and choose α, β to be some arbitrary (but nonzero) constant numbers. Then, as is easy to check,

$$\tilde{Q}(\mathbb{P}) = 4\pi, \qquad \tilde{Q}(R) = 4\pi, \tag{7.139}$$

and so we are left with having to evaluate

$$F = \int \mathrm{d}^2x \; \mathrm{Tr}\, \partial_+ R \partial_- R R. \tag{7.140}$$

We find that

$$\begin{aligned} F &= \int \mathrm{d}^2x \; \mathrm{Tr}\, \partial_+ R \partial_- R R \\ &= \int \mathrm{d}^2x \; \mathrm{Tr} \left[\partial_+ \left(\frac{hh^\dagger}{|h|^2} \right) \partial_- \left(\frac{hh^\dagger}{|h|^2} \right) \left(1 - \frac{hh^\dagger}{|h|^2} \right) \right] \\ &= \mathrm{Tr} \int \mathrm{d}^2x \, \partial_+ \left(\frac{hh^\dagger}{|h|^2} \right) \frac{hh^\dagger}{|h|^2} \partial_- \left(\frac{hh^\dagger}{|h|^2} \right) \\ &= \int \mathrm{d}^2x \frac{|\partial_+ h|^2}{|h|^2} = \int \mathrm{d}^2x \frac{|\alpha|^2 |P_+ f|^2}{|f|^2 \left[|\alpha|^2 + |\beta|^2 |f|^2 / |P_+^2 f|^2 \right]}, \end{aligned} \tag{7.141}$$

as $h^\dagger \partial_- h / |h|^2 = -(\partial_+ h)^\dagger h / |h|^2 = 0$. Then, if we take f given by (7.138), we find that

$$F = \int \mathrm{d}^2x \frac{2|\alpha|^2}{(1+x^2)^2 \left[|\alpha|^2 + |\beta|^2 (1+x^2)^4/4 \right]}. \tag{7.142}$$

It is clear that if $\beta \neq 0$, F is a function of only $\gamma = |\beta|/|\alpha|$, and that $F(\gamma) \to 0$ as $\gamma \to \infty$. However, as $S(Q_{ts}) - S(Q_{11}) = -\left[\epsilon^2 + \mu^2 + (\epsilon\mu - \epsilon^2)F\right]$, we see that this expression is negative if $F^2 + 4F - 4 < 0$. This condition can always be satisfied if we choose γ large enough. Observe that as $\gamma \to \infty$, our solution resembles more and more the noninstanton grassmannian solution.

We see that we have demonstrated that some nongrassmannian solutions have at least two modes of instability. These solutions, by an appropriate change of parameters, can be made to approach arbitrarily closely the non-instanton grassmannian solutions which are known to be unstable (they are already unstable in the grassmannian subspace). What about the grassmannian instanton solutions?

Our general discussion shows that they are unstable, but it is not clear whether there is more than one mode of instability. We consider this problem next.

7.6.3 *Stability of the Grassmannian Embeddings*

Let us restrict ourselves to the grassmannian embeddings of the $SU(N)$ chiral models. Then, as we know,

$$Q_0 = A(1 - 2\mathbb{P}), \tag{7.143}$$

where $\mathbb{P}$ is a grassmannian projector and A is a constant matrix, so chosen that $\det Q_0 = 1$. The above Q_0 satisfies the equation of motion of the chiral model (7.4), if

$$[\partial_+\partial_-\mathbb{P}\,,\mathbb{P}] = 0, \tag{7.144}$$

and, in particular, as we have discussed before, if $\mathbb{P}$ satisfies $\mathbb{P}\partial_+\mathbb{P} = 0$, we have an embedding of an instantonic solution. As all noninstanton grassmannian solutions are already unstable in the grassmannian subspace, let us consider only the instantonic embeddings.

The general field q in the neighbourhood of Q_0 can be written as

$$q = Q_0 \exp(iX), \tag{7.145}$$

where $X = X^\dagger$. In order that $q \in SU(N)$, we have to require that $\operatorname{Tr} X = 0$. Substituting $q = Q_0 \exp(iX)$ into the action, and calculating only to second order in X, we find that

$$\begin{aligned} S(q) - S(Q_0) &\sim \tfrac{1}{4}\int \mathrm{d}^2x\ \operatorname{Tr}\,[X_\mu X_\mu - Q_0\partial_\mu Q_0(X_\mu X - XX_\mu)] \\ &= \tfrac{1}{4}\int \mathrm{d}^2x\ \operatorname{Tr}\,[X_\mu X_\mu + 2(1-2\mathbb{P})\partial_\mu\mathbb{P}(X_\mu X - XX_\mu)] \\ &= \tfrac{1}{2}\int \mathrm{d}^2x\ [\operatorname{Tr}\partial_+X\partial_-X + 2\operatorname{Tr}(\partial_+\mathbb{P}\partial_-XX - \partial_-\mathbb{P}\partial_+XX)], \end{aligned} \tag{7.146}$$

where we have set $X_\mu = \partial_\mu X$ and used the fact that $\mathbb{P}\partial_+\mathbb{P} = 0$.

Let us now consider

$$X = \mathbb{P}K\mathbb{P} - \frac{1}{N}\operatorname{Tr}\mathbb{P}K, \tag{7.147}$$

where $K = vv^\dagger$. Then

$$\partial_+X = \partial_+\mathbb{P}K\mathbb{P} + \mathbb{P}K\partial_+\mathbb{P} - \frac{1}{N}\operatorname{Tr}\partial_+\mathbb{P}K, \tag{7.148}$$

and so

$$\begin{aligned} \operatorname{Tr}\partial_+X\partial_-X &= \operatorname{Tr}\partial_-\mathbb{P}\partial_+\mathbb{P}K\mathbb{P}K + \operatorname{Tr}\partial_+\mathbb{P}\partial_-\mathbb{P}K\mathbb{P}K \\ &\quad - \frac{1}{N}\operatorname{Tr}(\partial_-\mathbb{P}K)\operatorname{Tr}(\partial_+\mathbb{P}K), \end{aligned} \tag{7.149}$$

and

$$\operatorname{Tr}\partial_+\mathbb{P}\partial_-XX - \operatorname{Tr}\partial_-\mathbb{P}\partial_+XX = -\operatorname{Tr}\partial_-\mathbb{P}\partial_+\mathbb{P}KPK, \tag{7.150}$$

where, in deriving these expressions, we have used the equations of motion ($\mathbb{P}\partial_+\mathbb{P} = 0$, etc). Thus we find

$$S(q) - S(Q_0) = \tfrac{1}{2}\int \mathrm{d}^2x\{\mathrm{Tr}\,\partial_+\mathbb{P}\partial_-\mathbb{P}K\mathbb{P}K - \mathrm{Tr}\,\partial_-\mathbb{P}\partial_+\mathbb{P}K\mathbb{P}K \\ - (1/N)\,\mathrm{Tr}\,(\partial_-\mathbb{P}K)\,\mathrm{Tr}\,(\partial_+\mathbb{P}K)\}. \tag{7.151}$$

The last term is clearly negative, and, if we choose K so that we can use the divergence theorem and drop the surface term, the first two terms can be recast as

$$-\tfrac{1}{2}\int \mathrm{d}^2x\ \mathrm{Tr}\,\partial_-\mathbb{P}K\partial_+\mathbb{P}K. \tag{7.152}$$

Thus

$$S(q) - S(Q_0) = -\tfrac{1}{2}\int \mathrm{d}^2x\,|v^\dagger\partial_+\mathbb{P}v|^2\left(1+\frac{1}{N}\right), \tag{7.153}$$

which clearly shows that this is a negative mode. A little thought shows that, to justify dropping the surface term, we have to require that $|v^\dagger f|^2/|f|^2 \to 0$ at infinity. This is easy to impose; for example, if $N = 3$ and f is given by (7.138), we can take

$$v = \begin{Bmatrix} 0 \\ v_2 \\ v_3 \end{Bmatrix}. \tag{7.154}$$

We see that our negative mode of fluctuation depends on two parameters, v_2 and v_3. However, as is easy to check, we do not have a superposition principle for the independent parameters of v; we see that the way the parameters of v appear in the expression for the fluctuation prevents us from taking the number of parameters of v for the number of negatives modes. Are there further independent negative modes of fluctuations?

Let us, first of all, observe that, instead of our $X = \mathbb{P}K\mathbb{P} - (1/N)\,\mathrm{Tr}\,\mathbb{P}K$, we could take $X = (1-\mathbb{P})M(1-\mathbb{P}) - (1/N)\,\mathrm{Tr}\,\{(1-\mathbb{P})M\}$. As $Q_0\partial_\mu Q_0 = 2(1-\mathbb{P})\partial_\mu\mathbb{P} = 2(1-2R)\partial_\mu R$, where $R = 1 - \mathbb{P}$, it is clear that this X is also a negative mode of the fluctuation, provided that $M = ww^\dagger$ is so chosen that $|f|^2|w|^2 - |w^\dagger f|^2 \to 0$ at infinity. In our special case this corresponds to

$$w = \begin{Bmatrix} w_1 \\ 0 \\ 0 \end{Bmatrix}. \tag{7.155}$$

Can we now take a superposition of these two modes such that this superposition remains a negative mode of the fluctuation? To answer this question we consider

$$X = \mathbb{P}K\mathbb{P} + \alpha(1-\mathbb{P})M(1-\mathbb{P}) - \frac{1}{N}\,\mathrm{Tr}\,\mathbb{P}K - \frac{\alpha}{N}\,\mathrm{Tr}\,\{M(1-\mathbb{P})\}, \tag{7.156}$$

where again $K = vv^\dagger$ and $M = ww^\dagger$. Then a few lines of algebra, with the use of the equations of motion, give

$$\begin{aligned}\mathrm{Tr}\,\partial_+X\partial_-X &= \mathrm{Tr}\,\partial_-\mathbb{P}\partial_+\mathbb{P}K\mathbb{P}K + \mathrm{Tr}\,\partial_+\mathbb{P}\partial_+\mathbb{P}K\mathbb{P}K \\ &+ \alpha^2\,\mathrm{Tr}\,M\partial_+\mathbb{P}\partial_-\mathbb{P}M(1-\mathbb{P}) + \alpha^2\,\mathrm{Tr}\,M\partial_-\mathbb{P}\partial_+\mathbb{P}M(1-\mathbb{P}) \\ &- 2\alpha\,\mathrm{Tr}\,M\partial_+\mathbb{P}K\partial_-\mathbb{P} - \frac{1}{N}\mathrm{Tr}\,(K\partial_-\mathbb{P})\,\mathrm{Tr}\,(K\partial_+\mathbb{P}) \\ &- \frac{\alpha^2}{N}\mathrm{Tr}\,(M\partial_+\mathbb{P})\,\mathrm{Tr}\,(M\partial_-\mathbb{P}) + \frac{\alpha}{N}\mathrm{Tr}\,(K\partial_+\mathbb{P})\,\mathrm{Tr}\,(M\partial_-\mathbb{P}) \\ &+ \frac{\alpha}{N}\mathrm{Tr}\,(K\partial_-\mathbb{P})\,\mathrm{Tr}\,(M\partial_+\mathbb{P}).\end{aligned} \tag{7.157}$$

Also we find that

$$\begin{aligned}\mathrm{Tr}\,\partial_+\mathbb{P}\partial_-XX - \mathrm{Tr}\,\partial_-\mathbb{P}\partial_+XX &= -\mathrm{Tr}\,\partial_-\mathbb{P}\partial_+\mathbb{P}K\mathbb{P}K \\ -\alpha^2\,\mathrm{Tr}\,\partial_+\mathbb{P}\partial_-\mathbb{P}M(1-\mathbb{P})M &+ 2\alpha\,\mathrm{Tr}\,\partial_-\mathbb{P}M\partial_+\mathbb{P}K.\end{aligned} \tag{7.158}$$

Thus we have

$$\begin{aligned}S(q) - S(Q_0) &= \tfrac{1}{2}\int \mathrm{d}^2x\{\mathrm{Tr}\,\partial_+\mathbb{P}\partial_-\mathbb{P}K\mathbb{P}K - \mathrm{Tr}\,\partial_-\mathbb{P}\partial_+\mathbb{P}K\mathbb{P}K \\ &+ \alpha^2[\mathrm{Tr}\,M\partial_-\mathbb{P}\partial_+\mathbb{P}M(1-\mathbb{P}) - \mathrm{Tr}\,M\partial_+\mathbb{P}\partial_-\mathbb{P}M(1-\mathbb{P})] \\ &+ 2\alpha\,\mathrm{Tr}\,\partial_-\mathbb{P}M\partial_+\mathbb{P}K - \frac{1}{N}\mathrm{Tr}\,(K\partial_-\mathbb{P})\,\mathrm{Tr}\,(K\partial_+\mathbb{P}) \\ &- \frac{\alpha^2}{N}\mathrm{Tr}\,(M\partial_+\mathbb{P})\,\mathrm{Tr}\,(M\partial_-\mathbb{P}) + \frac{\alpha}{N}\mathrm{Tr}\,(K\partial_+\mathbb{P})\,\mathrm{Tr}\,(M\partial_-\mathbb{P}) \\ &+ \frac{\alpha}{N}\mathrm{Tr}\,(K\partial_+\mathbb{P})\,\mathrm{Tr}\,(M\partial_-\mathbb{P}\}.\end{aligned} \tag{7.159}$$

We see that the appearance of the mixed terms (proportional to α) prevents us from drawing a conclusion in full generality. Choosing K and M, as before, allows us to rewrite this expression as

$$
\begin{aligned}
&S(q) - S(Q_0) \\
&= -\tfrac{1}{2}\int d^2x |v^\dagger \partial_+ \mathbb{P}v|^2\Big(1+\frac{1}{N}\Big) - \tfrac{1}{2}\alpha^2 \int d^2x |w^\dagger \partial_+ \mathbb{P}w|^2\Big(1+\frac{1}{N}\Big) \\
&\quad + \tfrac{1}{2}\alpha \int d^2x \Big(2|w^\dagger \partial_+ \mathbb{P}v|^2 + \frac{1}{N}(v^\dagger \partial + \mathbb{P}v)(w^\dagger \partial_- \mathbb{P}w) \\
&\quad + \frac{1}{N}(w^\dagger \partial_+ \mathbb{P}w)(v^\dagger \partial_- \mathbb{P}v)\Big) \\
&= -\tfrac{1}{2}\int d^2x \Bigg(\frac{|v^\dagger P_+ f|^2 |v^\dagger f|^2}{|f|^4} \\
&\quad + \alpha^2 \frac{|w^\dagger P_+ f|^2 |w^\dagger f|^2}{|f|^4} - 2\alpha \frac{|v^\dagger f|^2 |w^\dagger P_+ f|^2}{|f|^4}\Bigg) \\
&\quad - \frac{1}{2N}\int d^2x |(v^\dagger \partial_+ \mathbb{P}v) - \alpha(w^\dagger \partial_+ \mathbb{P}w)|^2.
\end{aligned}
\tag{7.160}
$$

The last term is clearly negative, so we have to look in detail at the first term. As it is difficult to say anything definite about it in general, let us look at our special solution corresponding to f given by (7.138). Then

$$
P_+ f = \begin{Bmatrix} 2x_+ \\ \sqrt{2}(1-|x|^2) \\ -2x_- \end{Bmatrix} \frac{1}{1+|x|^2}. \tag{7.161}
$$

Thus the integrand of the first term is given by

$$
\frac{-4|x|^2}{(1+|x|^2)^6}\Big[|v|^4 + \alpha^2 |x|^4 |w|^4 - 2\alpha |v|^2 |w|^2\Big], \tag{7.162}
$$

where we have set $v_2 = 0$, and decided to drop the indices on v and w. (It can be checked that for $v_2 \neq 0$ our expression is not negative for all α.) However, as

$$
\int d^2x \frac{|x|^2}{(1+|x|^2)^6} = \int d^2x \frac{|x|^6}{(1+|x|^2)^6} \tag{7.163}
$$

we see that the first term gives

$$
-4(|v|^2 - \alpha|w|^2)^2 \int d^2x \frac{|x|^2}{(1+|x|^2)^2}, \tag{7.164}
$$

which again is nonpositive, and vanishes only for $|v|^2 = \alpha|w|^2$. It is easy to check that, when $|v|^2 = \alpha|w|^2$, $(v^\dagger \partial_+ \mathbb{P}v) - \alpha(w^\dagger \partial_+ \mathbb{P}w)$ does not vanish, thus showing that the total expression is negative definite. We see that for our special solution of the $SU(3)$ model we have found two independent negative modes of fluctuations. It is very difficult to say much more in general, as we have not been able to find a general method of generating such modes.

It is worth pointing out that, by analogy with a similar problem in a grassmannian case, some negative modes of the fluctuation operator are given by solutions of the associated background Dirac problem [12]. To see this we observe that the associated background Dirac problem reduces to equations

$$\partial_+ \psi_- + [A_+, \psi_-] = 0 \tag{7.165}$$

and

$$\partial_- \psi_+ + [A_-, \psi_+] = 0, \tag{7.166}$$

where $\psi_\pm$ denotes the helicity eigenstates of ψ (cf chapter 5) and $A_\pm = \frac{1}{2}g^\dagger \partial_\pm g$. Notice that, if ψ_- solves (7.165), then a solution of (7.166) is given by $\psi_+ = (\psi_-)^\dagger$.

We seek fluctuations which are hermitian, i.e. $X = X^\dagger$. Thus we can take for X a hermitian solution of (7.165), which, due to the comment above, is also a solution of (7.166). Such a fluctuation is clearly a negative mode, as

$$\begin{aligned} &\partial_- X\, \partial_+ X + [A_+, X] + [A_-, X] \\ &= -\partial_- X\, \partial_+ X = -(\partial_+ X)^\dagger \partial_+ X. \end{aligned} \tag{7.167}$$

However, it is not easy to find hermitian solutions of (7.165). When the chiral field g corresponds to a one-uniton configuration, i.e. $g = A(1-2P)$, $A_+ = P_+$; then for ψ_+ we can take $\psi_+ = \alpha 1 + \beta P$, which corresponds to the negative modes discussed before.

In fact we can generalise this observation further, and point out that, if a solution of the $U(N)$ chiral model g is given by

$$g = A(1 - P_1)(1 - P_2)\dots(1 - P_k), \tag{7.168}$$

for which all projectors P_l commute with each other (i.e. g describes a grassmannian field), any of the projectors P_l is a solution of (7.165), and so provides a negative mode of fluctuation. It would be interesting to find other solutions of (7.165).

7.6.4 *Interpolating Nature of Our Solutions*

In chapter 3 we discussed various properties of grassmannian solutions. We looked at action densities for some of these solutions, and found that, in general, they exhibit various well defined peaks which allow us to interpret them as describing extended structures with identifiable positions and sizes. This discussion can be extended also to solutions of the chiral models.

To do this let us consider some explicit examples of our solutions derived in the previous sections [13]. In section 7.3 we showed that the complete set of solutions of the $U(3)$ model is given by [7] the self-dual solutions (given by (7.23) and (7.24)),

$$\begin{aligned} Q_1 &= (1 - 2P_0) \\ Q_2 &= (1 - 2(P_0 + P_1)), \end{aligned} \tag{7.169}$$

and the two-uniton solutions (given by the last equation of (7.33))

$$Q_3 = (1 - 2(P_0 + P_1))(1 - 2P(V_2)), \tag{7.170}$$

where

$$V_2 = \alpha f + \beta P_+^2 f. \tag{7.171}$$

Looking at the form of this solution we observe that, when α goes to zero, the solution becomes the constant solution $Q_3 = 1$, but when β tends to zero, the solution approaches the non-instanton solution of the $\mathbb{C}P^2$ model, namely $Q_3 = (1 - 2P_1)$. To be completely specific let us take f in the form

$$f = \begin{pmatrix} (x_+ - 4)(x_+ + 4) \\ (x_+ - 5.5)(x_+ + 2.5) \\ (x_+ - 2.5)(x_+ + 5.5) \end{pmatrix}. \tag{7.172}$$

In this case, using (7.107), it is easy to evaluate the action corresponding to each solution. For both Q_1 and Q_2 the action is 2 (in units of 2π).

The action of the two-uniton solution depends on the expressions chosen for α and β. If we choose them both to be constant then, for large x_+, V_1 resembles a polynomial of the second order. The topological charge corresponding to $P(V_1)$ is therefore equal to 2, and the action of the two-uniton solution is 4. This can be seen explicitly in figure 7.1(a) in which we plot the Lagrangian density for this solution. For our parameters α and β we take 1000 and 1 respectively, and we observe that our plot exhibits four peaks, of which two are quite spiky and the remaining two are more spread out. Each peak contributes one unit to the action. Notice that the positions of the peaks appear to correspond to the zeros of the components of f.

As we have said before, the value of the action is quantised. So it must remain the same as we vary the values of α and β. On the other hand, when α is zero, we know that the solution is a constant, and so its action vanishes. So what happens as α vanishes? The answer to this question can be deduced from figures 7.1(a)–(e). We observe that, when α tends to zero (and the value of β is kept constant), the four peaks spread out and form a ring which moves out to infinity (the north pole of S^2) reaching it when α takes the value 0. This shows that we must be very careful

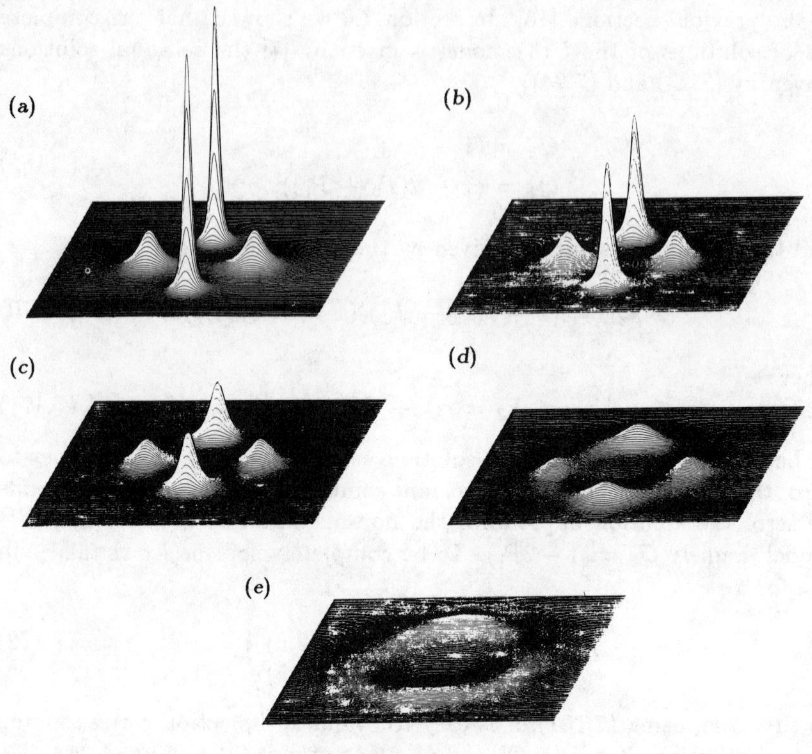

Figure 7.1

when we compute the value of the action, and that, without the inclusion of the extra contribution at infinity, the procedure of taking the limit of α vanishing does not commute with the integration of the Lagrangian density.

Next we consider nonconstant α and β. If we take $\beta = 1$ and $\alpha = \epsilon(x_+ - 0)\,(x_+ - 5 - 5i)$, the topological charge corresponding to $P(V_2)$ now equals 4, and the action of Q_3 is 6. We present the plot of the Lagrangian density in figures 7.2(*a*)–(*d*) and we observe that, indeed, the density exhibits six peaks. The two additional peaks are very spiky, and, as is easy to check, they are located around the zeros of α, namely 0 and $5 + 5i$. Choosing the vertical scale appropriately, it is easy to check that the spiky peaks are of finite height and their narrowness is real, not an artefact of our numerical approximation. All six peaks contribute one unit to the value of the action. If we now take the limit of ϵ going to zero, the peaks in the middle of the graph spread out and form a ring which absorbs the outside peak when it crosses it. We see that this behaviour of the central peaks very much resembles the behaviour found before.

We have studied different forms of the α and β functions, and in each

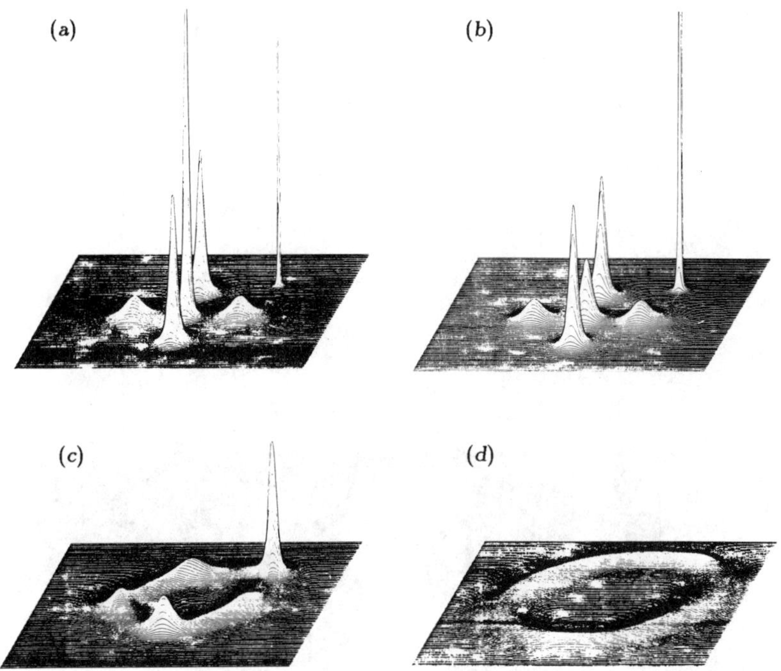

Figure 7.2

case we have found that the ring absorbs all the peaks which are on its way as it moves out. Here, at first sight, we appear to have a new paradox. Namely, when we take α in the form of a second order polynomial, $\alpha = a\,x_+^2 + bx_+ + c$, and take the limit of a going to zero, the action of the solution ought to decrease by one unit. However, this is not the case. In fact, we find, that in this limit, one of the two spiky peaks becomes more and more spiky ending up in the form of a delta function contribution, which again can be missed unless proper care is exercised in the calculation of the action. Actually the same is also true if β tends to zero; this time both the spiky peaks have to disappear.

Let us add that, if one of the functions α or β is not a polynomial but, say, is given by a trigonometric function of its argument, the topological charge of $P(V)$ is infinite. We can expect an infinite number of peaks in the Lagrangian density and, indeed, we find that this is the case. In figure 7.3 we exhibit the plot of the Lagrangian density for such a configuration corresponding to $\alpha = \epsilon \cos x_+$. Actually, as ϵ tends to zero the ring still forms and, as before, absorbs all those peaks that it crosses. A similar

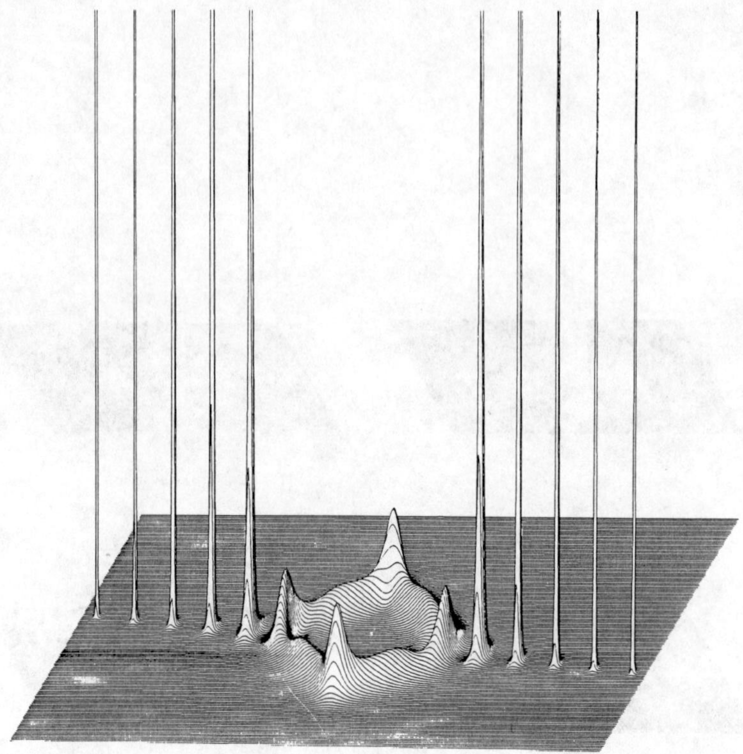

Figure 7.3

discussion can be also given for other solutions. In each case when chiral solutions are not grassmannian, the additional parameters on which these solutions depend provide us with variables which can be used to interpolate between different grassmannian solutions. In each case, as we vary these parameters, we can go from one grassmannian solution to another with the peak structure appropriately modified by the ring structure which is exhibited by the intermediate solutions.

References

[1] Fridling B and van de Ven A 1986 Renormalisation of generalised two-dimensional nonlinear sigma models *Nucl. Phys.* B **268** 719
de Alwis S 1985 The freedom of the Wess–Zumino–Witten model *Phys. Lett.* **164B** 67
Bakas I and McMullan D 1987 Alternative description of the Wess–Zumino-Witten system *Phys. Lett.* **189B** 141

[2] Uhlenbeck K 1985 Harmonic maps into Lie groups (classical solutions of the chiral model) *J. Diff. Geom.* to be published

[3] Eichenherr H and Forger M 1980 More about non-linear sigma models on symmetric spaces *Nucl. Phys.* B **164** 528; Erratum 1987 *Nucl. Phys.* B **282** 745

[4] Borchers M J and Garber W D 1980 Local theory of solutions for the $O(2k+1)$ σ model *Commun. Math. Phys.* **72** 77

[5] Piette B, Stokoe I and Zakrzewski W J 1988 On the stability of solutions of the $U(N)$ chiral model in two dimensions *Z. Phys.* C **37** 449

[6] Wood J C 1987 The explicit construction and parametrisation of all harmonic maps from the two-sphere to the unitary group *Proc. Lond. Math. Soc.* to be published

[7] Piette B and Zakrzewski W J 1988 General solutions of the $U(3)$ and $U(4)$ chiral σ models in two dimensions *Nucl. Phys.* B **300** 207

[8] Antoine J-P and Piette B 1988 Solutions of Euclidean σ models on non-compact Grassmann manifolds *J. Math. Phys.* **29** 1687

[9] Piette B and Zakrzewski W J 1987 General solutions of the $U(N)$ chiral σ Models in two dimensions *Durham University preprint DTP-29/87*

[10] Din A M and Zakrzewski W J 1980 Stability properties of classical solutions to non-linear σ models *Nucl. Phys.* B **168** 173

[11] Rawnsley J 1987 Private communication

[12] Burstall F 1987 Private communication

[13] Piette B and Zakrzewski W J 1988 Properties of classical solutions of the $U(N)$ chiral σ models in two dimensions *Nucl. Phys.* B **300** 223

8 Sigma Models with the Wess–Zumino Term

8.1 Lax-pair Problem

Let us return now to the grassmannian model discussed in chapter 3. It has often been said that the $\mathbb{C}P^{N-1}$ model is integrable [1]. What does this mean?

For a dynamical system of $2N$ dimensions ($p = (p_1, \ldots p_N)$, $q = (q_1, \ldots q_N)$) the term 'completely integrable' means that the system possesses N independent constants of motion $F_j(p, q)$, $j = 1, \ldots N$. Thus we can define N action-like variables and N corresponding angle variables, which, in consequence, vary linearly with time. In infinite dimensional systems such as field theories things are much more formal and the expression 'completely integrable' implies that an infinite number of new variables, corresponding to action-like variables of finite dimensional systems, can be found such that they represent constants of motion. But how do we find these constants of motion or even prove their existence?

Before we try to answer this question let us digress a little and say a few words about perhaps the most studied (1+1)-dimensional nonlinear equation, namely, the Korteweg–de Vries (KdV) equation on the real line:

$$q_t + \alpha q q_x + q_{xxx} = 0, \tag{8.1}$$

where q denotes a real field $q = q(t, x)$, and the subscripts t and x denote the derivatives with respect to these variables.

Although this equation has been studied for many years (for a review of this and many other nonlinear equations see [2]), the real breakthrough came in 1967 when Gardner *et al* [3] showed that this equation possesses an infinite number of conservation laws. Moreover, they found all solutions of (8.1) which tended sufficiently fast to a constant value as $|x| \to \infty$.

Shortly afterwards, Miura [4] examined the modified KdV equation

$$v_t - \beta v^2 v_x + v_{xxx} = 0, \tag{8.2}$$

and the associated conserved densities for both equations and discovered a transformation which related solutions of these two equations:

$$q = \frac{1}{\alpha}(-\beta v^2 + \epsilon\sqrt{6\beta}v_x), \quad \epsilon = \pm 1. \tag{8.3}$$

Rewriting this relation as an expression for v_x and then repeatedly substituting it into (8.2) (and differentiating it with respect to x) until all x derivatives are eliminated, leads to a pair of coupled partial differential equations

$$\begin{aligned} v_x &= \frac{\epsilon}{\sqrt{6\beta}}(\alpha q + \beta v^2) \\ v_t &= -\frac{\alpha\epsilon}{\sqrt{6\beta}}\left(\frac{1}{3}\beta v^2 q + q_{xx} + \sqrt{\frac{2\beta}{3}}\epsilon v q_x + \frac{\alpha q^2}{3}\right). \end{aligned} \tag{8.4}$$

However, exploiting the invariance of the KdV equation under the Galilean transformation

$$\begin{aligned} x' &= x + \alpha\lambda t \\ t' &= t \\ q' &= q + \lambda, \end{aligned} \tag{8.5}$$

where λ is an arbitrary parameter, allows us to linearise (8.4) by setting

$$v = -\epsilon\sqrt{\frac{6}{\beta}}\frac{\psi_x}{\psi}. \tag{8.6}$$

Then, after some manipulations (for more details see [2]) (8.4) becomes

$$\begin{aligned} \psi_{xx} + \frac{\alpha}{6}q\psi &= \frac{\alpha k^2}{6}\psi \\ \psi_t + \frac{\alpha}{3}(q + 2k^2)\psi_x + \left(f(k,t) - \frac{\alpha}{6}q_x\right)\psi &= 0, \end{aligned} \tag{8.7}$$

where $k^2 = \lambda$ and f is an arbitrary function of t and k.

If we now introduce the linear operators

$$\begin{aligned} L &= -\left(\frac{\partial^2}{\partial x^2} + \frac{\alpha}{6}q\right), \\ A &= -\left(4\frac{\partial^3}{\partial x^3} + \alpha q\frac{\partial}{\partial x} + \frac{\alpha}{2}q_x + f\right), \end{aligned} \tag{8.8}$$

the rescaling $k^2 \rightarrow -6k^2/\alpha$ allows us to rewrite (8.7) as

$$\begin{aligned} L\psi &= k^2\psi \\ \psi_t &= A\psi, \end{aligned} \tag{8.9}$$

and as it is easy to check, the compatibility of these equations for ψ implies

$$\partial_t L - [A, L] = 0. \tag{8.10}$$

Substituting the explicit form of the operators L and A we find that (8.10) reproduces (8.1). We see that we have rewritten (8.1) as a compatibility condition for ψ.

This last result was generalised to a large class of models by Lax [5], who showed that the existence of operators like A and L (called in the literature the Lax pair), is intrinsically tied to the integrability of the models as it holds for many systems which possess infinite numbers of conservation laws.

The idea of the existence of the Lax pair was then introduced by Zakharov and Mikhailov [6], in the context of field theory models, as their definition and the criterion of integrability of these models. As the $\mathbb{C}P^{N-1}$ is integrable what Lax-pair problem does it correspond to? What are the operators A and L in this case?

As we are interested in looking at this problem in Euclidean space we have to consider a generalised form of the Lax-pair problem, in which the equation for the auxiliary field ψ which involves the L operator is not an eigenvalue equation as in (8.9) but resembles the other equation (involving A). Thus, in our case, using the $x_\pm$ variables, the Lax-pair problem equations correspond to

$$\begin{aligned} \psi_+ &= A\psi \\ \psi_- &= B\psi, \end{aligned} \tag{8.11}$$

where the subscripts $\pm$ refer to the derivatives with respect to $x_\pm$ and the compatibility equation gives

$$A_+ - B_- + [A, B] = 0.$$

Let us now apply these ideas to the $\mathbb{C}P^{N-1}$ model, and let us use the projector formalism introduced in chapter 3. There we showed that we can describe grassmannian models by a complex unitary $N \times N$ matrix g, which is also hermitian. This last condition allows us to put $g = g_0(1-2\mathbb{P})$, where g_0 is a constant unitary matrix and $\mathbb{P}$ is a projector. Then the Lax-pair problem (also called the Hilbert–Riemann problem in some papers) takes the following form

$$\begin{aligned} \psi_+ &= \frac{g_+g^\dagger}{1+\lambda}\,\psi \\ \psi_- &= \frac{g_-g^\dagger}{1-\lambda}\,\psi, \end{aligned} \tag{8.12}$$

where, as we are considering a model involving a complex field, ψ is a complex field and where g and $g^\dagger$ denote our unitary matrices. The compatibility of this set of equations provides a set of equations for the matrices g.

However, in our case, the equations simplify as

$$\partial_-(1-2\mathbb{P})(1-2\mathbb{P}) = 2[\mathbb{P}_+, \mathbb{P}],$$

and so we find that they are given by

$$\begin{aligned} \psi_+ &= 2\frac{[\mathbb{P}_+, \mathbb{P}]}{1+\lambda}\psi \\ \psi_- &= 2\frac{[\mathbb{P}_-, \mathbb{P}]}{1-\lambda}\psi. \end{aligned} \tag{8.13}$$

It is easy to check that the compatibility of these equations is given by

$$[\partial_+\partial_-\mathbb{P}, \mathbb{P}] = 0,$$

i.e. the equations of motion of the $\mathbb{C}P^{N-1}$ model. (The simplest way to check this is to cross-differentiate the equations and then look at the residues of the $\lambda = \pm 1$ poles.)

It is interesting to solve the linear problem; i.e. to find ψ, which solves (8.13).

Consider first the purely instantonic case. Then

$$\mathbb{P} = \mathbb{P}^0 = \frac{ff^\dagger}{|f|^2}. \tag{8.14}$$

This projector satisfies

$$\partial_+\mathbb{P}^0 = \frac{(P_+f)f^\dagger}{|f|^2}, \qquad \partial_-\mathbb{P}^0 = \frac{f(P_+f)^\dagger}{|f|^2}, \tag{8.15}$$

and so

$$[\partial_+\mathbb{P}^0, \mathbb{P}^0] = \frac{P_+f(f)^\dagger}{|f|^2}, \qquad [\partial_-\mathbb{P}^0, \mathbb{P}^0] = \frac{f(P_+f)^\dagger}{|f|^2}. \tag{8.16}$$

Then, realising the importance of the projector $\mathbb{P}^0$, and seeking a solution $\psi = \psi(\lambda)$, which goes to 1 as $\lambda \to \infty$, we make an ansatz

$$\psi = 1 + a(\lambda)\mathbb{P}^0, \tag{8.17}$$

and find that the equations of motion are satisfied if

$$a(\lambda) = \frac{2}{\lambda - 1} = -\frac{2}{1-\lambda}.$$

For anti-instantons we proceed in an analogous way and find

$$z = \frac{g}{|g|}, \quad \partial_+ g = 0, \quad \mathbb{P} = \tilde{\mathbb{P}} = \frac{g(g)^\dagger}{|g|^2}, \qquad (8.18)$$
$$\psi = 1 - \frac{2}{1+\lambda}\tilde{\mathbb{P}}.$$

Next we look at our noninstanton solutions. Let us consider first $\mathbb{P}^1$ given by

$$\mathbb{P}^1 = \frac{(P_+ f)(P_+)^\dagger}{|P_+ f|^2}. \qquad (8.19)$$

A few lines of algebra then show that

$$\partial_+ \mathbb{P}^1 = \frac{(P_+^2 f)(P_+ f)^\dagger}{|P_+ f|^2} - \frac{(P_+ f)(f)^\dagger}{|f|^2}, \qquad (8.20)$$

and also

$$\partial_- \mathbb{P}^1 = \frac{P_+ f (P_+^2 f)^\dagger}{|P_+ f|^2} - \frac{f(P_+ f)^\dagger}{|f|^2}, \qquad (8.21)$$

which, of course, is a hermitian conjugate of $\partial_+ \mathbb{P}^1$. Hence, the commutators are given by

$$[\partial_+ \mathbb{P}^1, \mathbb{P}^1] = \frac{(P_+^2 f)(P_+ f)^\dagger}{|P_+ f|^2} + \frac{(P_+ f)(f)^\dagger}{|f|^2}, \qquad (8.22)$$

and

$$[\partial_- \mathbb{P}^1, \mathbb{P}^1] = -\frac{P_+ f (P_+^2 f)^\dagger}{|P_+ f|^2} - \frac{f(P_+ f)^\dagger}{|f|^2}. \qquad (8.23)$$

We can now show that ψ^1, which solves our linear problem, is given by

$$\psi^1 = 1 + \frac{4\lambda}{(\lambda-1)^2}\mathbb{P}^0 - \frac{2}{1-\lambda}\mathbb{P}^1. \qquad (8.24)$$

Notice the appearance in ψ^1 of the second-order pole at $\lambda = 1$.

Similarly, for

$$\mathbb{P}^k = \frac{(P_+^k f)(P_+^k f)^\dagger}{|P_+^k f|^2}, \qquad (8.25)$$

we find that the solution is given by

$$\psi^k = 1 + \frac{4\lambda}{(1-\lambda)^2}\sum_{l=0}^{k-1}\mathbb{P}^l - \frac{2}{1-\lambda}\mathbb{P}^k. \qquad (8.26)$$

Of course, we can use the completeness relation

$$\sum_{i=0}^{N-1} \mathbb{P}^i = 1, \tag{8.27}$$

to rewrite this expression in many alternative ways. In particular we can rewrite it as

$$\begin{aligned} \psi^k &= \left(\frac{1+\lambda}{1-\lambda}\right)\left(1 - \frac{2}{1-\lambda}\sum_{l=0}^{k-1}\mathbb{P}^l - \frac{2}{1+\lambda}\sum_{l=k+1}^{N-1}\mathbb{P}^l\right) \\ &= \left(\frac{1+\lambda}{1-\lambda}\right)^2\left(1 - \frac{4\lambda}{(1+\lambda)^2}\sum_{l=k+1}^{N-1}\mathbb{P}^l - \frac{2}{1+\lambda}\mathbb{P}^k\right), \end{aligned} \tag{8.28}$$

from which we see that our ψ^k can be written as a sum either of second-order poles at $\lambda = 1$ or at $\lambda = -1$, or of first-order poles at both $\lambda = 1$ and $\lambda = -1$. Observe that, for $k = N-1$, our solution becomes

$$\psi^{N-1} = \left(\frac{1+\lambda}{1-\lambda}\right)^2\left(1 - \frac{2}{\lambda+1}\mathbb{P}^{N-1}\right), \tag{8.29}$$

and we see that we have recovered our result for anti-instantons, (8.18).

Notice that our solutions are so normalised that $\psi(0) = g = 1 - 2\mathbb{P}$, where $\mathbb{P}$ is the projector which appears in the equations for ψ. Thus, having found ψ, we know the projector straightaway.

Let us now discuss briefly the Lax-pair problem for more general grassmannians. We recall that, in the projection formulation, the equations of motion are given by $[\partial_+\partial_-\mathbb{P}, \mathbb{P}] = 0$, where $\mathbb{P}$ is the corresponding projector.

It turns out that now there is a distinction as to whether our grassmannian solutions are generic or special. For the generic solutions, described by a set of M adjacent vectors from the sequence $\{e_1, e_2, \ldots, e_\beta, \ldots, e_N\}$, say,

$$z(\beta) = \{e_\beta, \ldots, e_{\beta+M-1}\}, \tag{8.30}$$

we have $\mathbb{P} = \sum_{\alpha=\beta}^{\beta+M-1} \mathbb{P}_\alpha$, where $\mathbb{P}_\alpha = e_\alpha e_\alpha^\dagger$. Then, as is easy to check, the solution is given by

$$\psi = 1 + \frac{4\lambda}{(1-\lambda)^2}\sum_{\alpha=1}^{\beta-1}\mathbb{P}_\alpha - \frac{2}{1-\lambda}\mathbb{P}. \tag{8.31}$$

We see a complete analogy with the $\mathbb{C}P^{N-1}$ case.

For the special solutions the situation is more complicated [7]. Let us discuss it here, as an illustration, only at the case of $M = 2$ grassmannians, and look at the solutions based on one vector f. Then defining $g_\beta = \partial_+^{\beta-1} f$ and considering $\beta_2 > \beta_1$, we can take

$$\mathbb{P} = \mathbb{P}_{\beta_1} + \mathbb{P}_{\beta_2}, \qquad \text{where} \qquad \mathbb{P}_\alpha = e_\alpha e_\alpha^\dagger. \tag{8.32}$$

This projector solves the equations of motion, and a few lines of algebra show that the solution of (8.13) is given by

$$\begin{aligned}\psi = {} & \frac{8\lambda(\lambda^2+1)}{(\lambda-1)^4}\sum_{\alpha=1}^{\beta_1-1} - \frac{2(3\lambda^2+1)}{(1-\lambda)^3}\mathbb{P}_{\beta_1} \\ & + \frac{4\lambda}{(\lambda-1)^2}\sum_{\alpha=\beta_1+1}^{\beta_2-1}\mathbb{P}_\alpha - \frac{2}{1-\lambda}\mathbb{P}_{\beta_2},\end{aligned} \tag{8.33}$$

which can be rewritten as

$$\begin{aligned}\psi \sim 1 + {} & \frac{4\lambda}{(\lambda-1)^2}\sum_{\alpha=1}^{\beta-1}\mathbb{P}_\alpha - \frac{2}{1-\lambda}\mathbb{P}_{\beta_1} \\ & - \frac{2}{1+\lambda}\mathbb{P}_{\beta_2} - \frac{4\lambda}{(1+\lambda)^2}\sum_{\alpha=\beta_2+1}^{N}\mathbb{P}_\alpha.\end{aligned} \tag{8.34}$$

We note that this solution exhibits poles of up to either the second order simultaneously at $\lambda = 1$ and $\lambda = -1$, or the fourth order at $\lambda = 1$ or alternatively at $\lambda = -1$. For $M > 2$ it is easy to find special types of solutions which lead to still higher-order poles at $\lambda = \pm 1$ in ψ. The order of these poles is related to the number of 'gaps' in the group of vectors used in the construction of the solution (see chapter 3).

We should point out that the traditional technique of the Lax-pair problem, as usually applied to nonlinear equations [8], involves an ansatz for ψ, which has only first-order poles in the λ plane. As we have found above, a more complicated ansatz can sometimes lead to different and interesting kinds of solutions.

We would also like to add that, in the traditional approach to the Lax-pair problem, one tries to determine new solutions from the old ones, and to determine at the same time new field configurations which solve their nonlinear equations of motion. This is done by setting

$$\psi^{\text{new}}(\lambda) = \chi(\lambda)\psi^{\text{old}}(\lambda) \tag{8.35}$$

and then considering the resultant equation for $\chi(\lambda)$. When applied to our case, it turns out that the resultant equations seem harder to solve than the original ones [7]!

Let us finish with one more comment about the general structure of our solutions to the $\mathbb{C}P^{N-1}$ model and the relations between the different solutions. Take a noninstanton solution (i.e. a solution that describes a mixture of instantons and anti-instantons), say, a solution described by $\mathbb{P}_l = e_l e_l^\dagger$, where $\hat{e}_p = (P_+^{p-1} f)$. Then recall that

$$\mathbb{P}(l) = \sum_{p=1}^{l} \mathbb{P}_l \tag{8.36}$$

describes an instantonic solution of $G(l, N)$, and that the solution of the corresponding Lax-pair problem corresponding to $\mathbb{P}(l)$ is given by

$$\psi^0(l) = 1 - \frac{2}{1-\lambda}\mathbb{P}(l) = 1 - \frac{2}{1-\lambda}\sum_{p=1}^{l}\mathbb{P}_p. \tag{8.37}$$

Then, considering such solutions for $l = k$ and for $l = k+1$, we observe that

$$\psi^0(k+1)\psi^0(k) = 1 + \frac{4\lambda}{(1-\lambda)^2}\sum_{p=1}^{k}\mathbb{P}_p - \frac{2}{1-\lambda}\mathbb{P}_{k+1} = \psi^1(k+1), \tag{8.38}$$

and we see that this last expression is the solution of the Lax-pair problem corresponding to the projector $\mathbb{P}^1$ considered before. This formulation brings out, in perhaps the clearest possible way, the interdependent nature of our solutions.

8.2 The Wess–Zumino Term

To introduce the Wess–Zumino term let us follow Witten [9] and consider a simple mechanical problem.

Consider a particle of mass m constrained to move on an arbitrary two-dimensional sphere of radius 1. Its Lagrangian is therefore $L = \frac{1}{2}m\int \mathrm{d}t\, \dot{x}_i^2$, and the equations of motion are given by

$$m\ddot{x}_i + m x_i \left(\sum_k \dot{x}_k^2\right) = 0, \tag{8.39}$$

taken together with the constraint $\sum_i x_i^2 = 1$. This Lagrangian and the equations of motion possess separate symmetries $t \to -t$ and $x_i \to -x_i$. If we want an equation invariant only under the combined transformation

$$t \to -t, \qquad x_i \to -x_i \tag{8.40}$$

the simplest choice is

$$m\ddot{x}_i + m x_i \left(\sum_k \dot{x}_k^2 \right) = \alpha\, \epsilon_{ijk}\, \dot{x}_j x_k, \tag{8.41}$$

where α is a constant number.

The derivation of this equation from a Lagrangian is difficult as there is no obvious term whose variation equals the right-hand side (since $\epsilon_{ijk} x_j x_k \dot{x}_i = 0$). However, as pointed out by Witten [9], there is a solution. Let us observe that the right-hand side of the expression above can be understood as the Lorentz force acting on an electric charge, due to its interaction with a magnetic monopole located at the centre of the sphere. If we now introduce a vector potential $\underline{A}$ such that

$$\underline{\nabla} \wedge \underline{A} = \frac{\underline{r}}{r^3}, \tag{8.42}$$

then the action for our problem is given by

$$S = \int (\tfrac{1}{2} m \dot{x}_i^2 + \alpha A_i \dot{x}_i) \mathrm{d}t. \tag{8.43}$$

We see that A_i contains a Dirac string and the spurious symmetry has been eliminated.

It is easy to see that our extra term provides, as claimed, an additional contribution to the equations of motion. To see this we observe that the variation of the additional term gives us

$$\alpha\, \epsilon^{ijk} \epsilon^{\alpha\beta} \int x_i\, \partial_\alpha x_j\, \partial_\beta \delta x_k\, \mathrm{d}^2\sigma, \tag{8.44}$$

where ϵ^{ijk} and $\epsilon^{\alpha\beta}$ stand for antisymmetric tensors of its arguments. Then, as $\vec{x}$ lies on a unit sphere in three dimensions the contribution of $\epsilon^{\alpha\beta}\epsilon^{ijk}\delta x_i \partial_\alpha x_j\, \partial\beta x_k$ vanishes, and so the remaining contribution gives the required term in the equations of motion.

Let us observe that the value of α in the expressions above is arbitrary. However, as Witten [9] has pointed out, the value of α is quantised when we consider the quantum version of the model.

To see this, observe that in the quantum consideration of the problem, we look at

$$\int \mathrm{d}x_i(t)\, e^{-S}. \tag{8.45}$$

If we now consider the calculation of the simplest object, i.e. the partition function, we are interested in closed orbits γ on the $x_i x_i = 1$ sphere, and so we will have to consider the contribution of $\exp(i\alpha \int_\gamma A_i \mathrm{d}x^i)$. However, we can use Stokes' theorem, and reexpress this contribution as

$$\exp\left(i\alpha \int_\gamma A_i \mathrm{d}x^i\right) = \exp\left(i\alpha \int_D F_{ij} \mathrm{d}\Sigma^{ij}\right), \tag{8.46}$$

where D is a disc whose boundary is γ, and where F_{ij} denote the magnetic field strength. Of course our curve γ and our disc D lie on our two-dimensional surface of the sphere in three dimensions. However, γ is a boundary of two discs; so to speak — the inside D_1 and the outside D_2 of γ, such that $D_1 \cup D_2 = S^2$. Notice that the orientation of γ as seen from D_1 is opposite to that seen from D_2. So, if in the previous expression by D we meant D_1, we also find that

$$\exp\left(i\alpha \int_\gamma A_i \mathrm{d}x^i\right) = \exp\left(-i\alpha \int_{D_2} F_{ij}\, \mathrm{d}\Sigma^{ij}\right). \tag{8.47}$$

To get the same answer independently of which disc we choose we must require that

$$\exp\left(i\alpha \int_{D_1 \cup D_2} F_{ij}\, \mathrm{d}\Sigma^{ij}\right) = 1. \tag{8.48}$$

However, as $D_1 \cup D_2 = S^2$ and $\int_{S^2} F_{ij}\, \mathrm{d}\Sigma^{ij} = 4\pi$, we see that we must require that α is an integer or half integer. Thus we see that we have recovered Dirac's quantisation condition for the product of electric and magnetic charges (here set to 1).

We can summarise our results as follows: in a classical theory we can include our additional term with an arbitrary coefficient; however, if we want to quantise our model, the value of this coefficient can no longer be arbitrary, but must belong to a discrete set of values. The inclusion of this term naturally involves a space with one extra dimension and such that the 'physical space' is its boundary.

So far we have discussed only a simple mechanical analogue of the Wess–Zumino term. In the next section we will introduce such a term for the chiral models. We will then be able to show that there is an intimate connection between the solutions of the chiral models with the Wess–Zumino term and the solutions of the Lax-pair problem discussed in the previous section in the case of the grassmannian solutions.

8.3 *U(N)* Model with the Wess–Zumino Term

In this section we will study the $U(N)$ sigma models with the Wess–Zumino term. What is the form of this extra term? By analogy with the case discussed in the previous section, we would expect that the Wess–Zumino term be given by a three-dimensional integral, which is locally (but not globally) a total divergence. A little thought shows that the action of the $U(N)$ sigma models with the Wess–Zumino term (which from now on we will refer to as the WZ-σ models) is given by

$$S = \int \mathrm{d}x^2 \, \tfrac{1}{4} \operatorname{Tr}(\partial_\mu g^\dagger \partial_\mu g) + \tfrac{1}{6} i \lambda \int \mathrm{d}x^3 \operatorname{Tr}(\tilde{g}^\dagger \partial_\mu \tilde{g} \, \tilde{g}^\dagger \partial_\nu \tilde{g} \, \tilde{g}^\dagger \partial_\rho \tilde{g}) \epsilon^{\mu\nu\rho}, \quad (8.49)$$

where λ is a purely imaginary parameter, and g and $\tilde{g}$ are unitary matrices: $g^\dagger g = \tilde{g}^\dagger \tilde{g} = 1$. Moreover, in the second term, the unitary matrix g has been extended to $\tilde{g}$, a function of three variables (x, y and t), where the additional variable t satisfies $0 \leq t \leq 1$. Following Witten [12], we choose the boundary conditions of this extension to be such that $\tilde{g}(x, y, 1) = g(x, y)$ and $\tilde{g}(x, y, 0) = K$, where K is a constant unitary matrix. More details of our continuation will be given later on.

The equations of motion of the model can easily be shown not to depend on our extension but to be given by

$$(1 - \lambda)\partial_+(g^\dagger \partial_- g) + (1 + \lambda)\partial_-(g^\dagger \partial_+ g) = 0, \quad (8.50)$$

where we have already changed our variables to the more convenient $x_\pm = x \pm y$ introduced before, and where g is required to satisfy the unitarity constraints given above. Clearly, when $\lambda = 0$, the equations reduce to the well studied equations of the model without the Wess–Zumino term.

As we will discuss relations between the solutions of both models, we adopt the convention that the fields and the solutions of the model with the Wess–Zumino term will be denoted by g, and those of the model without this term by Q. Clearly, the solutions of the model without the Wess–Zumino term satisfy

$$\partial_+(Q^\dagger \partial_- Q) + \partial_-(Q^\dagger \partial_+ Q) = 0, \quad (8.51)$$

together with the constraint $Q^\dagger Q = 1$.

In the next sections we will present explicit solutions to (8.50) and we will discuss some of their properties. To do this, in the next section we will show how to relate the solutions of (8.50) to the solutions of the Lax-pair system for the corresponding sigma model without the Wess–Zumino term. We will then describe the construction of these solutions and study their properties. We will compute the value of the action for some of these solutions, and then prove that all the solutions of the WZ-σ model have the same number of negative modes as the corresponding solutions of the model without the Wess–Zumino term (which we studied in the previous chapter), and, as a consequence, that they are all unstable.

8.4 Solutions of the WZ-σ Model

Despite the fact that one of the original motivations for the inclusion of the Wess–Zumino term in the Lagrangian has been to reduce some of it symmetries [12], many of these symmetries remain [13, 14]. If we only consider the classical version of the theory and its equations of motion, the functional space of the solutions of the equations of motion exhibits many symmetries [14] which can be exploited in the construction of these solutions. For example, it is well known that if Q is a solution of (8.51) then $Q^\dagger$ is also its solution. This symmetry can be extended to a property of g's, i.e. for all λ's: if $g(\lambda)$ is a solution of (8.50) then so is $g(-\lambda)^\dagger$.

As we have already mentioned, the two-dimensional σ models are known to be integrable [6, 15], as we can associate with them a Lax-pair formalism, and so they possess an infinite number of conservation laws. Recently [14] it has been shown that the WZ-σ models are also integrable. In what follows we will show that this is not unexpected, as we will demonstrate that the solutions of both models are related, and can be derived from each other.

To do this, let us consider the equations of the Lax-pair problem for the $U(N)$ σ model:

$$\partial_\pm \Psi = \Psi \frac{2A_\pm}{1\pm\lambda}, \tag{8.52}$$

where

$$A_\pm = \frac{Q^\dagger \partial_\pm Q}{2}. \tag{8.53}$$

The unitarity of Q implies some conditions on Ψ [16], namely that

$$\Psi^{\dagger-1}(\lambda) = \Psi(-\lambda^*). \tag{8.54}$$

As a consequence, we see that if λ is imaginary, then Ψ is a unitary matrix, and as Q satisfies (8.51), $g = \Psi$ satisfies (8.50). This means that the solutions of the Lax-pair problem for the σ model have provided us with a simple way of constructing solutions of the WZ-σ model.

Note that we could have chosen to construct the Lax-pair problem for our system in a different way; namely, we could have chosen

$$\partial_\pm \varphi = \frac{2B_\pm}{1\pm\lambda}\varphi, \tag{8.55}$$

where

$$B_\pm = \frac{\partial_\pm Q Q^\dagger}{2}. \tag{8.56}$$

Once again, as $Q^\dagger$ is a solution of (8.51), it is easy to show that $\varphi(-\lambda)$ satisfies (8.50) for the imaginary values of λ.

Before we proceed any further, let us show that there is a simple relationship between the solutions of (8.52) and those of (8.55). To see this we observe that we can always take $\varphi = QX^{-1}$ for some matrix X. Inserting this expression into (8.55), and multiplying it from the left by $Q^\dagger$ and from the right by X, we find that

$$X_\pm = X \frac{2\,A_\pm}{1 \pm 1/\lambda}, \tag{8.57}$$

thus showing that

$$\varphi(\lambda) = Q\,\Psi^{-1}(\lambda^{-1}). \tag{8.58}$$

In fact, we can consider $\varphi^\dagger$ as a solution of (8.52), in which Q is replaced by $Q^\dagger$.

In addition, we can show how any solution of the WZ-σ model can be transformed into a solution of the usual σ model. Assuming that g is a solution of (8.50), we define $C_\pm = (1 \pm \lambda) g^\dagger \partial_\pm g$. It is then straightforward to show that

$$\partial_+ C_- + \partial_- C_+ = 0, \qquad C_-^\dagger = -C_+. \tag{8.59}$$

These two equations imply [6, 16] that we can set

$$C_\pm = Q^\dagger \partial_\pm Q, \tag{8.60}$$

where $Q^\dagger Q = 1$, and then Q is a solution of (8.51), which, however, in general does depend on λ. We see that each solution of the WZ-σ model can be related to a solution of the usual σ model, with the relation provided by the solutions of the Lax-pair equations for the σ model.

We may now ask if any solution g of (8.50) is automatically also a solution of one of the Lax-pair equations (8.52) or (8.55). This question is equivalent to asking whether $C_\pm$ does not depend on λ. As such, this question is not well posed, as, in reality, we have here a family of solutions parametrised by λ.

Take, for example, $Q(\beta)$, a family of solutions of the $U(N)$ σ model where β is some parameter. Computing $\Psi(\beta)$ provides us with a family of solutions of the WZ-σ model. If we now decide to choose $\beta = \lambda$, then $C_\pm = (1 \pm \lambda) g^\dagger \partial_\pm g$ will depend on λ, in agreement with the claim above.

The question we should be asking is then: given a solution $g_0(\lambda_0)$ of the WZ-σ model for a fixed value λ_0 of λ, is it possible to extend it to a function $g'(\lambda)$, such that g' is a solution of both (8.50) and of either the Lax pair (8.52) or (8.55) for all values of λ (in which $A_\pm$ or respectively $B_\pm$ are independent of λ), and in addition $g'(\lambda_0) = g_0(\lambda_0)$. To do this we simply solve (8.52) for $A_\pm = g_0^\dagger \partial_\pm g_0\,(1 \pm \lambda_0)$ or (8.55) for $B_\pm = \partial_\pm g_0\, g_0^\dagger (1 \pm \lambda_0)$. Then $\Psi_0(\lambda)$ or $\varphi(-\lambda)$ is the required solution.

8.5 Solutions of the Lax-pair Equations for the $U(N)$ σ Model

Before constructing any solutions of the Lax-pair equations (8.52) or (8.55) given in the previous section, let us briefly recall Uhlenbeck's [17] construction of solutions of the $U(N)$ sigma model discussed in the previous chapter.

There we showed that, due to Uhlenbeck's theorem [17], any solution Q_l of (8.51) can be factorised as

$$Q_l = K(1 - 2R_1)(1 - 2R_2)\dots(1 - 2R_l), \tag{8.61}$$

where K is a constant matrix, l is an integer (the uniton number), and R_i's are projectors which can be constructed by induction, as they satisfy

$$\begin{aligned} R_l A_-^{l-1}(1 - R_l) &= 0 \\ (1 - R_l)(\partial_- R_l + A_-^{l-1} R_l) &= 0, \end{aligned} \tag{8.62}$$

where

$$A_-^l = \frac{Q_l^\dagger \partial_- Q_l}{2}. \tag{8.63}$$

From this construction we can deduce, that if we define (by induction)

$$K_l = K_{l-1}(1 - aR_l), \tag{8.64}$$

where a is a complex number (which does not depend on l) and R_i's are the projectors which satisfy the equations given above, then

$$K_l^{-1} = (1 - bR_l)\, K_{l-1}^{-1}, \tag{8.65}$$

where b is a complex number, which satisfies

$$a + b - ab = 0. \tag{8.66}$$

As a result we can now prove that

$$\begin{aligned} K_l^{-1}\partial_- K_l &= bA_-^l \\ K_l^{-1}\partial_+ K_l &= aA_+^l. \end{aligned} \tag{8.67}$$

We prove this by induction. The result is trivially valid for the constant factor and

$$\begin{aligned} K_l^{-1}\partial_- K_l =& b[\partial_- R_l(1 - aR_l) + A_-^{l-1} \\ & - bR_l A_-^{l-1} - aA_-^{l-1}R_l + abR_l A_-^{l-1}R_l] \\ =& b[A_-^{l-1} + \partial_- R_l - a((1 - R_l)\partial_- R_l + A_-^{l-1}R_l - R_l A_-^{l-1})] \\ =& bA_-^l. \end{aligned} \tag{8.68}$$

A similar calculation proves also the other equation for K_l.

As an immediate consequence, we see that if we choose, following the discussion for the grassmannian instantonic case,

$$a = \frac{2}{1 + \lambda}, \qquad b = \frac{2}{1 - \lambda}, \tag{8.69}$$

the condition (8.66) is satisfied, and

$$\Psi = K_l \tag{8.70}$$

is a solution of (8.52) corresponding to Q_l.

Moreover, as this solution is unique up to a multiplication by a constant matrix, we see that we have thus solved (8.52) completely.

We would like to point out that our expression for K_l agrees with the expression discussed in the first section of this chapter. Although there we discussed only the grassmannian solutions, we know that these solutions are also solutions of the chiral model, and, when considered as solutions of the $U(N)$ models, are in fact contained in the general expression for some l and some choice of projectors R_i. Thus we see that the interesting dependence on λ observed in the grassmannian case discussed in the first section comes from the product of factors a and the properties of our projectors R_i.

Let us point out at this stage that our solutions have already been given in Uhlenbeck's paper [17]. In that paper Uhlenbeck, using a different but related parameter λ, introduced auxiliary functions E_λ, which she then used to prove her factorisation theorem for the ordinary chiral model. These functions E_λ correspond, in fact, to our functions Ψ.

We finish this section by pointing out that, due to the relation (8.58), the solutions to (8.55) are given by

$$\varphi = Q_l K_l^{-1}, \tag{8.71}$$

where the parameter b appearing in K_l^{-1}, as defined in (8.65), is given by

$$b = \frac{-2\lambda}{1-\lambda}. \tag{8.72}$$

8.6 Properties of the Solutions

As we have seen, the solutions of the $U(N)$ WZ-σ model can be obtained very easily from the solutions of the $U(N)\,\sigma$ model. It would then be interesting to see whether the properties of both sets of solutions are similar. In particular, is the action of the solutions of the model with the Wess–Zumino term also quantised, and can it be related to the topological charges of the projectors R_i's out of which the solutions are constructed? Another interesting question would be to determine whether the solutions are also unstable, or if, on the contrary, the Wess–Zumino term stabilises them.

Before we can answer these questions, we have to find a convenient way of computing explicitly the contribution of the Wess–Zumino term. As this term is given by a three-dimensional integral, we know [12, 14], as we have already said in the introduction, that we have to extend our field $g(x, y)$ to a field $\tilde{g}(x, y, t)$, defined over the three-dimensional space locally parametrised by x, y and t, with $0 \leq t \leq 1$ and such that $\tilde{g}(x, y, 1) = g(x, y)$ and $\tilde{g}(x, y, 0) = 1$.

This extension shows that our cylinder, with an infinitely large base described by x, y and 'height' t, is being mapped by $\tilde{g}$ into a full two sphere in $U(N)$, the surface of which is nothing else but the field $g(x, y)$. To perform this explicitly, let us first consider a one-uniton-like field configuration $g_1 = (1 - a R_1)$, where $a = 2/(1 + \lambda)$. For our extension we can now use

$$\begin{aligned} \tilde{g}(x_+, x_-, t) &= (1 - a R_1)^t \\ &= (1 - a(t) R_1) = \exp(i\alpha t R_1), \end{aligned} \tag{8.73}$$

where

$$a(t) = (1 - e^{i\alpha t}), \tag{8.74}$$

with

$$i\alpha = \ln(1 - a) = \ln \frac{(\lambda - 1)}{(\lambda + 1)}. \tag{8.75}$$

The extension for other configurations is given by induction:

$$\begin{aligned} \tilde{g}_l(x_+, x_-, t) &= \tilde{g}_{l-1}(x_+, x_-, t)\,(1 - a R_l)^t \\ &= K_l(a(t)). \end{aligned} \tag{8.76}$$

To extend solutions corresponding to φ, we apply the same method by raising each factor of both Q and Ψ to the power t.

To study the properties of our solutions of the WZ-σ model, let us first of all calculate explicitly the values of their action. We start by considering the solution of (8.50), constructed by solving (8.52) for the one-uniton field configuration $Q = (1 - 2R_1)$, namely:

$$g = (1 - a R_1), \tag{8.77}$$

where $a = 2/(1 + \lambda)$ and the projector R_1 satisfies $\partial_- R_1 R_1 = 0$. Note that, had we solved (8.55) instead of (8.52), we would have obtained the same solution. It is now easy to show that for this solution

$$\begin{aligned} S &= \int \mathrm{d}x^2\, \mathrm{Tr}\,(\partial_- R_1\, \partial_+ R_1)\, |a|^2 \\ &\quad - \lambda \int \mathrm{d}x^3\, i\alpha |a(t)|^2\, \mathrm{Tr}\,[(\partial_+ R_1\, \partial_- R_1 - \partial_- R_1\, \partial_+ R_1) R_1] \\ &= \left(|a|^2 + \int_0^1 \mathrm{d}t\, i\alpha\lambda |a(t)|^2\right) \int \mathrm{d}x^2\, \mathrm{Tr}(\partial_- R_1\, \partial_+ R_1). \end{aligned} \tag{8.78}$$

Writing $\lambda = ik$ and using the relations between a, α and λ, it is easy to perform the t integration and obtain

$$S = 4h(k) \int \mathrm{d}x^2 \, \mathrm{Tr}\,(\partial_- R_1 \, \partial_+ R_1), \tag{8.79}$$

where

$$h(k) = 1 - k \tan^{-1}(k^{-1}). \tag{8.80}$$

It is important to note that in (8.74) α is defined only up to the addition of an integer multiple of 2π and, as a consequence, the value of (8.78) is not uniquely defined. This well known fact [12, 14] comes from the possibility of choosing different extensions of g, which in our case correspond to different choices of the arbitrary constant in α. In (8.79) and in (8.80) this freedom is hidden in the choice of the phase of $\tan^{-1}$ in $h(k)$. If we fix this phase by choosing $0 \le \tan^{-1}(k^{-1}) \le \pi$, then $h(k)$ is monotonic in k. As k varies from $-\infty$ to $+\infty$, then $h(k)$ decreases from ∞ to 0, taking the value 1 for $k = 0$.

The computation of the value of the action for more general solutions is far more complicated. If we consider, for example, the solution constructed from a two-uniton solution Q_2 (where we have taken $K = 1$) by using the Lax pair (8.52), we have

$$g = (1 - a\,R_1)(1 - a\,R_2), \tag{8.81}$$

where $a = 2/(1+\lambda)$. Then when $\lambda = 0$, the action of the solution is given by

$$S_0 = 4 \int \mathrm{d}x^2 [\tilde{Q}(R_1) + \tilde{Q}(R_2)], \tag{8.82}$$

where

$$\tilde{Q}(R_i) = \mathrm{Tr}\,[\partial_+ R_i \, R_i \, \partial_- R_i - \partial_- R_i \, R_i \, \partial_+ R_i]] \tag{8.83}$$

is the topological charge density corresponding to R_i, which can be computed explicitly for any given solution, as was shown in the previous chapter. For nonvanishing values of λ we can use the Uhlenbeck equations, and it is then easy to show that

$$\begin{aligned} S = {} & h(k)\, S_0 - i\alpha\lambda \int \mathrm{d}x^3 |a(t)|^2 \\ & \times \mathrm{Tr}\,(\partial_+ R_2 \, \partial_- R_2 \, R_1 - \partial_- R_2 \, \partial_+ R_2 \, R_1 - \partial_- R_1 \, \partial_+ R_2). \end{aligned} \tag{8.84}$$

If R_2 is a basic (or antibasic) uniton discussed in the previous chapter, and so satisfies $R_2 \partial_- R_1 = 0$ (resp. $\partial_- R_1 (1 - R_2) = 0$), we have

$$S = h(k)\, S_0 - i\alpha\lambda \int_0^1 \mathrm{d}t \, |a(t)|^2 \int \mathrm{d}x^2 \partial_+ \partial_- \, Tr(R_2 \, R_1). \tag{8.85}$$

However, as

$$\int \mathrm{d}x^2 \partial_+ \partial_- Tr(R_1\, R_2) \;=\; 0 \tag{8.86}$$

for any two projectors, we conclude that

$$S \;=\; h(k)\, S_0, \tag{8.87}$$

which generalises the result obtained for the one-uniton-like configuration in the sense that the action of the solution of the model with the Wess–Zumino term is given by the action of the corresponding solution of the σ model multiplied by $h(k)$.

We now consider a solution constructed from the two-uniton solution Q_2 of the σ model, but derived by using the second Lax-pair equations (8.55). We find

$$g \;=\; (1 \;-\; 2\, R_1)(1 \;-\; 2\, R_2)(1 \;-\; b\, R_2)(1 \;-\; b\, R_1), \tag{8.88}$$

where b takes the form as in (8.72). The action of this solution can be shown to be equal to the action of our previous solution (8.81), and so is given by (8.84), or (8.87) when our unitons are basic or antibasic.

Unfortunately, so far we have been unable to find a corresponding result for the more general case. However, it is easy to show that if the action of Q, a solution of the σ model without the Wess–Zumino term, is finite, the action of g, the corresponding solution of the WZ-σ model, is also finite. To prove this, we notice that the action (8.49) takes the form

$$\begin{aligned} S \;=\; & \frac{4}{(1-\lambda)(1+\lambda)} \int \mathrm{d}x^2\, \mathrm{Tr}[A_-^{l\dagger} A_-^l] \\ & - \lambda \int \mathrm{d}x^3\, \mathrm{Tr}\Big[|a(t)|^2 (A_-^{l\dagger} A_-^l \;-\; A_-^l A_-^{l\dagger}) \tilde{g}_l^\dagger \partial_t \tilde{g}_l\Big], \end{aligned} \tag{8.89}$$

where A_-^l is given by (8.63). Using our expression for $\tilde{g}$, it is easy to show by induction that

$$\tilde{g}_l^\dagger \partial_t \tilde{g}_l \;=\; i\alpha \sum_{j=1}^{l} \tilde{R}_j, \tag{8.90}$$

where

$$\tilde{R}_j \;=\; (1 - a(t)R_l)^\dagger \ldots (1 - a(t)R_{j+1})^\dagger R_j (1 - a(t)R_{j+1}) \ldots (1 - a(t)R_l), \tag{8.91}$$

for $j \leq l-1$ and $\tilde{R}_l = R_l$. Then, as

$$0 \leq \int \mathrm{d}x^2\, \mathrm{Tr}(A_-^{l\dagger} A_-^l \tilde{R}_j) \leq \int \mathrm{d}x^2\, \mathrm{Tr}(A_-^{l\dagger} A_-^l) \tag{8.92}$$

for all R_j, we see that (8.89) must be finite, if Q_l is a finite action solution of the σ model.

To study the stability properties of our solutions g, we follow the procedure of the previous chapter. We take $g' = ge^{\epsilon X}$, where X is an antihermitian matrix and, as the first-order terms in ϵ vanish because of the equations of motion, the lowest nonvanishing terms of $\delta S = S(g') - S(g)$ are given by

$$\begin{aligned}
\delta S =& \frac{\epsilon^2}{4}\Bigg\{ \int \mathrm{d}x^2 \,\mathrm{Tr}\,[\partial_\mu X^\dagger \partial_\mu X - g^\dagger \partial_\mu g (X \partial_\mu X - \partial_\mu X X)] \\
&+ 2i\lambda \int \mathrm{d}x^3 \,\mathrm{Tr}\left[\tilde{g}^\dagger \partial_\nu \tilde{g} \tilde{g}^\dagger \partial_\alpha \tilde{g} \left(\frac{X^2}{2} \tilde{g}^\dagger \partial_\mu \tilde{g} + \tilde{g}^\dagger \partial_\mu \tilde{g} \frac{X^2}{2} + \partial_\mu \frac{X^2}{2} \right.\right. \\
&\left. - X \tilde{g}^\dagger \partial_\mu \tilde{g} X - X \partial_\mu X \right) + (\partial_\mu X + \tilde{g}^\dagger \partial_\mu \tilde{g} X - X \tilde{g}^\dagger \partial_\mu \tilde{g}) \\
&\left. \times (\partial_\nu X + \tilde{g}^\dagger \partial_\nu \tilde{g} X - X \tilde{g}^\dagger \partial_\nu \tilde{g}) \tilde{g}^\dagger \partial_\alpha \tilde{g} \right] \epsilon^{\mu\nu\alpha} \Bigg\} \\
=& \frac{\epsilon^2}{4}\Bigg(\int \mathrm{d}x^2 \,\mathrm{Tr}\,[\partial_\mu X^\dagger \partial_\mu X - g^\dagger \partial_\mu g (X \partial_\mu X - \partial_\mu X X)] \\
&+ 2i\lambda \int \mathrm{d}x^3 \,\mathrm{Tr}\,[\partial_\mu X \partial_\nu X \tilde{g}^\dagger \partial_\alpha \tilde{g} \\
&- \tfrac{1}{2} (\partial_\mu X X - X \partial_\mu X) \tilde{g}^\dagger \partial_\nu \tilde{g} \tilde{g}^\dagger \partial_\alpha \tilde{g}] \epsilon^{\mu\nu\alpha} \Bigg) \\
=& \frac{\epsilon^2}{4}\Bigg(\int \mathrm{d}x^2 \,\mathrm{Tr}\,[\partial_\mu X^\dagger \partial_\mu X \; - \; g^\dagger \partial_\mu g (X \,\partial_\mu X \; - \; \partial_\mu X \,X)] \\
&+ i\lambda \int \mathrm{d}x^3 \partial_\mu \,\mathrm{Tr}[(X \partial_\nu X \; - \; \partial_\nu X X) \tilde{g}^\dagger \partial_\alpha \tilde{g}] \epsilon^{\mu\nu\alpha} \Bigg).
\end{aligned} \tag{8.93}$$

Next we integrate by parts, and using the complex variables $x_\pm$, we obtain

$$\begin{aligned}
\delta S =& \frac{\epsilon^2}{2} \int \mathrm{d}x^2 \,\mathrm{Tr}\,[\partial_- X^\dagger \partial_+ X \; + \; \partial_+ X^\dagger \partial_- X \; - \; (1 \; + \; \lambda)\, g^\dagger \partial_+ g \\
&\times (X \,\partial_- X \; - \; \partial_- X \,X) \; - \; (1 \; - \; \lambda)\, g^\dagger \partial_- g \,(X \,\partial_+ X \; - \; \partial_+ X \,X)].
\end{aligned} \tag{8.94}$$

As we have seen before, g is related to a solution Q of (8.51) by

$$g^\dagger \partial_\pm g \; = \; \frac{Q^\dagger \partial_\pm Q}{1 \pm \lambda}, \tag{8.95}$$

and so, as a consequence, we end up with

$$\delta S \; = \; \frac{\epsilon^2}{4} \int \mathrm{d}x^2 \,\mathrm{Tr}\,[\partial_\mu X^\dagger \partial_\mu X \; - \; Q^\dagger \partial_\mu Q \,(X \,\partial_\mu X \; - \; \partial_\mu X \,X)], \tag{8.96}$$

which is nothing else but the stability equation for the σ model without the Wess–Zumino term for the solution Q studied in the previous chapter. In other words, the negative fluctuation modes for g are exactly the same as the negative modes for Q and, as we showed in the last chapter that all the solutions of the σ model are unstable, so are the solutions of the WZ-σ model.

Let us finish this chapter by pointing out that our solutions (8.70) are solutions of the chiral model with the Wess–Zumino term. However, as we discussed in chapter 3, the grassmannian models can be considered as special cases of the corresponding chiral models in which the basic unitary matrix-valued fields g are also hermitian: $g = g^\dagger$. Moreover, in the construction of multi-uniton solutions to the chiral model discussed in chapter 7, we have found that one-uniton solutions do satisfy this condition and so are, in fact, also solutions of the grassmannian model. This is also true for those multi-uniton solutions for which the corresponding projectors commute. However, these properties do not generalise to the solutions of the chiral model. To see this observe that

$$(G(\lambda))^\dagger = G(\lambda^*) \neq G(\lambda), \tag{8.97}$$

even for the solution corresponding to one uniton. This is not surprising, as one cannot add the Wess–Zumino term to grassmannian models; to have models with such a term the sigma models must be valued in the group and not in a coset space.

References

[1] Eichenherr H and Forger M 1981 Higher local conservation laws for nonlinear sigma models on symmetric spaces *Commun. Math. Phys.* **82** 227
Goldschmidt Y Y and Witten E 1980 Conservation laws in some two-dimensional models *Phys. Lett.* **91B** 392
Brezin E, Itzykson C, Zinn-Justin J and Zuber J-B 1979 Remarks about the existence of non-local charges in two-dimensional models *Phys. Lett.* **82B** 442
de Vega H 1979 Field theories with an infinite number of conservation laws and Backlund transformations in two dimensions *Phys.Lett.* **87B** 233
Abdalla M C 1985 Integrability of chiral nonlinear sigma models with a Wess–Zumino term *Phys. Lett.* **152B** 215
Eichenherr H 1980 Infinitely many conserved local charges for the $\mathbb{C}P^{N-1}$ models *Phys. Lett.* **90B** 121

[2] Dodd R K, Eilbeck J C, Gibbon J D and Morris H C 1985 *Solitons and Nonlinear Wave Equations* (New York: Academic)
Newell A C 1985 *Solitons in Mathematics and Physics* Society for Industrial and Applied Mathematics

[3] Gardner C S, Green J M, Kruskal M D and Miura M R 1967 Method for solving the Korteweg–de Vries equation *Phys. Rev. Lett.* **19** 1095

[4] Miura M R 1968 Korteweg–de Vries equation and generalizations. I. A remarkable explicit nonlinear transformation *J. Math. Phys.* **9** 1202

[5] Lax P D 1968 Integrals of nonlinear equations of evolution and solitary waves *Commun. Pure Appl. Math.* **21** 467

[6] Zakharov V E and Mikhailov A V 1978 Relativistically invariant two dimensional models in field theory which are integrable by means of the inverse scattering problem method *Zh. Eksp. Teor. Fiz.* **74** 1953 (1979 *Sov. Phys.–JETP* **47** 1017

[7] Din A M, Horvath Z and Zakrzewski W J 1984 The Riemann–Hilbert problem and finite action $\mathbb{C}P^{N-1}$ solutions *Nucl. Phys.* B **233** 269

[8] Ueno K and Nakamura Y 1982 The hidden symmetry of chiral fields and the Riemann–Hilbert problem *Phys. Lett.* **117B** 208
Wu Y-S 1983 The group theoretical aspects of infinitesimal Riemann–Hilbert transform and hidden symmetry *Commun. Math. Phys.* **90** 461
Newman E T 1986 Gauge theories, the holonomy operator and the Riemann–Hilbert problem *J. Math. Phys.* **27** 2797

[9] Witten E 1983 Global aspects of current algebra *Nucl. Phys.* B **223** 422

[10] Wess J and Zumino B 1971 Consequences of anomalous Ward identities *Phys. Lett.* **37B** 95

[11] Polyakov A and Wiegmann P 1983 Theory of nonabelian Goldstone bosons in two dimensions *Phys. Lett.* **131B** 121; 1984 Goldstone fields in two dimensions with multivalued actions *Phys. Lett.* **141B** 223

[12] Witten E 1984 Non-Abelian bosonisation in two dimensions *Commun. Math. Phys.* **92** 455

[13] Braaten E, Curtright T L and Zachos C K 1985 Torsion and geometrostatis in nonlinear sigma models *Nucl. Phys.* B **260** 630

[14] Forger M and Zizzi P 1987 Twisted chiral models with Wess–Zumino term and strings *Nucl. Phys.* B **287** 131

[15] Goldschmidt Y Y and Witten E 1980 Conservation laws in some two dimensional models *Phys. Lett.* **91B** 392

[16] Harnad J, Saint-Aubin Y and Shnider S 1980 Backlund transformation for non-linear sigma models with values in Riemannian symmetric spaces *Commun. Math. Phys.* **92** 329

[17] Uhlenbeck K 1985 Harmonic maps into Lie groups (classical solutions of the chiral model) *J. Diff. Geom.* to be published

9 More General Models; Anomalies

9.1 Introduction

The chiral models discussed in the last chapter are only one possible generalisation of the $\mathbb{C}P^{N-1}$ sigma models. Various other generalisations are posssible and have been discussed by many people. Most of these studies have been concerned with the cases when the target manifolds were either group spaces or some coset spaces. Various properties of such models have been studied, and many classical solutions derived and their properties examined in detail. Many such models have been shown to be integrable; that is, their equations of motion can be shown to correspond to the integrability problem for an auxiliary matrix value linear problem (the Lax-pair problem). In this case one can show the existence of an infinite number of conservation laws and so the dynamics of the system is quite constrained. Of course, the integrability does not imply that the equations are easy to solve (although, in principle, the Lax-pair equations reduce the problem to a problem in algebra).

Recently, motivated by some observations from the geometric properties of the strings, interest has shifted towards sigma models with the Wess–Zumino term [1], which we have already discussed in the previous chapter. The Wess–Zumino term was first introduced by Wess and Zumino [2] who, in the context of the chiral theory of pions, showed that it represents the effects of flavour anomalies. More recently it was reintroduced by Polyakov and Wiegmann [3] and by Witten [4], who studied various effects associated with its presence in different models. In addition, Witten [1] pointed out that, as the Wess–Zumino term breaks reflection symmetries on the manifold, this term should be included in any low-energy approximation to QCD based on Skyrme ideas [4]. We will discuss these ideas in the next chapter; here we want to look at various generalisations of our σ models and interpretations of their properties from a more geometrical point of view. In this, we are motivated by interesting papers by Gates *et al*, Curtright and Zachos, Howe and Sierra, and by Braaten *et al* [5, 6], who managed to show that the additional contribution to the equations of motion from the Wess–Zumino term can be interpreted as the contribution from the 'torsion' of the target manifold space.

We begin this chapter by presenting some of the ideas contained in these papers. In doing so we follow the paper by Braaten *et al* [6] very closely and so we concentrate our attention on the case when the target manifold is given by S^3, although most of our discussion will be more general. To consider the model for which the target manifold is given by S^3, we look first at the conventional $O(4)$ nonlinear sigma model of Gell-Mann and Levy [7], as the scalar fields of this model are restricted to lie on the unit 3-sphere S^3 by the contraint of nonlinearity.

9.2 $O(4)$ Sigma Model

Consider the $O(4)$ sigma model. To describe it we can use four real fields ϕ^i, where $i = 0, \ldots, 3$, and such that

$$(\phi^0)^2 + (\phi^1)^2 + (\phi^2)^2 + (\phi^3)^2 = 1; \tag{9.1}$$

i.e. the condition that the vector $\vec{\phi}$ lies on a sphere. For the action we take

$$S_1 = \frac{1}{2\lambda^2} \int \mathrm{d}^2x\, \partial_\mu \phi^i \partial^\mu \phi^i. \tag{9.2}$$

Here we can consider this model in either Minkowski or Euclidean space. Notice that the transformation $\vec{\phi} \to \lambda\vec{\phi}$ shows that λ can be identified with the inverse radius of the 3-sphere. However, we can 'solve the constraint' and use only ϕ^1, ϕ^2 and ϕ^3. Then, using either of the two possible solutions,

$$\phi^0_\pm = \pm\sqrt{1 - \phi^a\phi^a} \qquad a = 1, 2, 3, \tag{9.3}$$

allows us to write

$$S_1 = \frac{1}{2\lambda^2} \int \mathrm{d}^2x\, g_{ab}(\phi)\, \partial_\mu \phi^a \partial^\mu \phi^b, \tag{9.4}$$

where the metric on the field manifold is given by

$$g_{ab}(\phi) = \delta_{ab} + \frac{\phi_a\phi_b}{1 - \vec{\phi}\vec{\phi}}. \tag{9.5}$$

Now we can use various ideas from differential geometry. In particular,

$$g(\phi) = \det g_{ab} = \frac{1}{1 - \vec{\phi}\vec{\phi}}, \tag{9.6}$$

and so

$$\left[g_{ab}(\phi)\right]^{-1} = g^{ab}(\phi) = \delta^{ab} - \phi^a\phi^b. \tag{9.7}$$

From this we can calculate the Levi-Civita connection and the Riemann curvature tensor. We find

$$\Gamma^a_{bc} = \tfrac{1}{2}g^{ad}\{\partial_b g_{cd} + \partial_c g_{bd} - \partial_d g_{bc}\} = \phi^a g_{bc}(\phi), \tag{9.8}$$

and

$$\begin{aligned} R_{abcd} &= g_{ae}\left\{\partial_c\Gamma^e_{db} - \Gamma^f_{cb}\Gamma^e_{df} - \partial_d\Gamma^e_{cb} + \Gamma^f_{db}\Gamma^e_{cf}\right\} \\ &= g_{ac}(\phi)g_{bd}(\phi) - g_{ad}(\phi)g_{bc}(\phi). \end{aligned} \tag{9.9}$$

Observe that our action, when written in terms of 4 ϕ^i, is invariant under $O(4) \equiv O(3) \times O(3)$ transformations. The present action is clearly invariant under $O(3)$ transformations; of course, the remaining invariance is still there, but it is hidden. In fact, it is easy to check that, infinitesimally, our action is invariant under

$$\delta\phi^a = \epsilon^{abc}\phi^b\eta^c + (1 - \vec{\phi}^2)^{-1/2}\kappa^a, \tag{9.10}$$

where η^a and κ^a are independent parameters of our transformations. Of course, from a geometrical point of view our transformations can be viewed as isometries of the metric.

However, we can add to our Lagrangian a further contribution corresponding to the already familiar Wess–Zumino term. To do this we go over to three dimensions and consider

$$\epsilon^{\mu\nu\kappa}\,\epsilon^{ijkl}\,\phi^i\partial_\mu\phi^j\partial_\nu\phi^k\partial_\kappa\phi^l. \tag{9.11}$$

A little thought shows that this is an interesting expression from a topological point of view. It represents the topological charge density in three dimensions (i.e. it is associated with the maps $S^3 \to S^3$) and so, locally, it is given by a total divergence. This expression, when properly normalised, contributes an integer multiple of 2π to the action in three dimensions and it does indeed represent the Wess–Zumino term.

To obtain more insight into the form of (9.11) and its role in two dimensions, let us rewrite it in the form which uses only the three independent variables ϕ^i, $i = 1, 2, 3$. Using (9.3) to eliminate ϕ^0, and exploiting the fact that $\epsilon^{0abc} = \epsilon^{abc}$, we find that we obtain

$$\frac{\epsilon^{abc}\,\epsilon^{\mu\nu\kappa}}{\left(1 - \vec{\phi}^2\right)^{1/2}}\,\partial_\mu\phi^a\partial_\nu\phi^b\partial_\kappa\phi^c \tag{9.12}$$

times an overall factor which depends on the choice of the sign in (9.3). However, using the expression for g (9.6), we see that we can rewrite the action corresponding to (9.11) as

$$S^{(3)} = \int \mathrm{d}^3x\, S_{abc}\,\epsilon^{\mu\nu\kappa}\partial_\mu\phi^a\partial_\nu\phi^b\partial_\kappa\phi^c, \tag{9.13}$$

where

$$S_{abc} = \pm f g^{1/2} \epsilon^{abc}, \tag{9.14}$$

for some choice of a constant f.

Observe that S_{abc} is a covariantly constant tensor,

$$D_a S_{bcd} = \partial_b S_{bcd} - \Gamma^e_{ab} S_{ecd} - \Gamma^e_{ac} S_{bed} - \Gamma^e_{ad} S_{bce} = 0. \tag{9.15}$$

This can be verified by an explicit enumeration.

This very strong result shows that S_{abc} has a vanishing generalised curl, to which the Levi-Civita terms do not contribute as $\Gamma^a_{bc} = \Gamma^a_{cb}$. Thus S_{abc} is a *closed* three-form and so may be represented *locally* on the field manifold as a curl of a second-rank antisymmetric tensor e_{ab}:

$$S_{abc} = \eta\, \partial_{[a} e_{bc]}. \tag{9.16}$$

However, because of the sign arbitrariness, S_{abc} is *not exact*, and so cannot be represented globally as a curl of a single e_{ab}. Moreover, e_{ab} is defined only up to a curl:

$$e_{ab} \to e'_{ab} = e_{ab} + \partial_{[a}\xi_{b]}, \tag{9.17}$$

as $\partial_{[a}\, \partial_{[b}\, \xi_{c]]} = 0$.

This closure property allows us to make a connection between the three-dimensional topological charge density and an additional contribution to the two-dimensional action of the chiral model. Following the discussion of the previous chapter we identify our two-dimensional space with the boundary of a three-dimensional space that appears in our three-dimensional integrals. Then we can apply the divergence theorem and find

$$\begin{aligned} S^{(3)} &= \eta \int \mathrm{d}^3x\, \partial_{[a} e_{bc]}\, \epsilon^{\mu\nu\lambda}\, \partial_\mu \phi^a \partial_\nu \phi^b \partial_\lambda \phi^c \\ &= \eta \int \mathrm{d}^3x\, \partial_\lambda \left[\epsilon^{\mu\nu\lambda} (e_{ab}\, \partial_\mu \phi^a \partial_\nu \phi^b)\right] \\ &= \eta \int \mathrm{d}^2x\, \epsilon^{\mu\nu}\, e_{ab} \partial_\mu \phi^a \partial_\nu \phi^b, \end{aligned} \tag{9.18}$$

where we have used the fact that $\partial_\mu e_{ab} = \partial_c e_{ab} \partial_\mu \phi^c$. Moreover, the antisymmetry of $\epsilon^{\mu\nu\lambda}$ kills all other terms and provides a justification for the replacement $\partial_a e_{bc} \to \partial_{[a} e_{bc]}$.

We recall that e_{ab} is defined only up to a curl; however, changing it by a curl only introduces a surface contribution in two dimensions, which vanishes:

$$\int \mathrm{d}^2x\, \epsilon^{\mu\nu}\, \partial_{[a}\xi_{b]} \partial_\mu \phi^a \partial_\nu \phi^b = \int \mathrm{d}^2x\, \partial_\mu (\epsilon^{\mu\nu} \xi_b \partial_\nu \phi^b) = 0. \tag{9.19}$$

Next we determine our potential e_{ab}. We try an ansatz:

$$e_{ab} = \epsilon_{abc}\phi^c f(\vec{\phi}^2). \tag{9.20}$$

A few lines of algebra then show that as $g = \det g_{ab}(\phi) = (1 - |\vec{\phi}|^2)^{-1}$,

$$\partial_{[a}\, e_{bc]} = \pm g^{1/2} A\, \epsilon_{abc} \tag{9.21}$$

is satisfied, provided

$$\dot{f} + \frac{3f}{2x} = \frac{\pm A}{x\sqrt{1-x}}, \tag{9.22}$$

where $x = \vec{\phi}^2$. This equation is solved by

$$f(x) = \pm\frac{A}{2x^{3/2}}\,(\sin^{-1}\sqrt{x} - \sqrt{x}\sqrt{1-x}), \tag{9.23}$$

where the $\pm$ sign refers to the solution corresponding to the choice of ϕ_0 in (9.3) on the upper/lower hemisphere of S^3. As $\sin^{-1}\sqrt{x} \sim \sqrt{x} + \frac{1}{6}x\sqrt{x} + \ldots$ as $x \to 0$, we see that $f(0) = \pm\frac{1}{3}A$. The coefficient A is arbitrary at this stage; however, it should be so chosen that the complete action is well defined quantum mechanically — that is, it does not depend on the choice of the hemisphere. This is only possible if the three-dimensional action is given by an integer multiple of 2π.

We can now proceed to study various properties of the combined action $S^{(1)} + S^{(3)}$ in order to gain some insight into the significance of the e_{ab} term. To begin with, let us not use the specific form of our metric and of e_{ab}, but instead regard them as some differentiable, symmetric and antisymmetric covariant tensors on the field manifold.

The action is given by

$$S = \frac{1}{2\lambda^2}\int d^2x\,\{g_{ab}(\phi)\delta^{\mu\nu} + \tfrac{2}{3}p e_{ab}(\phi)\epsilon^{\mu\nu}\}\partial_\mu\phi^a\partial_\nu\phi^b, \tag{9.24}$$

where p is some constant. The classical equations of motion are given by

$$D_\mu\partial^\mu\phi^a = \left(\delta^{ab}\partial_\mu + \Gamma^a{}_{bc}\partial_\mu\phi^c - \frac{2p}{c}S^a{}_{bc}\epsilon_{\mu\nu}\partial^\nu\phi^c\right)\partial^\mu\phi^b = 0, \tag{9.25}$$

where $\Gamma^a{}_{bc}$ is defined as before, and where $S^a{}_{bc} = g^{ad}S_{dbc}$.

We see that S_{abc} plays a role similar to that of the Levi-Civita connection and so we can incorporate it into our connection. Due to its antisymmetry in bc we identify it with the *torsion* on the field manifold and, with this identification, we will refer to e_{ab} as the *torsion potential*. Thus we see that the addition of the Wess–Zumino term can be interpreted as the introduction of the torsion on the field manifold.

We can now define the full connection to be

$$\mathcal{F}_{abc} = \Gamma_{abc} - S_{abc}, \qquad \mathcal{F}^a{}_{bc} = g^{ad} F_{dbc} \tag{9.26}$$

and define the corresponding covariant derivatives using this connection:

$$\begin{aligned} \mathcal{D}_a V_b &\equiv \partial_a V_b - \mathcal{F}^c{}_{ab} V_c = D_a V_b + S^c{}_{ab} V_c, \\ \mathcal{D}_a V^b &\equiv \partial_a V^b + \mathcal{F}^b{}_{ac} V^c = D_a V^b - S^b{}_{ac} V^c. \end{aligned} \tag{9.27}$$

Considering a parallel transport of vectors, we can now define a *generalised curvature tensor* $\mathcal{R}$. It is given by

$$[\mathcal{D}_a, \mathcal{D}_b] V_c = \mathcal{R}_c{}^d{}_{ab} V_d + 2S^d{}_{ab} \mathcal{D}_d V_c. \tag{9.28}$$

Notice the additional term, which displays the standard 'translation' contribution proportional to the torsion tensor. Here $\mathcal{R}$ is defined by analogy with R, and is given by

$$\begin{aligned} \mathcal{R}_{abcd} &= g_{ae}\{\partial_c \mathcal{F}^a{}_{db} - \mathcal{F}^f{}_{cb} \mathcal{F}^e{}_{df} - \partial_d \mathcal{F}^e{}_{cb} + \mathcal{F}^f{}_{db} \mathcal{F}^e{}_{cf}\} \\ &= R_{abcd} + D_c S_{abd} - D_d S_{abc} + S_{fac} S^f{}_{db} - S_{fad} S^f{}_{cb}. \end{aligned} \tag{9.29}$$

So far these results are general. In the $O(4)$ case,

$$D_c S_{abd} = D_d S_{abc} = 0 \tag{9.30}$$

and so we are left with

$$\mathcal{R}_{abcd} = R_{abcd} + S_{fac} S^f{}_{db} - S_{fad} S^f{}_{cb}. \tag{9.31}$$

Next we calculate R_{abcd} and the other two terms in the $O(4)$ case. We find that

$$R_{abcd} = g_{ac} g_{bd} - g_{ad} g_{bc}, \tag{9.32}$$

and that the contribution of $S_{fac} S^f{}_{db} - S_{fad} S^f{}_{cb}$ is also proportional to the same expression, thus showing that

$$\mathcal{R}_{abcd} = (1 - \alpha^2)\, R_{abcd}, \tag{9.33}$$

where α is proportional to the constant p in (9.24). It follows that, if we choose this constant so that $\alpha = \pm 1$, the generalised curvature vanishes. Then the manifold is said to be 'parallelised'. The meaning of this term comes from the parallel transport of vectors. According to (9.28), for $\alpha = \pm 1$, if we perform a pair of independent infinitesimal $\mathcal{D}$ transportations of a vector, and compare the result with that obtained by interchanging the order of the two transportations, in general the two resultant vectors will be parallel but displaced from each other, due to the translation term, i.e. they will be 'parallel at a distance'.

The importance of the result involving parallelisibility can be seen in its impact on the renormalisability of the model. Under renormalisation the geometry of the model changes. The metric follows a trajectory in the space of field manifolds, the parameter of which is the length scale. As Braaten *et al* [6] have shown, the one-loop on-shell ultraviolet divergences of the theory can be eliminated by the addition of two counterterms to the action. These counterterms are obtained by simply renormalising the metric and the torsion potentials. The necessary one-loop corrections are given by [6]

$$\begin{aligned} \frac{1}{\lambda^2} g^{(1)}_{ab} &= \frac{1}{2\pi(2-d)} \mathcal{R}_{(ab)} \\ \frac{2\eta}{3\lambda^2} e^{(1)}_{ab} &= \frac{1}{2\pi(2-d)} \mathcal{R}_{[ab]}, \end{aligned} \tag{9.34}$$

where the brackets on the indices of $\mathcal{R}$ in the first and the second equations above denote their symmetrisation and antisymmetrisation respectively. Here

$$\mathcal{R}_{ab} = \mathcal{R}^c{}_{acb} = R_{ab} - S^{cd}{}_a S_{cdb} + D^c S_{cab}, \tag{9.35}$$

and we have used dimensional regularisation. From these counterterms we can calculate the scale dependence of the renormalised metric g_{ab} and torsion e_{ab}.

If we continue to d dimensions in such a way that the field ϕ is dimensionless, but the bare metric and torsion potentials $g^{(0)}_{ab}$ and $e^{(0)}_{ab}$ both have mass dimension $(d-2)$, then each of them satisfies

$$f^{(0)}_{ab} = M^{d-2}\Big\{ f_{ab} + \frac{1}{d-2} \check{f}^{(1)}_{ab} + \cdots \Big\}, \tag{9.36}$$

where M is the renormalisation group mass scale, and we have not written the second and higher loop corrections.

Requiring that the bare quantities are independent of M gives

$$M\frac{\mathrm{d}f^{(0)}}{\mathrm{d}M} = 0 \quad \rightarrow \quad M\frac{\mathrm{d}}{\mathrm{d}M} f_{ab} = -\check{f}^{(1)}_{ab}. \tag{9.37}$$

Thus we see that we obtain

$$\begin{aligned} M\frac{\mathrm{d}}{\mathrm{d}M}\Big\{\frac{1}{\lambda^2} g_{ab}\Big\} &= \mathcal{R}_{(ab)\,1/2\pi} \\ M\frac{\mathrm{d}}{\mathrm{d}M}\Big\{\frac{2\lambda}{\lambda^3} e_{ab}\Big\} &= \mathcal{R}_{[ab]\,1/2\pi}. \end{aligned} \tag{9.38}$$

In our case $\mathcal{R}_{abcd} = (1-\alpha^2)R_{abcd}$, and so

$$\mathcal{R}_{ab} = (1-\alpha^2)R_{ab} = 2(1-\alpha^2)g_{ab}. \tag{9.39}$$

Thus, as $g_{[ab]} = 0$, we find that the correction to e_{ab} vanishes, and for g_{ab} we have

$$M\frac{\mathrm{d}}{\mathrm{d}M}\left\{\frac{1}{\lambda^2}g_{ab}\right\} = 2(1-\alpha^2)\frac{1}{2\pi}g_{ab}. \tag{9.40}$$

We see that the torsion term is not renormalised (a fact that we might have expected given the topological nature of this term). Moreover, since the renormalisation of the metric is proportional to the metric itself, we observe that we may incorporate all the information into a single statement about the renormalisation of the scale factor λ:

$$\frac{\mathrm{d}}{\mathrm{d}\ln M}\lambda^2 = \frac{1}{\pi}(1-\alpha^2). \tag{9.41}$$

However, as $\alpha = \lambda^2 N/2\pi$, we can rewrite this as

$$\frac{\mathrm{d}}{\mathrm{d}\ln M}\lambda^2 = -\frac{1}{\pi}\lambda^4\left[1-\left(\frac{\lambda^2 N}{2\pi}\right)^2\right]. \tag{9.42}$$

We can now integrate (9.42), and obtain

$$4(s-s_0) = \frac{1}{z} + \tfrac{1}{2}\ln(1-z) - \tfrac{1}{2}\ln(1+z), \tag{9.43}$$

where $z = \lambda^2 N/2\pi$ and $s = \ln M/N$. The plot of $z(s)$ is sketched below (with $z(0) = \frac{1}{2}$)

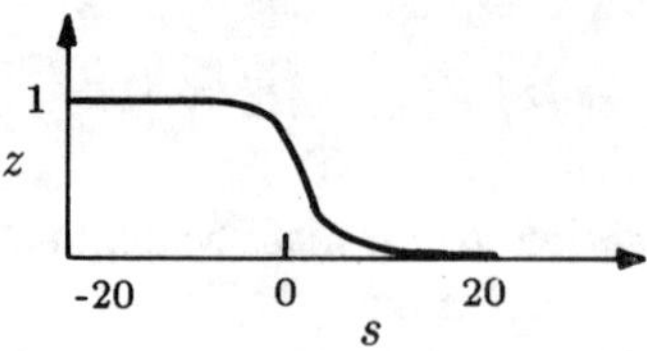

and, as $s = \ln M/N$, we observe that the theory possesses the ultraviolet fixed point at $\lambda = 0$ and the infrared one at $\lambda^2 = 2\pi/N$.

We can visualise the renormalisation process geometrically by exploiting the equivalence of the coupling constant $\lambda = 1/f$ with the inverse radius of the sphere S^3. We find that the radius increases with energy, and as a result the local curvature tends to zero; in the infrared limit the radius decreases with decreasing energy, but it stops decreasing as it reaches its minimum critical radius $\sqrt{N/2\pi}$. It can be shown that this infrared fixed point is not an artefact of the one-loop approximation [6]. Moreover, Witten [8] has presented some arguments suggesting that at this fixed point the theory is equivalent to a free field theory. We will return to this point in the last chapter when we discuss Witten's bosonisation.

Notice that the nontrivial fixed point arises when the manifold is parallelised, i.e. when $\mathcal{R}_{abcd} = 0$. It seems reasonable to expect this equivalence to hold also at the full quantum level, and it has been argued [8] that the S-matrix is trivial for this model. Moreover, it can be shown [6] that such infrared fixed points arise in a large class of parallelisable models.

9.3 Supersymmetric Extensions

Our results obtained in the previous section can be easily extended to include fermions; in particular, we have in mind extensions which lead to theories which exhibit supersymmetry. Such extensions were discussed by various people [9]; here we will follow closely the original paper of Braaten *et al* [6].

To construct supersymmetric extensions, we could proceed as we did in chapter 4, using the constrained fields; however, it is more natural to use unconstrained fields and include their metric, etc. Hence our fields $\phi^a(x,y)$ become superfields $\Phi^a(x,y,\theta_1,\theta_2)$.

As in chapter 5, we expand our superfields

$$\Phi^a(x,y,\theta_1,\theta_2) = \phi^a(x,y) + \theta_i\Psi_i^a(x,y) + \theta_1\theta_2 F^a(x,y), \tag{9.44}$$

where all the fields are real (this resembles the $O(3)$ formulation of the $\mathbb{C}P^1$ model), and introduce our supercovariant derivatives:

$$\partial' = \partial_\theta + i\partial\!\!\!/\theta. \tag{9.45}$$

Then

$$\partial'\Phi^a = \Psi^a + [\gamma_5 F^a + i\partial\!\!\!/\phi^a]\theta - i\theta_1\theta_2\partial\!\!\!/\gamma_5\Psi^a, \tag{9.46}$$

where Ψ is a spinorial field:

$$\Psi = \begin{pmatrix}\Psi_1\\ \Psi_2\end{pmatrix}. \tag{9.47}$$

In a similar way we find that

$$(\partial'\Phi)^\dagger = (\Psi^a)^\mathsf{T} + \theta^\mathsf{T}(-\gamma_5 F^a - i\partial\!\!\!/\phi^a) - i\theta_1\theta_2\Psi^\mathsf{T}\gamma_5\overleftarrow{\partial\!\!\!/}. \tag{9.48}$$

The manifestly supersymmetric action that contains the general bosonic action with an arbitrary metric g_{ab} and torsion e_{ab} is then given by

$$S = \frac{1}{2\lambda^2}\int \mathrm{d}^2x\,\mathrm{d}^2\theta[g(\Phi)_{ab} - \alpha e(\Phi)_{ab}](\partial'\Phi^a)^\dagger(1+\gamma_5)\partial'\Phi^b. \tag{9.49}$$

Next we expand $g(\Phi)_{ab}$ and $e(\Phi)_{ab}$ as second-order polynomials in θ_1 and θ_2, and obtain

$$g_{ab}(\Phi) = g_{ab}(\phi) + \frac{\partial g_{ab}}{\partial \phi^c}(\theta_i \Psi_i^c + \theta_1\theta_2 F^c)$$

$$+\frac{1}{2}\frac{\partial^2 g_{ab}}{\partial\phi^c\partial\phi^d}(\theta_i\Psi_i^c)(\theta_j\Psi_j^d) = g_{ab}(\phi) + \frac{\partial g_{ab}}{\partial\phi^c}(\theta_1\Psi_1^c + \theta_2\Psi_2^c) \tag{9.50}$$

$$+\theta_1\theta_2\left(\frac{\partial g_{ab}}{\partial\phi^c}F^c - \frac{1}{2}\frac{\partial^2 g_{ab}}{\partial\phi^c\partial\phi^d}(\Psi_1^c\Psi_2^d + \Psi_1^d\Psi_2^c)\right),$$

and an analogous expression for $e_{ab}(\Phi)$. Note that the reality of our fields simplifies the calculation somewhat, as at each stage either g_{ab} or e_{ab} projects out of $(1+\gamma_5)$ only one term.

Next we integrate over θ_1 and θ_2. This involves very tedious calculations; notice, however, that, due to the rules of integration over grassmannian variables discussed in chapter 1, only terms proportional to $\theta_1\theta_2$ give a nonvanishing contribution , and the result of this integration is the total coefficient of $\theta_1\theta_2$. This coefficient consists of several terms. Among them we find $g_{ab}\partial_\mu\phi^a\partial_\nu\phi^b$, $g_{ab}(\Psi^a)^\top \not\partial\Psi^b$, $g_{ab}F^aF^b$, $\alpha e_{ab}\epsilon^{\mu\nu}\partial_\mu\phi^a\partial_\nu\phi^b$, $-i\alpha e_{ab}(\Psi^a)^\top\gamma_5 \mathrm{d}\Psi^b$ and several further terms which involve derivatives of e_{ab} or g_{ab}. Let us observe that the expressions involving second-order derivatives give us terms involving four Ψ fields, and that first-order derivatives terms give us contributions quadratic in Ψ.

What about our auxiliary fields F^a? It is easy to check that they appear in $g_{ab}F^aF^b$, and also in expressions involving first derivatives of g_{ab} or e_{ab} together with two Ψ fields. In fact, detailed calculations [6] show that their total contribution to the Lagrangian density is given by

$$g_{ab}F^aF^b - \Gamma_{abc}F^a(\Psi^b)^\top\Psi^c + \rho S_{abc}F^a(\Psi^b)^\top\gamma^5\Psi^c, \tag{9.51}$$

where $S_{abc} = \partial_{[a}\, e_{bc]}$, and where $\Gamma^a{}_{bc}$ is the usual Levi-Civita connection discussed in the last section.

As the auxiliary fields appear quadratically in the action, without spacetime derivatives, they can be eliminated and replaced by their equations of motion:

$$F^a = \tfrac{1}{2}(\Gamma^a{}_{bc} - \rho S^a{}_{bc})(\Psi^b)^\top(1+\gamma_5)\Psi^c. \tag{9.52}$$

This shows that the auxiliary field F^a is a full connection contracted with a spinor bilinear. Notice that, due to the symmetry of the Ψ spinors,

$$F^a = \tfrac{1}{2}\Gamma^a{}_{bc}(\Psi^b)^\top\Psi^c - \tfrac{1}{2}\rho S^a{}_{bc}(\Psi^b)^\top\gamma_5\Psi^c. \tag{9.53}$$

Substituting this expression into the Lagrangian allows us to rewrite it in terms of expressions involving only ϕ^a and Ψ^a fields. We obtain

$$\begin{aligned} S = \frac{1}{2\lambda^2} \int & d^2x (g_{ab}\partial_\mu\phi^a\partial^\mu\phi^b + i g_{ab}(\Psi^a)^\top \not{D}\Psi^b \\ & + \alpha e_{ab}\epsilon^{\mu\nu}\partial_\mu\phi^a\partial_\nu\phi^b - \tfrac{1}{2}\alpha\partial_\mu(i e_{ab}(\Psi^a)^\top\gamma_5\gamma^\mu\Psi^b) \\ & + \tfrac{1}{2}\mathcal{R}_{abcd}(\Psi^a)^\top(1+\gamma_5)\Psi^c(\Psi^b)^\top(1+\gamma_5)\Psi^d, \end{aligned} \tag{9.54}$$

where

$$(D_\mu\Psi)^a = [\delta^{ab}\partial_\mu + \Gamma^a{}_{bc}\partial_\mu\phi^c - S^a{}_{bc}\epsilon_{\mu\nu}\partial^\nu\phi^c]\Psi^b. \tag{9.55}$$

We observe that the spinor analogue of our torsion term is locally a total divergence and so may be discarded, if the spinor fields vanish at ∞, as is normally assumed. Notice that, if we require that the model to possess topological properties, and in particular its bosonic part to reproduce the results of the previous section, then the tensor $e_{ab}(\Phi)$ must be the same function of Φ as in the purely bosonic case.

Observe also that, when the bosonic space is parallelised, the quartic spinor interaction vanishes; this suggests that the parallelism of the ϕ manifold is responsible for the infrared renormalisation group fixed point both for the supersymmetric and the purely bosonic models.

9.4 Anomaly in Two Dimensions

Quantum mechanics is not a theory by itself; it relies on classical mechanics. To construct a quantum system with consistent dynamics we construct it first with consistent and nontrivial classical dynamics and then quantise it.

As an example of this procedure let us consider a quantum theory involving gauge fields. Then we have a field strength tensor $F_{\mu\nu}$, and if the system involves also some matter fields J_μ which can be treated as sources for the gauge fields, then classically the dynamics of the system is described by

$$D^\mu F_{\mu\nu} = J_\nu. \tag{9.56}$$

However, for the consistency of classical dynamics, the sources must satisfy

$$D^\nu J_\nu = 0. \tag{9.57}$$

This will be true classically, if the gauge fields are coupled in correctly, and the matter dynamics admits a group of continuous symmetry transformations, and J_μ is the associated Noether current.

However, when sources include chiral fermions, we can have an anomaly; the conservation of the current may be destroyed, and we may have a *quantum mechanical violation* of the classical symmetry. Of course, the existence of the anomaly is not a necessity; the fermions may be in an anomaly-free representation or it may be possible to cancel the anomaly. This last option is available in many sigma models via the addition of the Wess–Zumino term. We will discuss an example of such a cancellation at the end of this chapter.

That such a phenomenon should happen ought not to surprise us; some theories are conformally invariant (our sigma models in two dimensions, gauge theories in four dimensions, etc), and this invariance is often broken by quantum effects and, associated with them, processes of regularisation and renormalisation. Naive calculations often give infinite answers, and the procedures which assign meaning to these answers introduce a mass scale and so break the conformal invariance. However, not all two-dimensional models have their conformal invariance broken at the quantum level. For example, strings in consistent backgrounds and some supersymmetric models remain conformally invariant after quantisation.

Should we be concerned if the theory has an anomaly? Does the theory still make sense? Before we attempt to answer these questions, let us see whether there are any further consequences of such an anomaly.

First of all we observe that, if it is the source current (i.e. the current which couples to the gauge field) which is not conserved, then the fermion determinant $Z(A)$, which arises in the theory from integrating out the fermion fields, is also not gauge-invariant. To see this observe that, as $Z(A) = \det(\partial\!\!\!/ + A\!\!\!/)$, then

$$\frac{\delta}{\delta A_\mu}[-i \ln Z(A)] = \langle J^\mu \rangle \,, \tag{9.58}$$

and so

$$\begin{aligned} Z(A_\mu + D_\mu\theta) - Z(A_\mu) &\sim \int \mathrm{Tr}\; D_\mu\theta \frac{\delta}{\delta A_\mu} Z(A_\mu) \\ &= iZ(A_\mu)\,\mathrm{Tr}\int \theta D^\mu \langle J_\mu \rangle \neq 0, \end{aligned} \tag{9.59}$$

which shows that the gauge invariance is broken.

Moreover, in addition, there are problems with Gauss's law,

$$G^a = \nabla_i(A)E_i^a + i\bar{\Psi}\lambda^a\Psi = 0, \tag{9.60}$$

which we have written in the Weyl gauge ($A_0 = 0$). G^a, which are, in fact, the generators of time-independent gauge transformations, involve constraints between canonical variables, and so cannot be set to zero quantum-mechanically (as an operator identity). Instead we impose

$$G^a(\underline{r})\,|\mathrm{phys}\rangle = 0, \tag{9.61}$$

which is consistent, as

$$[G^a(\underline{r}),\, G^b(\underline{r}')] = f^{abc}G^c(\underline{r})\delta(\underline{r} - \underline{r}'). \tag{9.62}$$

However, when fermions are present, we get an extra term in this commutation relation (corresponding to a Schwinger term). Its existence can be demonstrated in various ways — for example, by looking in detail at the properties of the phase of a physical state or by performing a careful analysis of the fermionic quantities. In fact we will perform such an analysis in the last chapter and we will demonstrate the appearance of a Schwinger term in a closely related case of the fermionic current algebra.

So far we have discussed a little the consequences of the anomaly; but, in fact, we have not proved its existence. Let us do this now in the case of two dimensions.

Let us recall (chapter 5) that we can take for our γ matrices $\gamma^0 = \sigma^1$, $\gamma^1 = i\sigma^2$ and then we have $\gamma_5 = -i\sigma^3$ (we consider the problem in Minkowski space). Then, as $\epsilon^{\mu\nu}\gamma_\nu = i\gamma^\mu\gamma_5$, where $\epsilon^{01} = 1 = -\epsilon_{01}$, we find that, if we define the following vector and axial currents

$$j^\mu = \Psi^\dagger\gamma^\mu\Psi, \qquad j_5^\mu = i\Psi^\dagger\gamma^\mu\gamma_5\Psi, \tag{9.63}$$

these two currents are related by

$$j_5^\mu = \epsilon^{\mu\nu}j_\nu. \tag{9.64}$$

This result is very specific to two dimensions, but not so specific as to mislead us about the existence of the anomaly (the anomaly exists in higher dimensions too, but the proof is more involved).

Classically, both currents are conserved. Let us consider them now in quantum theory. To do this we look at the vacuum correlation function of two vector currents. Due to the Poincaré invariance, its general form is

$$\begin{aligned}\langle j^\mu(x)j^\nu(y)\rangle &= g^{\mu\nu}\Pi_1(x-y) - \frac{\partial^\mu\partial^\nu}{\Delta}\Pi_2(x-y)\\ &+\Big(\frac{\partial^\mu\epsilon^{\nu\alpha}\partial_\alpha}{\Delta} + \frac{\partial^\nu\epsilon^{\mu\alpha}\partial_\alpha}{\Delta}\Big)\Pi_3(x-y),\end{aligned}$$

where $\Delta = \partial_\mu\partial^\mu$, and where Π_i are some functions of their arguments. From this we find that

$$\begin{aligned}\langle j_5^\mu(x)j^\nu(y)\rangle &= \epsilon^{\mu\alpha}\langle j_\alpha(x)j^\nu(y)\rangle\\ = \epsilon^{\mu\nu}\Pi_1(x-y) - \epsilon^{\mu\alpha}&\frac{\partial_\alpha\partial^\nu}{\Delta}\Pi_2(x-y) - \Big(g^{\mu\nu} - 2\frac{\partial^\mu\partial^\nu}{\Delta}\Big)\Pi_3(x-y).\end{aligned}$$

If we now impose the condition of the conservation of the vector current j_μ,

$$\partial_\mu\langle j^\mu(x)j^\nu(y)\rangle = 0 \tag{9.65}$$

we find that this requires that $\Pi_1 = \Pi_2$ and $\Pi_3 = 0$. Similarly the condition of the conservation of the axial current

$$\partial_\mu\langle j_5^\mu(x)j^\nu(y)\rangle = 0 \tag{9.66}$$

gives $\Pi_1 = \Pi_3 = 0$. It is clear that we cannot impose both conditions simultaneously, as this would require all three functions Π_i to vanish, resulting in a trivial theory. So what should we do? Clearly we can preserve one invariance, but the other one will be broken. Which one should we keep? It turns out that from a mathematical point of view this is optional; however, if the theory involves gauge fields which couple to vector or axial vector currents which are source currents we should preserve the conservation of those currents, as otherwise this would lead to the breakdown of gauge invariance. For further discussion see, e.g. a very pedagogical article by Jackiw [10].

Can we see this effect in a functional integral setting? Here we can use the ideas of Fujikawa [11], who observed that this effect can be related to the nontrivial transformation properties of the fermionic integration measure. To discuss these, let us consider an abelian vector theory in which the fermionic part of the Lagrangian is given by

$$L_f = i\bar{\Psi}\gamma^\mu\partial_\mu\gamma - i\bar{\Psi}\gamma^\mu A_\mu\gamma. \tag{9.67}$$

Then, under a local axial transformation given by

$$\gamma \to e^{i\alpha\gamma_5}\gamma, \qquad \bar{\Psi} \to \bar{\Psi}e^{i\alpha\gamma_5}, \tag{9.68}$$

where $\alpha = \alpha(x_\mu)$, the fermionic part of the Lagrangian L_f changes to

$$L_f \to L_f - \bar{\Psi}\gamma^\mu\gamma_5\partial_\mu\alpha = L_f - J_5^\mu\partial_\mu\alpha, \tag{9.69}$$

where $J_5^\mu = \bar{\Psi}\gamma^\mu\gamma_5\Psi$, and where we have used $\{\gamma^\mu, \gamma_5\} = 0$. Hence, integrating by parts, we find that $S \to S + \int \alpha\partial_\mu J_5^\mu$, and so we see that, had the action been invariant, we would have concluded that $\partial_\mu J_5^\mu = 0$.

However, in the quantum version of the theory we must consider the effects of the functional integration over all Ψ and $\bar{\Psi}$, and so we have to check whether there is no additional contribution from the change of the integration measure

$$D[\bar{\Psi}]\,D[\Psi] \to D[\bar{\Psi}']\,D[\Psi'] = D[\bar{\Psi}]\,D[\Psi]\,J, \tag{9.70}$$

where J is the Jacobian of the change of variables. To check for the appearance of such a Jacobian we have to define our fermionic measure with some care. We expand

$$\Psi = \sum a_n\Psi_n, \qquad \bar{\Psi} = \sum \bar{b}_n\Psi_n^\dagger, \tag{9.71}$$

where Ψ_n forms a convenient complete set of orthogonal functions. The coefficients a_n, $\bar{b}_n$ of the expansion above are grassmann valued and we can define the functional measure by

$$D[\bar{\Psi}]\,D[\Psi] = \prod_{nm} da_n\,d\bar{b}_m. \tag{9.72}$$

For our functions Ψ_n it is convenient to take the eigenfunctions of $\not{D}$, i.e.

$$\not{D}\Psi_n = \lambda_n \Psi_n, \tag{9.73}$$

which satisfy

$$\int \Psi_n^\dagger \Psi_m \, \mathrm{d}^d x = \delta_{mn}, \qquad \sum_n \Psi_n^\dagger(x)\Psi_n(y) = \delta^d(x-y). \tag{9.74}$$

Then as

$$\Psi \to \Psi' = e^{i\alpha\gamma_5}\Psi, \tag{9.75}$$

we find that, infinitesimally (i.e. for small α), the coefficients a_n change to

$$a_n' = a_n + i\int \mathrm{d}^d x \, \Psi_n^\dagger(x)\alpha(x)\gamma_5\Psi(x) = a_n + \sum_m C_{mn} a_m, \tag{9.76}$$

where

$$C_{mn} = i\int \mathrm{d}^d x \, \Psi_n^\dagger(x)\alpha(x)\gamma_5\Psi_m(x). \tag{9.77}$$

We see that our change of variables gives

$$\prod_n \mathrm{d}a_n = (\det C)\prod_n \mathrm{d}a_n', \tag{9.78}$$

and so we find that our Jacobian is given by

$$(\det C)^2 = \exp(2\,\mathrm{Tr}\ln C) = \exp\left(2i\int \mathrm{d}^d x \alpha(x)\sum_n \Psi_n^\dagger(x)\gamma_5\Psi_n(x)\right). \tag{9.79}$$

Thus we have to consider

$$2\sum_n \Psi_n^\dagger(x)\gamma_5\Psi_n(x). \tag{9.80}$$

However, it is easy to see that this expression diverges, and so has to be regularised. We can do this by setting

$$\sum_n \Psi_n^\dagger(x)\gamma_5\Psi_n(x) = \lim_{\substack{y\to x \\ M\to\infty}} \sum_n \Psi_n^\dagger(y)\gamma_5 e^{-\lambda_n^2/M^2}\Psi_n(x), \tag{9.81}$$

which, using the completeness of $\Psi_n(x)$, can be rewritten as

$$\lim_{\substack{y\to x \\ M\to\infty}} \mathrm{Tr}\left[\gamma_5 \exp\left(-\frac{\not{D}^2}{M^2}\right)\delta^d(x-y)\right].$$

However, as

$$\delta^d(x-y) = \int \frac{\mathrm{d}^d k}{(2\pi)^d} e^{ik(x-y)}, \tag{9.82}$$

and

$$\not{D}^2 = \gamma_\mu \gamma_\nu D^\mu D^\nu = D^2 + \tfrac{1}{4}[\gamma_\mu, \gamma_\nu] F^{\mu\nu}, \tag{9.83}$$

we see that we can rewrite it further as

$$\lim_{M\to\infty} \int \frac{\mathrm{d}^d k}{(2\pi)^d} \operatorname{Tr}\left[\gamma_5 \exp\left(-\frac{1}{4}[\gamma_\mu, \gamma_\nu]\frac{F^{\mu\nu}}{M^2}\right) \exp\left(-\frac{k^2}{M^2}\right)\right] \tag{9.84}$$

(the remaining terms vanish as $x \to y$).

Next we change the integration variables from k to k/M, and obtain

$$\lim_{M\to\infty} M^d \int \frac{\mathrm{d}^d (k/M)}{(2\pi)^d} \operatorname{Tr}\left[\gamma_5 \exp\left(-\frac{1}{4}[\gamma_\mu, \gamma_\nu]\frac{F^{\mu\nu}}{M^2}\right) \exp\left(-\frac{k^2}{M^2}\right)\right]. \tag{9.85}$$

The final step involves the expansion of $\exp(-\frac{1}{4}[\gamma_\mu, \gamma_\nu]F^{\mu\nu}/M^2)$ and the calculation of the leading term of

$$\operatorname{Tr}\left[\gamma_5 \exp\left(-\frac{1}{4}[\gamma_\mu, \gamma_\nu]\frac{F^{\mu\nu}}{M^2}\right)\right] \tag{9.86}$$

as $M \to \infty$. In two dimensions we find that

$$\operatorname{Tr}(\gamma_5[\gamma_\mu, \gamma_\nu]) = 4i\epsilon_{\mu\nu} \tag{9.87}$$

and so we obtain, for the Jacobian,

$$(\det C)^2 = \exp\left(-2\int \mathrm{d}^2 x\, \alpha(x)\epsilon_{\mu\nu}F^{\mu\nu}(x)\right), \tag{9.88}$$

thus showing that, in two dimensions,

$$\partial_\mu j_5^\mu = 2\epsilon_{\mu\nu}F^{\mu\nu}. \tag{9.89}$$

Notice that this result shows that, when we use Fujikawa's procedure, we automatically make the choice of preserving the conservation of the vector current, and so associate our anomaly with the nonconservation of the axial current.

Let us add, in passing, that in four dimensions the discussion is very similar, and the leading term is given by

$$\operatorname{Tr}(\gamma_5 \sigma_{\mu\nu}F^{\mu\nu}\sigma_{\rho\lambda}F^{\rho\lambda}), \tag{9.90}$$

where $\sigma_{\mu\nu} = [\gamma_\mu, \gamma_\nu]$. Then, using the properties of γ_μ, we can show [11] that

$$\partial_\mu j_5^\mu = 4F^{\mu\nu} * F_{\mu\nu}. \tag{9.91}$$

We observe that the right-hand terms in each case are the topological charge densities (we have not kept all the overall factors). As expected, the result is dimension-dependent, and we see that the Fujikawa method provides a convenient regularisation of the fermionic integrals resulting in the nonconservation of the axial vector current. This is a useful regularisation if the theory is based on vector gauge invariance, which remains unbroken. In elementary particle physics such invariance characterises strong and electromagnetic interactions, hence the Fujikawa regularisation is very convenient and useful to use.

However, the weak interactions are not governed by vector couplings, and so it is interesting to contemplate also the possibility of preserving the conservation of the axial vector current, while allowing the vector current to be nonconserved.

In two dimensions a large part of the discussion can be made very explicit, and the results have been known for a long time. In fact, in 1962 Schwinger [12] calculated explicitly

$$\Delta_1(A_\mu) = \det(\partial\!\!\!/ + ieA\!\!\!/) \tag{9.92}$$

in two dimensions. His method, based on the point-splitting regularisation, allowed him to find that

$$\ln \Delta_1(A_\mu) = \frac{e^2}{2\pi} \int \mathrm{d}^2x\, \mathrm{d}^2y A^\mu(x) \left(g_{\mu\nu} a_1 - \frac{\partial_\mu \partial_\nu}{\Delta} \right) \delta^2(x-y) A^\nu(y), \tag{9.93}$$

where $\Delta = \partial_\mu \partial^\mu$, and where a_1 is an arbitrary constant which can be chosen at will and depends on the regularisation of the theory.

If we regularise the theory in such a way that the vector current is conserved, we have $a_1 = 1$, as with this choice

$$\partial_\mu j^\mu = \partial_\mu \frac{\delta \ln \Delta_1(A_\mu)}{\delta A_\mu} = \partial_\mu \left(A^\mu - \frac{\partial^\mu \partial_\nu}{\Delta} A^\nu \right) = 0.$$

To see that $a_1 = 1$ corresponds to the preservation of gauge invariance, we observe that

$$\int \mathrm{d}^2x\, A^\mu(x) \left(g_{\mu\nu} - \frac{\partial_\mu \partial_\nu}{\Delta} \right) A^\nu(x) = - \int \mathrm{d}^2x\, F^{\mu\nu} \frac{1}{\Delta} F_{\mu\nu}, \tag{9.94}$$

which is explicitly gauge invariant.

The axial current is given by

$$j_5^\mu = \epsilon^{\mu\nu} A_\nu - \epsilon^{\mu\nu} \frac{\partial_\nu \partial_\alpha}{\Delta} A^\alpha, \tag{9.95}$$

and so is not conserved as its divergence is given by

$$\partial_\mu j_5^\mu = \epsilon^{\mu\nu}\partial_\mu A_\nu = \tfrac{1}{2}\epsilon^{\mu\nu}F_{\mu\nu}. \tag{9.96}$$

In a similar way we can calculate [10] the axial and chiral determinants, i.e. $\det(\not\partial + e\gamma_5 \not A)$, and the expressions derived from

$$L_+ = \bar\Psi_-(i\not\partial - e\not A)\Psi_+ \quad \text{and} \quad L_- = \bar\Psi_-(i\not\partial - e\not A)\Psi_-, \tag{9.97}$$

where the 'chiral fermions' are given by $\Psi_\pm = \frac{1}{2}(1 \pm i\gamma_5)\Psi$. Of course, as is well known, $L = L_+ + L_-$. On the other hand, both Weyl determinants involving Ψ_+ and Ψ_- can be evaluated explicitly. Let us call

$$\Delta_\pm = \{\det[(\not\partial + \not A)(1 \pm i\gamma_5)]\}^{1/2}, \tag{9.98}$$

and observe that $\Delta_\pm$ depends only on $A^\pm = \frac{1}{2}(A^0 \pm A^1)$. If we then regularise each determinant in a way which preserves this property, we find that the vector determinant (regularised in a way which preserves the conservation of the vector current) is related to them by

$$\ln \Delta(A) = \ln(\Delta_+\Delta_-) + \int \mathrm{d}^2x\, A^+A^-. \tag{9.99}$$

Thus we see that the determinant for the Dirac fermions is not a product of the determinants for the Weyl fermions of each chirality. The extra term depends on the 'coupling between A^+A^-', even though there is no such term in the classical Lagrangian.

Another interesting property of the anomaly in two dimensions is that it provides a mass term for the gauge field (or, strictly speaking, for the only degree of freedom in the gauge field). To see this observe that, if we introduce

$$\tfrac{1}{2}\epsilon^{\mu\nu}F_{\mu\nu} = *F, \tag{9.100}$$

then $\partial_\mu j_5^\mu = *F$. On the other hand $\partial_\mu F^{\mu\nu} = ej^\nu$, and so $\epsilon_{\nu\alpha}\partial_\mu F^{\mu\nu} = -2\partial_\alpha * F$. Then, using the relation (9.64) between the currents, we find that $*F$ satisfies

$$\Delta * F + \tfrac{1}{2}e * F = 0 \tag{9.101}$$

and so we see that $*F$ describes a massive field.

Hence we see that in two dimensions we can have 'chiral fermions' interacting with gauge fields giving rise to massive vector bosons and consequently deconfined femions. Of course, the use of the term 'vector bosons' is not really appropriate as in two dimensions we do not have any spin. We see that the *gauge invariance* is *broken* (in a sense that the mass appears), not explicitly or spontaneously but just quantum mechanically. So far no similar effect has been found in four dimensions; it would be very exciting if we could explain the origin of the Z_0 and $W_\pm$ mass in this way!

Before we proceed further let us make a few general comments. First of all let us add that often [13] a different approach to the study of the anomaly is used. We consider

$$W(A_\mu) \;=\; \operatorname{Tr}\ln(i\not{D} - im), \tag{9.102}$$

where we have introduced a mass term to cure the possible infrared problems that we have been ignoring, and then we expand $W(A_\mu)$ formally in the power series in A_μ:

$$\begin{aligned} W(A_\mu) &= \operatorname{Tr}\ln\{(i\not\partial - im)[1-(i\not\partial - im)^{-1}\not{A}]\} \\ &= \operatorname{Tr}\ln(i\not\partial - im) + \operatorname{Tr}\ln[1-(i\not\partial - im)^{-1}\not{A}] \\ = -\sum_{n=1}^{\infty}\frac{1}{n}\int \mathrm{d}^2y_1\cdots &\int \mathrm{d}^2y_n A_{\mu_1}(y_1)\cdots A_{\mu_n}(y_n)\int \mathrm{d}^2p_1\frac{e^{ip_1(y_1-y_2)}}{(2\pi)^2} \\ \cdots\int \mathrm{d}^2p_n &\frac{e^{ip_n(y_n-y_1)}}{(2\pi)^2}\operatorname{Tr}\{G(p_1)\gamma_{\mu_1}G(p_2)\gamma_{\mu_2}\ldots G(p_n)\gamma_{\mu_n}\}, \end{aligned}$$

where

$$G(p) \;=\; \frac{-\not{p}+m}{p^2+m^2}. \tag{9.103}$$

This corresponds to an expansion in the number of external A_μ fields. Clearly the contribution with one such field ($n=1$) vanishes (due to the property of traces and the symmetrical integration), and the first nontrivial term corresponds to two A_μ fields ($n=2$):

The expression describing this contribution can be easily regularised, and we recover our previous expression. Moreover, it must be possible to find a regularisation which would eliminate all the higher order terms, as the result of Schwinger clearly proves. Thus the anomaly in two dimensions is associated with the graph. (In physically more relevant four dimensions the anomaly is associated with the diagram; the also diverges, but this divergence is cured by the renormalisation of the theory).

In general, if we have *source currents* then they have to be covariantly conserved (so far we have discussed only the abelian case; in the nonabelian case we need full conservation). Otherwise the gauge invariance would be lost and the theory would have to be rejected. However, as Wess and

Zumino have shown [2], the anomalous term satisfies certain constraints — called the 'Wess–Zumino consistency constraints'. A clearly presented discussion of the consequences of these constraints can be found in the previously mentioned article by Jackiw [10].

Here let us restrict ourselves to just a few comments. If we consider only the Noether currents which are classically conserved, then the appearance of the anomaly in the theory is not a problem. The theory has to be regularised in a way which does not violate the gauge principle. The way the anomaly terms appear is independent of whether the currents are source currents or just Noether currents. In practice this process is tied up with the regularisation of loop diagrams, and so there is some arbitrariness involved. The applied renormalisation should eliminate all the infinities of the theory in a way which preserves as much a possible the original symmetry structure of the theory.

In currently fashionable theories of elementary particles (QCD) we have a vector theory, and the axial singlet current is not conserved; instead we find

$$\partial_\mu j_5^\mu \sim \operatorname{Tr} F_{\mu\nu} * F^{\mu\nu}. \tag{9.104}$$

The anomaly is locally the total divergence, and so can be associated with the redefinition (renormalisation) of the angle θ, which is a parameter of the theory, and can be associated with the fact that the true vacuum of the theory is the so-called 'θ-vacuum' (for more details see most books on QCD, e.g. [14]). The theory remains invariant under gauge transformations (as long as they do not change the topological number). The axial nonsinglet current is also not conserved, as

$$D_\mu j_{5a}^\mu \sim \operatorname{Tr} \tau_a F_{\mu\nu} * F^{\mu\nu}. \tag{9.105}$$

This time the anomaly is not a total divergence; however, as the current is not an observable (i.e. no field couples to it), its nonconservation is innocuous.

The weak-interaction part of the Lagrangian corresponds to a chiral theory. The appropriate chiral currents satisfy

$$D_\mu j_\pm^\mu = \pm A, \tag{9.106}$$

where the anomaly term A is partly determined by the Wess–Zumino consistency conditions. The non-vanishing value of A breaks the gauge invariance, and so A must vanish, as otherwise the theory would be inconsistent as a quantum theory, since it would have problems satisfying unitarity, etc.

However, the value of A depends on the representation in which the fermion fields are placed and on the group itself. For some groups it vanishes automatically (such groups are 'safe'), for others it is set to zero by a judicious choice of representations for the fermions. In the theory describing weak interactions we set it to zero by the latter procedure. In fact this is achieved by requiring an equal number of quarks and leptons, a 'prediction' which is successful phenomenologically.

It has been suggested [10] that it may be possible to construct a consistent theory by adding to it a Wess–Zumino term; the total Lagrangian will be gauge invariant quantum mechanically, as the noninvariant part of the Lagrangian associated with the anomaly will be cancelled by the explicitly noninvariant piece given by a Wess–Zumino term (here by the Wess–Zumino term we mean any term which satisfies the Wess–Zumino consistency conditions). And although, so far, no realistic realisation for this procedure has been found in higher dimensional gauge theories, this prescription can be shown to be operational in the cancellation of the sigma model anomalies.

But what are the anomalies of the sigma models? What are the symmetries of these models which are broken by these anomalies? If we consider a general sigma model in which the target manifold is M then, following the procedure discussed in the first two sections of this chapter, we can introduce ϕ_i as local coordinates on M with the corresponding metric being given by $g_{ij}(\phi)$. Of course, the choice of these coordinates is nonunique and the action

$$S[\phi, g_{ij}(\phi)] = \int \mathrm{d}x^2\, g_{ij}(\phi)\, \partial_\mu \phi^i \partial_\mu \phi^j \tag{9.107}$$

is invariant under the general coordinate transformation

$$\phi^i \to \phi^{i\prime},\ g_{ij}(\phi) \to g'_{ij}(\phi') = g_{kl}(\phi) \frac{\partial \phi^k}{\partial \phi^{i\prime}} \frac{\partial \phi^l}{\partial \phi^{j\prime}}. \tag{9.108}$$

It is this invariance that may be broken when chiral fermions are present. However, when the complete Lagrangian includes also an appropriate Wess–Zumino term in the classical action the anomaly due to the fermions can be cancelled by the non-reparametrisation invariant contribution of the Wess–Zumino term.

The origin and the existence of the σ model anomaly and its cancellation by the Wess–Zumino term was first dicussed by Moore and Nelson and Bagger *et al* [15] and recently reviewed in detail by Hull in his lectures at the Vancouver Super Field Theory Workshop [16]. Hence instead of reproducing all the details of the proof of this cancellation we will restrict ourselves only to sketching the main ideas and send the interested reader to the excellent review article by Hull [16]. Mooreover, in our discussion we will follow his review and keep his notation.

To introduce the sigma model anomaly we have to consider coupling to the sigma model of various complex fermions of definite chirality. Such fermions, as we discussed in chapter 5, are eigenstates of $\gamma^3 = \gamma^1\gamma^2$, with the right-handed and the left-handed fermions corresponding to the eigenvalues $+1$ and -1 of γ^3 respectively. This time, however, to keep in agreement with Hull, we work in Minkowski space. Then, if we consider a sigma model involving r right-handed fermions $\lambda_{\rm R}^A$ ($A = 1,\ldots,r$) and s left-handed fermions $\xi_{\rm L}^M$ ($M = 1,\ldots,s$) the generalised Dirac action for these fermions is given by $S_{\rm F} = S_{\rm FL} + S_{\rm FR}$, where

$$\begin{aligned} S_{\rm FR} &= i\int \mathrm{d}x^2\, \bar{\lambda}_{\rm R}^A\, \gamma^\mu\, D_\mu \lambda_{\rm R}^A \\ S_{\rm FL} &= i\int \mathrm{d}x^2\, \bar{\xi}_{\rm R}^M\, \gamma^\mu\, \tilde{D}_\mu \xi_{\rm R}^M\,, \end{aligned} \tag{9.109}$$

where the covariant derivatives, which are in general different, are given by

$$\begin{aligned} D_\mu \lambda_{\rm R}^A &= \partial_\mu \lambda_{\rm R}^A + \hat{A}_\mu^{AC} \lambda_{\rm R}^C \\ \tilde{D}_\mu \xi_{\rm L}^M &= \partial_\mu \xi_{\rm L}^M + \hat{B}_\mu^{MN} \xi_{\rm L}^N\,, \end{aligned} \tag{9.110}$$

for some connections $\hat{A}_\mu^{AC}$ and $\hat{B}_\mu^{MN}$, which are functions of the our local coordinates on M, i.e. ϕ^i.

The generalised Dirac actions (9.109) possess a local $O(r) \times O(s)$ symmetry

$$\begin{aligned} \delta\lambda_{\rm R}^A &= \epsilon\Lambda^{AB}\lambda_{\rm R}^B,\, \delta A_\mu^{AB} = -\epsilon(\partial_\mu\Lambda^{AB} + A_\mu^{CA}\Lambda^{BC} - A_\mu^{CB}\Lambda^{AC}) \\ \delta\xi_{\rm L}^M &= \epsilon\theta^{MN}\xi_{\rm L}^N\,,\, \delta B_\mu^{MN} = -\epsilon(\partial_\mu\theta^{MN} + B_\mu^{PM}\theta^{NP} - B_\mu^{PN}\theta^{MP}), \end{aligned} \tag{9.111}$$

where $\Lambda^{AB} = -\Lambda^{BA}$, $\Lambda^{AB}\Lambda^{BC} = \delta^{AC}$, taken together with $\Lambda^{AB} = \Lambda^{AB}(\phi^i)$, $\partial_\mu\Lambda^{AB} = \partial_i\Lambda^{AB}\partial_\mu\phi^i$ and similarly for θ^{MN}. This symmetry corresponds to the gauge symmetry of gauge theories.

Next we observe that the path integral over the chiral fermions

$$\exp(iS_{\rm eff}[A,B]) = \int \mathrm{d}[\lambda_{\rm R}]\,\mathrm{d}[\bar{\lambda}_{\rm R}]\,\mathrm{d}[\xi_{\rm L}]\,\mathrm{d}[\bar{\xi}_{\rm L}]\, \exp(i[S_{\rm FR} + S_{\rm FL}]) \tag{9.112}$$

is formally identical to that in a chiral gauge theory. The effective action $S_{\rm eff}$ defined by the above expression is not gauge invariant under (9.111); in analogy with gauge theories it can be shown that as a result of this transformation we obtain the chiral anomaly

$$\delta S_{\rm eff} = \int \mathrm{d}x^2(\Omega[A,\Lambda] - \Omega[B,\theta]) \tag{9.113}$$

for some Ω. The computation of Ω is now well understood. As Ω depends on the fields A and B, which can be treated as gauge fields, first, one looks at the consistency conditions satisfied by Ω [2] and then solves them using the so-called descent equations [17]. In this approach one considers a three-form

$$\omega_3 = \mathrm{Tr}\,[A \wedge F - \tfrac{1}{3} A \wedge A \wedge A], \tag{9.114}$$

where $A = A^{AB}_\mu \, \mathrm{d}x^\mu$ and $F = \mathrm{d}A + A \wedge A$. This three-form is not gauge invariant under (9.111) but its variation under this transformation is an exact form, i.e. there exists a two-form ω_2 such that

$$\delta_\Lambda \omega_3 = \mathrm{d}\omega_2. \tag{9.115}$$

Then it is easy to check that

$$\omega_2 = \mathrm{Tr}\,[\Lambda \wedge \mathrm{d}A] \tag{9.116}$$

and it can be shown that the anomaly is given (up to the addition of trivial terms) by the contribution of two such forms corresponding respectively to the fields A and B:

$$\delta S_{\mathrm{eff}} = -\frac{1}{4\pi q} \int (\omega_2[A, \Lambda] - \omega_2[B, \theta]). \tag{9.117}$$

Writing this expression explicitly we find

$$\begin{aligned} \delta S_{\mathrm{eff}} = -\frac{1}{2\pi} \int \mathrm{d}x^2 \, \epsilon^{\mu\nu} \partial_\mu \phi^i \partial_\nu \phi^j [\Lambda^{AB} (\partial_i A^{AB}_j - \partial_j A^{AB}_i) \\ - \theta^{MN} (\partial_i B^{MN}_j - \partial_j B^{MN}_i)]. \end{aligned} \tag{9.118}$$

The anomaly can be trivial (i.e. can be cancelled by the addition of a finite local counterterm); this would happen if

$$\mathrm{d}\omega_3(A) = \mathrm{d}\omega_3(B), \tag{9.119}$$

which would hold, for instance, if $r = s$ and $A_i = B_i$, so that the theory was left–right symmetric.

If the anomaly is nontrivial it can be cancelled by the addition of a suitable Wess–Zumino term to the classical action. To add such a term we follow the procedure of the second section of this chapter and consider a torsion term, $b_{ij}(\phi)$, an antisymmetric tensor on M. Then the corresponding Wess–Zumino term is given by (up to normalisation) by

$$S_{\mathrm{WZ}} = q \int b = q \int \epsilon^{\mu\nu} \, b_{ij}(\phi) \partial_\mu \phi^i \partial_\nu \phi^j \, \mathrm{d}^2 x. \tag{9.120}$$

To complete the cancellation we have now to choose our torsion term $b_{ij}(\phi)$ conveniently. This problem was discussed by Hull and Witten [18], who showed that this can be achieved if we choose b_{ij} in such a way that the term b in (9.120) transform under $O(r) \times O(s)$ as

$$\delta b = \frac{1}{4\pi q}[\omega_2(A, \Lambda) - \omega(B, \theta)]. \tag{9.121}$$

It this case the variation of the Wess–Zumino term precisely cancels the anomalous variation (9.117).

Moreover, it can be shown [15, 18] that such a torsion term always exists and that its choice is consistent with the quantisation condition for the coefficient q of the Wess–Zumino term.

Finally, let us say a few words about what would happen had the anomaly not been cancelled. Clearly, the theory would not be invariant under (9.111). However, such an invariance is needed for the global definition of the theory. If it were broken it would be possible to define an effective action only in one coordinate patch, but would not be possible to extend this definition to the whole space. In this sense the sigma model anomaly can be understood as a topological obstruction preventing a global definition of an effective action.

References

[1] Witten E 1983 Global aspects of current algebra *Nucl. Phys.* B **223** 422

[2] Wess J and Zumino B 1971 Consequences of anomalous Ward identities *Phys. Lett.* **37B** 95

[3] Polyakov A and Wiegmann P 1983 Theory of nonabelian Goldstone bosons in two dimensions *Phys. Lett.* **131B** 121; 1984 Goldstone fields in two dimensions with multivalued actions *Phys. Lett.* **141B** 223

[4] Witten E 1984 Non-abelian bosonisation in two dimensions *Commun. Math. Phys.* **92** 455

[5] Gates S J, Hull C M and Rocek M 1984 Twisted multiplets and new supersymmetric non-linear σ models *Nucl. Phys.* B **248** 157
Curtright T J and Zachos C K 1984 Geometry, topology and supersymmetry in non-linear σ models *Phys. Rev. Lett.* **59** 1799
Howe P S and Sierra G 1984 Two-dimensional supersymmetric non-linear σ models with torsion *Phys. Lett.* **148B** 451

[6] Braaten E, Curtright T L and Zachos C K 1985 Torsion and geometrostatis in nonlinear sigma models *Nucl. Phys.* B **260** 630

[7] Gell-Mann M and Levy M 1960 The axial vector current in beta decay *Nuovo Cimento* **16** 705

[8] Witten E 1984 Non-Abelian bosonisation in two dimensions *Commun. Math. Phys.* **92** *455*
Di Vecchia P and Rossi P 1984 On the equivalence of the Wess–Zumino action and the free Fermi theory in two dimensions Phys. Lett. **140B** *344*

[9] Di Vecchia P, Knizhnik V G, Petersen J L and Rossi P 1985 A supersymmetric Wess–Zumino Lagrangian in two dimensions *Nucl. Phys.* B **253** 701

Abdalla E and Abdalla M C B 1985 Supersymmetric extension of the chiral model and the Wess–Zumino term in two dimensions *Phys. Lett.* **152B** 59
Abdalla M C B 1985 Integrability of chiral non-linear sigma models with a Wess–Zumino term *Phys. Lett.* **152B** 215

[10] Jackiw R 1985 Topological investigations of quantized gauge theories *Current Algebra and Anomalies* ed S B Trieman *et al* (Singapore: World Scientific) p 211

[11] Fujikawa K 1980 Path integral for gauge theories with fermions *Phys. Rev.* D **21** 2848

[12] Schwinger J 1962 Gauge invariance and mass II *Phys. Rev.* **128** 2425

[13] Nielsen N K, Rothe K D and Schroer B 1979 Fermionic Green function and functional determinant in QCD_2 *Nucl. Phys.* B **160** 330
Andrianov A and Bonora L 1984 Finite-mode regularization of the fermion functional integral I and II *Nucl. Phys.* B **233** 232, 247
Mignaco J A and Rego Monteiro M A 1986 A four-dimensional model with the fermionic determinant exactly evaluated *Phys. Lett.* **175B** 77

[14] Cheng Ta-Pej and Li Ling-Fong 1984 *Gauge Theory of Elementary Particle Physics* (Oxford: Oxford University Press)

[15] Moore G and Nelson P 1984 Anomalies in non-linear sigma models *Phys. Rev. Lett.* **53** 1519; 1985 The aetiology of sigma model anomalies *Commun. Math. Phys.* **100** 83
Bagger J, Nemeschansky D and Yankielowicz S 1985 Anomaly constraints on non-linear sigma models *Nucl. Phys.* B **262** 478

[16] Hull C M 1987 Lectures on non-linear sigma models *Super Field Theories* ed H C Lee *et al* (New York: Plenum) p 77

[17] Zumino B, Wu Y S and Zee Z 1984 Chiral anomalies, higher dimensions and differential geometry *Nucl. Phys.* B **239** 477

[18] Hull C M and Witten E 1985 Supersymmetric sigma model and the heterotic string *Phys. Lett.* **160B** 398

10 Skyrme Models

10.1 The Skyrme Idea

QCD is believed to be the correct model to describe strong interactions of elementary particles at energy scales below that corresponding to grand unification [1]. If we assume that the quark fields are massless, then QCD exhibits global symmetry under the full chiral transformation group

$$G = SU(N_{\rm f})_{\rm L} \times SU(N_{\rm f})_{\rm R}, \tag{10.1}$$

where $N_{\rm f}$ is the number of flavours (in practice $N_{\rm f} = 3$ or 4), and $SU(N_{\rm f})_{\rm L,R}$ act separately on the left and right quark fields (previously called $\Psi_{\pm}$). However, it is believed that this symmetry is spontaneously broken down to the diagonal (vector) $SU(N_{\rm f})$ symmetry, and so the vacuum states of the theory are in a one-to-one correspondence with different points of the $SU(N_{\rm f})$ manifold. Thus the low-energy dynamics of QCD can be described by introducing a field $U(x_\mu)$, which, under an $SU(N_{\rm f})_{\rm L} \times SU(N_{\rm f})_{\rm R}$ transformation described by unitary matrices (A, B), transforms as

$$U(x_\mu) \rightarrow U'(x_\mu) = AU(x)B^{-1}. \tag{10.2}$$

The effective Lagrangian for U must have the above-mentioned G symmetry, and to describe correctly the low-energy limit it must have the smallest possible number of derivatives. Thus, in order to have any reasonable dynamics, the lowest order terms (in the number of derivatives) must be given by

$$L_0 = \tfrac{1}{16}F^2 \int \mathrm{d}^4x \ \mathrm{Tr}\, \partial_\mu U \partial^\mu U^{-1}. \tag{10.3}$$

The perturbative expansion of U around $U = 1$ is given by

$$U = 1 + \frac{2i}{F} \sum_{a=1}^{N_{\rm f}^2-1} \tau^a \Pi^a + \ldots, \tag{10.4}$$

where τ^a (normalised by $\mathrm{Tr}\,\tau^a\tau^b = 2\delta^{ab}$) are the generators of $SU(N_f)$, and where Π^a are dynamical Goldstone boson fields. In physical applications $N_f = 2$ or 3, Π^a fields describe pseudoscalar mesons (π, K), and F is phenomenologically fixed to be given by $F \sim 190$ MeV.

This effective Lagrangian is known to incorporate all the relevant symmetries of QCD and we can derive from it (in the tree approximation) all the current algebra theorems governing the low-energy limit of Goldstone boson S-matrix elements.

Of course quarks carry in addition a colour quantum number (associated with the colour $SU(N)$ gauge group), and 't Hooft, Witten and others [2] have argued that, as $N \to \infty$, the lowest lying mesons and their interactions are governed by the above effective theory — thus providing theoretical justification for treating (10.3) as an effective theory of physical mesons.

Moreover, Witten has also pointed out that baryons behave as if they were solitons in this effective large-N meson field theory [3]. This observation of Witten's resurrected an old idea of Skyrme's, who many years ago studied the possibility of treating the ordinary proton and neutron as solitons in a nonlinear sigma model [4].

The recent observation of Witten allows us to check in what sense baryons can be treated as solitons, and it has produced a new stimulus to phenomenology. Moreover, the first suggestions made by Skyrme, later elaborated on by Finkelstein and Rubinstein [5], showed that solitons in a nonlinear sigma model in four dimensions could, in principle, be fermions. Witten's recent observations [6] show that for $N_f > 2$ they have to be fermions. The term responsible for this fact is the familiar Wess–Zumino term, which has to be included in the effective action, and its appearance in the action can be associated with the elimination of unwanted symmetries in the Lagrangian (and consequently allowing certain processes to take place) or with the consideration of anomalies in the more fundamental theory (QCD) from which the effective theory is meant to be derived in the $N \to \infty$. Thus we see that a large number of modern ideas fit together quite well and are interdependent.

Let us discuss briefly Witten's ideas. First of all we will concentrate on the effective meson theory and ignore the Wess–Zumino term. The effective Lagrangian, for the effective meson theory derived from QCD, is bound to be very complicated. At low energies it is given by (10.3). However, (10.3) is only an approximation, and in fact it does not possess any stable solitonic solutions. So if we want to consider fermions/baryons as solitons of the theory, clearly we must go beyond (10.3).

Let us consider further terms that we may want to add to (10.3). Of course, we are considering our theory at low energies only, and as we are talking about an effective theory we should envisage expansions in $\partial_\mu U$.

Thus Skyrme [4] chose to consider

$$L = \frac{F^2}{16}\,\mathrm{Tr}\,\partial_\mu U \partial^\mu U^{-1} + \frac{1}{32e^2}\,\mathrm{Tr}\,[(\partial_\mu U)U^{-1}, (\partial_\nu U)U^{-1}]^2 + \ldots. \tag{10.5}$$

The last term, called the 'Skyrme term', can be rewritten as

$$-\frac{1}{16e^2}\,\mathrm{Tr}\,(\partial_\mu U \partial^\mu U^\dagger \partial_\nu U \partial^\nu U^\dagger - \partial_\mu U \partial_\nu U^\dagger \partial^\mu U \partial^\nu U^\dagger). \tag{10.6}$$

This is not the only four-derivative term that we can write down. However, it can be shown that this is the unique four-derivative term that leads to a positive Hamiltonian (moreover, it is also the unique term with four derivatives which is of the second order in time derivatives).

If we now want to study meson interactions for physical mesons we have to introduce additional terms which are responsible for the meson masses. Thus we add to the Lagrangian density

$$L_m = \tfrac{1}{8} m_\pi^2 F^2\,\mathrm{Tr}\,U. \tag{10.7}$$

However, let us concentrate instead on the solitonic properties of the model. As the energy is given by

$$E = \int \mathrm{d}^3x \left(\frac{F^2}{16}\,\mathrm{Tr}\,(\partial_i U^\dagger \partial_i U) - \frac{1}{32e^2}\,\mathrm{Tr}\,\{[\partial_i U U^\dagger, \partial_j U U^\dagger]^2\} \right) \tag{10.8}$$

we see that, in order that a given configuration corresponds to a finite energy lump, we must require that

$$U \to \text{constant} \qquad \text{as} \quad |x| \to \infty. \tag{10.9}$$

Without any loss of generality we can take this constant to be the unit matrix.

However, since $U \to 1$ as $|x| \to \infty$, we see that we can take the physical space at constant time to be S^3 instead of R^3 (i.e. we have 'compactified' the space). As $U \in SU(N_\mathrm{f})$ and $\Pi_3(SU(N_\mathrm{f})) = Z$ for $N_\mathrm{f} \geq 2$, we see that the model has an infinite number of solitonic sectors characterised by an appropriate winding number B (the space of U's splits into an infinite number of classes). This integer-valued topological index is given by

$$\begin{aligned} B &= \frac{1}{24\pi^2}\epsilon^{ijk} \int \mathrm{d}^3x\,\mathrm{Tr}\,(\partial_i U U^\dagger \partial_j U U^\dagger \partial_k U U^\dagger) \\ &= \frac{1}{24\pi^2}\epsilon^{ijk} \int \mathrm{d}^3x\,\mathrm{Tr}\,(U^\dagger \partial_i U \partial_j U^\dagger \partial_k U). \end{aligned}$$

Observe that B can be thought of as representing the charge of the conserved current J_μ, given by

$$J_\mu = \frac{1}{24\pi^2}\epsilon^{\mu\nu\lambda\rho}\,\mathrm{Tr}\,(\partial_\nu U^\dagger \partial_\lambda U U^\dagger \partial_\rho U U^\dagger). \tag{10.10}$$

This current (we will shortly identify it as a baryonic current) is conserved for purely algebraic reasons:

$$\partial_\mu J^\mu = 0. \tag{10.11}$$

To show this observe that

$$J^\mu \sim \epsilon^{\mu\nu\lambda\rho}\,\mathrm{Tr}\,A_\nu A_\lambda A_\rho, \tag{10.12}$$

where $A_\mu = U^\dagger \partial_\mu U$, and that

$$\partial_\mu A_\nu - \partial_\nu A_\mu = -[A_\mu, A_\nu]$$

and so

$$\partial^\mu J_\mu \sim \epsilon^{\mu\nu\lambda\rho}\,\mathrm{Tr}\,A_\mu A_\nu A_\lambda A_\rho, \tag{10.13}$$

which vanishes by the cyclic symmetry of the trace and the antisymmetry of $\epsilon^{\mu\nu\lambda\rho}$.

Skyrme, and following him the other previously mentioned physicists, decided to interpret the index B as the baryon number, in which case the lowest energy state in the $B = 1$ sector should be identified with the nucleon. The lowest *classical* energy E_0 is attained by the static 'spherically symmetric' ansatz, given by

$$U_0(\underline{r}) = \exp\{iF_0(r)\tilde{\underline{r}}\cdot\underline{\tau}\}, \tag{10.14}$$

(the $SU(2)$ case), where $\tilde{\underline{r}} = \underline{r}/r$. The winding number B can be calculated for $U_0(\underline{r})$, and is given by

$$B = \frac{1}{\pi}[F_0(0) - F_0(\infty)]. \tag{10.15}$$

So we can achieve $B = 1$ with $F_0(0) = \pi$ and $F_0(\infty) = 0$. The function $F_0(r)$ and the energy E_0 must be determined numerically. This was done by Adkins *et al* [7], who found that the variation of F shown in the figure below minimises E_0 subject to the conditions mentioned before.

Before we go further, let us observe that our 'spherically symmetric' ansatz is so called because it satisfies

$$-i(\underline{r}\wedge\underline{\nabla})_i U_0 + [\tfrac{1}{2}\tau_i, U_0] = 0 \tag{10.16}$$

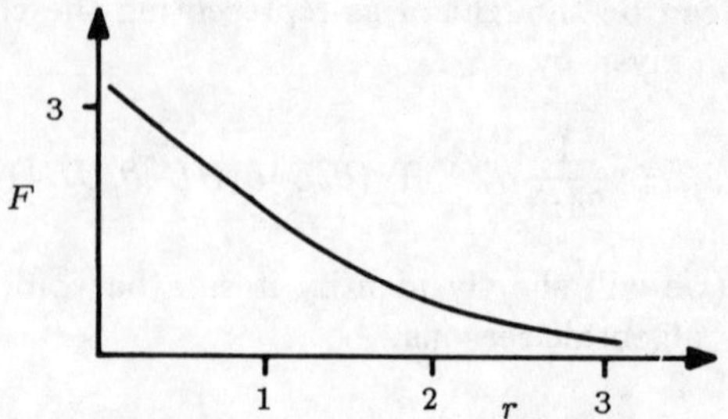

i.e. it involves the combined spatial and flavour rotations.

In the case of higher $SU(N_f)$ $N_f > 2$ symmetry groups we proceed by embedding $SU(2)$ into $SU(N_f)$ using only three matrices, which correspond to a $SU(2)$ subgroup (like λ_i, $i = 1, 2, 3$ for Gell-Mann's $SU(3)$ matrices).

So far our discussion has been totally classical. To proceed to quantisation we observe that our 'spherically symmetric' ansatz is not invariant under the unbroken $SU(N_f)$ flavour group. However, we can argue that the action of $SU(N_f)$ on the ansatz naturally gives rise to 'collective coordinates'. To introduce them we observe that if U_0 is a soliton solution, so is $U = AU_0A^{-1}$, where $A \in SU(2)$ and is a constant matrix. Moreover, the energy of U is the same as that of U_0. Thus the simplest quantum corrections come from treating A as a function of time and considering it as a quantum mechanical variable — a sort of collective coordinate. Clearly A parametrises the zero modes; promoting A to a dynamical variable $A(t)$, and quantising it, amounts to interchanging the infinite number of quantum variables $U(\underline{x}, t)$ for the three quantum degrees of freedom describing A. Of course, in addition, U is also described by an infinite number of classical coordinates — $F_0(r)$, which should be quantised in a fuller treatment.

Observe that the replacement

$$U(\underline{r}, t) \rightarrow A(t)U_0(\underline{r})A^{\dagger}(t) \tag{10.17}$$

corresponds to a global $SU(N_f)$ rotation ($N_f = 3$). But such a rotation takes

$$\exp\{i\underline{\tilde{n}} \cdot \underline{\tau} G(x, t)\}, \tag{10.18}$$

and sends it to

$$\exp\{iA\underline{\tilde{n}} \cdot \underline{\tau} A^{\dagger} G(x, t)\}, \tag{10.19}$$

which shows that under this rotation $\underline{\tilde{n}}$ changes but remains on the sphere $\underline{\tilde{n}} \cdot \underline{\tilde{n}} = 1$. So when we try to diagonalise the quantum Hamiltonian of the A variables, we see that picking out the energy associated with the collective motion of $\underline{\tilde{n}}$ is analogous to separating out the spherical harmonics in the nonrelativistic quantum mechanics of a particle in a central potential. In fact, this was the approach adopted by Bohr in his semiclassical quantisation of the hydrogen atom (he quantised only the rotational degrees of freedom).

Thus, substituting $U = A(t)U_0A^{-1}(t)$ into (10.5), where U_0 is our previous expression, we find

$$L = -E_0(F_0) + \Lambda(F_0)\,\mathrm{Tr}\,(\dot{A}\dot{A}^\dagger), \tag{10.20}$$

where $E_0(F_0)$ is the previously calculated quantity, and $\Lambda(F_0)$ is an explicit (but rather complicated) function of F_0. It is clear that the collective motion has added a rotational energy analogous to $\frac{1}{2}I\omega^2$, where $I \sim \Lambda(F_0)$ and $\mathrm{Tr}\,\dot{A}\dot{A}^\dagger = \omega^2$. To quantise we set $A = a_0 1 + i\underline{a}\cdot\underline{\tau}$, where $a_0^2 + \underline{a}^2 = 1$. Then

$$p_i = \frac{\partial L}{\partial \dot{a}_i} = 4\Lambda\dot{a}_i \qquad \sum_{i=0}^{3} a_i^2 = 1, \tag{10.21}$$

and so we see that

$$H = E_0 - \frac{1}{8\Lambda}\sum_{i=0}^{3}\left(\frac{\partial^2}{\partial a_i^2}\right) \tag{10.22}$$

together with the constraint $\sum_{i=0}^{3} a_i^2 = 1$.

Because of the constraint, the operator $-\sum_{i=0}^{3}\partial^2/\partial a_i^2$ can be interpreted as the Laplacian on the three sphere. The wavefunctions (by analogy with usual spherical harmonics) are traceless symmetric polynomials in the a_i's (i.e. $-\nabla^2(a_0+ia_1)^l = l(l+2)(a_0+ia_1)^l$). Then we find that

$$H = E(F_0) + \frac{l(l+1)}{2\Lambda(F_0)} \qquad l = \tfrac{1}{2}, \tfrac{3}{2}, \ldots, \tag{10.23}$$

when acting on states of definite isospin I and spin J with $\underline{I}^2 = \underline{J}^2 = \underline{L}^2$.

In this procedure we have to decide what to do about the symmetry $A \to -A$, as both A and $-A$ lead to the same U. This symmetry is removed by the addition of the Wess–Zumino term, whose effects we will discuss later on. Here we would like to decide whether we want to impose

$$\Psi(A) = \Psi(-A) \qquad \text{or} \qquad \Psi(A) = -\Psi(-A) \tag{10.24}$$

on the wavefunctions, which correspond to quantisation of the soliton as a boson or a fermion respectively. If we choose the latter, the wavefunctions become polynomials of odd degree in the a_i's. Thus the nucleons, of $I = J = \frac{1}{2}$, correspond to wavefunctions linear in a_i's, while the Δ's, of $I = J = \frac{3}{2}$, correspond to cubic polynomials. The model has two parameters — F and e (see (10.8)), and they should be fixed by the determination of the masses of N and of Δ. When this is done [7] we find that $F \sim 129$ MeV, i.e.

about 30% lower than the experimental value of 186 MeV. Adkins *et al* used the values of F and e determined in this way to calculate a number of static properties of baryons, such as the charge radii, magnetic moments, etc. The obtained results were in reasonable agreement with the experimental data (the accuracy being of the order of 20%). These results were used further by other groups, who considered effects associated with going from $SU(2)$ to $SU(3)$, studied various improvements and also considered the problem of the two-nucleon potential [8]. Results are encouraging though inconclusive.

At the same time the Skyrme model remains a good phenomenological tool for studying various properties of baryons and their interactions. In particular, in a recent paper [9], Brodsky *et al* use it to estimate the valence quark contribution to the proton spin, when chiral symmetry and $SU(3)$ symmetry are broken. They find that the model predicts this contribution to be very small, in agreement with the recent EMC data on the spin-dependent proton structure.

We should add a few words about the stability of our soliton (often called the skyrmion). We recall that F_0 is determined by the minimisation of $E_0(F_0)$ and not of the effective quantum Hamiltonian given by $H = E_0(F_0) + l(l+1)/\Lambda(F_0)$. As a result of this, it is easy to show that small oscillations about the skyrmion are not stable, but reveal modes that grow rather than oscillate harmonically. This is sometimes compensated for by increasing $E_0(F_0)$ at the expense of $\Lambda(F_0)$, thus suggesting that the soliton should undergo 'centrifugal stretching' to compensate for the additional centrifugal potential energy induced by $A(t)$. Such attempts may be reasonable, but they suffer from arbitrariness. Clearly, this problem deserves more attention.

Adkins and Nappi [10] have also considered the effects of the pion masses (after all, in the real world the $SU(2)_{\mathrm{L}} \times SU(2)_{\mathrm{R}}$ symmetry is not exact but only correct to within 5–10%). They have found that their results (apart from F) change little and that the changes are in the right direction. This reassures them as to the validity of their approximations.

Finally there is the question of the Wess–Zumino term, which we have only touched on briefly before. Here we follow quite closely the discussion given by Witten [11]. Let us return to our Lagrangian, which is given by

$$L = \frac{F^2}{16} \int \mathrm{d}^4x \ \mathrm{Tr}\, \partial_\mu U \partial^\mu U^\dagger + \text{ quartic terms.} \tag{10.25}$$

This Lagrangian is invariant under $U \to U^\top$. If we expand

$$U = 1 + \frac{2i}{F} \sum_{a=1}^{8} \lambda_a \Pi^a, \tag{10.26}$$

where λ^a are Gell-Mann $SU(3)$ matrices we see that this symmetry corresponds to $\Pi^+ \leftrightarrow \Pi^-$. Of course, in addition, we have the invariance under $\underline{r} \to -\underline{r}$, $t \to t$, $U \to U$ and finally also under $U \to U^\dagger$. This last symmetry corresponds to $\Pi^a \leftrightarrow -\Pi^a$.

If this were a true symmetry of the strong interactions, then the total number of mesons in any reaction would be conserved modulo two. However, this is not the case. The physical world seems to be (approximately) invariant under

$$\underline{r} \to -\underline{r}, \qquad t \to -t, \qquad U \to U^\dagger, \tag{10.27}$$

but not under each transformation separately (a nontrivial example is the existence of $K^+K^- \to \pi^+\pi^0\pi^-$). So we see that we have to eliminate the additional symmetries which ought not to be there. To do that, let us first of all see how we would eliminate them from the equations of motion. For simplicity let us temporarily ignore the quartic terms. The equation of motion corresponding to L is

$$\partial_\mu(U^\dagger \partial^\mu U) = 0. \tag{10.28}$$

Let us try to add a suitable term to this expression. As we want to violate the symmetry ($\underline{r} \to -\underline{r}$, $t \to -t$, $U \to U$) but still preserve the Lorentz symmetry we see that the additional term must include $\epsilon^{\mu\nu\lambda\rho}$. If we now restrict ourselves to the smallest number of derivatives, we get

$$\partial^\mu(U^\dagger\partial_\mu U) + \lambda\,\epsilon^{\mu\nu\lambda\rho}(U^\dagger\partial_\mu U)(U^\dagger\partial_\nu U)(U^\dagger\partial_\lambda U)({}^\dagger\partial_\rho U) = 0. \tag{10.29}$$

It can be now shown that this expression violates the simple parity transformation but is covariant under the QCD parity transformation (10.27). However, a little thought shows that our extra term cannot be derived from a conventional Lagrangian. An obvious guess would lead to an expression based on

$$\epsilon^{\mu\nu\lambda\rho}\,\mathrm{Tr}\,(U^\dagger\partial_\mu U\, U^\dagger\partial_\nu U\, U^\dagger\partial_\lambda U\, U^\dagger\partial_\rho U),$$

but this expression vanishes by the antisymmetry of ${}^{\mu\nu\lambda\rho}$ and the cyclic property of the trace. However, a Lagrangian from which this term can be derived does exist, but it requires a five-dimensional space as it corresponds to the Wess–Zumino term discussed in chapter 8. There we discussed the three-dimensional case; here we need its five-dimensional version.

To discuss it we consider our four-dimensional space as a boundary of a five-dimensional disc Q. We expect the action, corresponding to the Wess–Zumino term, to be given by an integral over Q. By analogy with the cases discussed in chapter 8, we take

$$S = \int_{\mathrm{Im}\,Q} \omega_{ijklm}\, \mathrm{d}\Sigma^{ijklm}, \tag{10.30}$$

where ω_{ijklm} is some function of U. Again, we have the quantisation condition

$$\int_{S^5} \omega_{ijklm}\, \mathrm{d}\Sigma^{ijklm} = 2\pi \text{ integer}, \tag{10.31}$$

where $S^5 = \mathrm{Im}Q \cup \mathrm{Im}Q'$. A little thought and the analogy with the expressions discussed before shows that

$$\mathrm{d}\Sigma^{ijklm}\,\omega_{ijklm} = \alpha\, \mathrm{d}\Sigma^{ijklm} \times \mathrm{Tr}\left(U^\dagger \frac{\partial U}{\partial y^i} U^\dagger \frac{\partial U}{\partial y^j} U^\dagger \frac{\partial U}{\partial y^k} U^\dagger \frac{\partial U}{\partial y^l} U^\dagger \frac{\partial U}{\partial y^m}\right),$$

where y^i are local coordinates on $\mathrm{Im}Q$ and where α is some constant factor.

It is interesting to look at our expression for U close to identity. Thus we take

$$U^\dagger \partial_i U = \frac{2i}{F}\partial_i A + O(A^2), \tag{10.32}$$

where $A = \sum_a \lambda^a \pi^a$ and so we have

$$\omega_{ijklm}\, \mathrm{d}\Sigma^{ijklm} = \alpha\, \mathrm{d}\tilde{\Sigma}^{ijklm}\, \mathrm{Tr}\,(\partial_i A \partial_j A \partial_k A \partial_l A \partial_m A) + O(A^6), \tag{10.33}$$

where $\mathrm{d}\tilde{\Sigma}^{ijklm}$ is the measure on Q (notice that the Jacobian factors cancel). The derived expression can be rewritten as

$$\mathrm{d}\tilde{\Sigma}^{ijklm}\, \partial_i \{\mathrm{Tr}\,(A \partial_j A \partial_k A \partial_l A \partial_m A)\}, \tag{10.34}$$

and so we see that to order A^5 (and in fact also to all higher orders) our expression is a total divergence. Hence,

$$\begin{aligned}\int_{\mathrm{Im}Q} \omega_{ijklm}\, \mathrm{d}\Sigma^{ijklm} &= \alpha \int_Q \mathrm{d}\tilde{\Sigma}^{ijklm}\, \partial_i \{\mathrm{Tr}\,(A \partial_j A \partial_k A \partial_l A \partial_m A)\} \\ &= \alpha' \int \mathrm{d}^4 x\, \mathrm{Tr}\,(A \partial_\mu A \partial_\nu A \partial_\alpha A \partial_\beta A)\epsilon^{\mu\nu\alpha\beta} + \ldots .\end{aligned}$$

In fact our complete expression is nothing else but the Wess–Zumino term determined by various people [12]. What our discussion has shown is that its coefficient is quantised. Moreover, it is easy to see that the additional term we have written has the correct symmetry properties. Thus we have found that, in the hypothetical world of massless K and π mesons, this effective Lagrangian describes the low-energy limit of $\mathrm{K}^+\mathrm{K}^- \to \pi^+\pi^0\pi^-$, and that the low-energy limit of the amplitude for this reaction must be an integer (in appropriate units).

What is the value of this integer? Can it vanish? Of course it cannot vanish, as this reaction is observed experimentally. Witten [6] has given many arguments suggesting that, if we write our additional term in the effective action as $L_{\text{WZ}} = n\Gamma$, where Γ is the Wess–Zumino term, then $n = N_c$, where N_c is the number of colours in the theory (observe that this is not inconsistent with our derivation of the effective Skyrme Lagrangian, which was obtained in the limit of $N_c \to \infty$). The simplest way to see this is to consider $\pi^0 \to 2\gamma$.

To consider such a process we have to introduce gauge fields into the model. This can be done in a straightforward way by the minimal coupling substitutions

$$\partial_\mu U \to D_\mu U, \tag{10.35}$$

where $D_\mu = \partial_\mu + ieA_\mu$. Then Γ changes also appropriately:

$$\Gamma \to \tilde{\Gamma} = \Gamma + \text{ corrections.} \tag{10.36}$$

This last change is quite complicated, but luckily it has been discussed by many people who have studied anomalies, etc [13]. If we use their result, we find that if $n = N_c = 3$ we obtain an expression for $\pi^0 \to 2\gamma$, which agrees with the calculation based on the QCD triangle diagram. In a similar way we can calculate $K^+K^- \to \pi^0\pi^+\pi^-$ and again we have agreement.

We should point out that if the number of flavours is 2, then there is no additional Wess–Zumino term; consequently skyrmions can be quantised as either bosons or fermions. However, as soon as $N_f > 2$ we have this term, and so skyrmions have to be fermions and this result is independent of the value of N_f. The correct value of the coefficient implies its quantisation as a fermion.

To determine whether a soliton (skyrmion) is a fermion or not, we have to compare the amplitudes for two processes; in one of them the soliton remains at rest for a long time T. The amplitude is $\exp(-E_0T)$, where E_0 is the soliton's energy. In the second process the soliton is adiabatically rotated through a 2π angle in the course of this long time T. The usual term,

$$L_0 = \frac{F^2}{16}\,\text{Tr}\,\partial_\mu U^\dagger \partial^\mu U \tag{10.37}$$

does not distinguish between these two processes, as L_0 is quadratic in time derivatives, and so

$$\int dt\ \text{Tr}\left(\frac{\partial U}{\partial t}\frac{\partial U^\dagger}{\partial t}\right) \to 0, \tag{10.38}$$

as $T \to \infty$. However, the anomalous term Γ is linear in time derivatives and distinguishes between the two processes. For the soliton at rest we find $\Gamma = 0$, but for a rotating soliton we find $\gamma = \pi$. Thus the amplitude of the rotating soliton is not $\exp\{-iE_0T\}$ but $(-1)^n \exp\{-iE_0T\}$, and so if n is odd the soliton behaves as a fermion.

To see which one we are dealing with, we take our ansatz for the skyrmion and a particular form of $A(t)$ (which corresponds to the rotation), and then put this expression into our expression for Γ. We find that

$$\Gamma = \pi \tag{10.39}$$

(the calculation is quite involved). A similar, although more lengthy, calculation has to be done for $N_{\mathrm{f}} > 3$.

Finally, the success of 'skyrmion physics' has led some people to consider the idea of putting vector mesons into the effective Lagrangian approach of Skyrme (see for instance recent papers by Brihaye *et al* [14] and others [15]). They have speculated that maybe such fields can be generated dynamically within this approach. The idea is that quantum effects generate quantum gauge fields associated with a symmetry of the Lagrangian. In particular, if we consider $G = SU(N_{\mathrm{f}})_{\mathrm{R}} \times SU(N_{\mathrm{f}})_{\mathrm{L}}$ and $H = SU(N_{\mathrm{f}})_V$, then we can hope that the mechanism of the $\mathbb{C}P^{N-1}$ model will become operative and some, or maybe even all, of the gauge fields associated with the unbroken symmetry group H will have their kinetic terms generated quantum mechanically. They can be then interpreted as the familiar vector mesons of the low-energy hadronic world (those that appear in various vector-meson dominance approximations, etc). Recently a group of Italian physicists [15] suggested that the appearance of vector mesons, which can be interpreted as a manifestation of such a 'hidden symmetry', also leads to the emergence of the Skyrme term in the limit of a large value of the hidden gauge fields' coupling constant. This interesting idea requires further elaboration.

Let us finish this section with a note of warning. All arguments based on quantum effects calculated within the effective Lagrangian framework are slightly suspicious. Such Lagrangians are not renormalisable and are not meant to be used as a field theory; they are to be used only in the tree approximation. In the skyrmion case there is also the question of how many terms to keep (in the effective expansion in the number of derivatives), and whether such an expansion makes sense. Recently the Oxford group [16] looked at this in some detail. They paid particular attention to all the approximations involved; in particular, as to whether we can devise a scheme which would allow all the terms involving more derivatives than four to be ignored. Their results are discouraging, but their analysis is a little incomplete. Clearly more is still to be said on this topic.

10.2 $\mathbb{C}P^1$ Skyrmions

So far, in our studies of grassmannian models, we have considered their properties in two Euclidean dimensions. The main motivation for this comes from their analogy to four-dimensional Yang–Mills theories and our interest in their classical and quantum properties.

However, in recent years, an increasing amount of attention has been paid to various field theory models which possess non-trivial topological properties, and so admit extended solutions. Unfortunately, in most such models only the static solutions are known explicitly. This is the case, for instance, for the genuine skyrmions, which we discussed in the previous section, or for nonabelian monopoles. As the problem of going beyond the static field configurations requires the solution of the full equations of motion, which is rather difficult, it has been suggested that static field configurations be interpreted as describing static extended objects and that the time evolution of the fields be approximated by the time evolution of these objects; this is in fact what is often done for the real skyrmions which we discussed in the previous sections.

The evolution of extended objects, in general, causes a distortion of their shape, and even though the problem may not be tractable in full generality, some observations can be made about the evolution of the system in the limit of small velocities. For example, Manton [17] noted that in the case of nonabelian monopoles, their initial motion follows the geodesics in the space of parameters of the static extended solutions. This observation has led to a partial understanding of the evolution of a system of monopoles, and provided information about the way they scatter off each other [18].

These remarks apply also to grassmannian models in (2+1) dimensions. As we said before, our Euclidean two-dimensional solutions can be thought of as describing extended structures, and, in the previous chapters, we showed that we can introduce the concepts of position and size for these structures. Moreover, they can be regarded as static solutions of the full (2+1)-dimensional version of the model. Following Wilczek and Zee [19], we will call these structures '$\mathbb{C}P^{N-1}$ skyrmions'. The evolution of these skyrmions can be studied under the assumption that all time dependence of our field variables resides only in the skyrmion parameters [20]. This amounts to a study of the dynamics of slowly moving skyrmions which may be thought of as approximations to genuine time-dependent classical solutions of the full theory.

Let us consider first the $\mathbb{C}P^1$ model in (2+1) dimensions. Its action is given by

$$S = \int_0^T \mathrm{d}t \int \mathrm{d}^2x \, (D^\mu z)^\dagger D_\mu z, \tag{10.40}$$

where, as in previous chapters, $z = (z_1, z_2)$ is a complex two-vector satisfying $|z|^2 = 1$, and $D_\mu z$ is the familiar covariant derivative of z given by

$$D_\mu z = \partial_\mu z - (z^\dagger \partial_\mu) z, \tag{10.41}$$

and the index μ runs from 0 to 2.

The time-independent solutions of the equations of motion are of course our Euclidean solutions; these solutions are given by the instanton and anti-instanton configurations discussed in chapter 3.

Let us recall that the general k-instanton configuration is given by (see the discussion in chapter 3)

$$z = \frac{f}{|f|}, \tag{10.42}$$

with the holomorphic vector f given by

$$\begin{aligned} f_1 &= \lambda_1 (x_+ - a_1)\ldots(x_+ - a_k) \\ f_2 &= \lambda_2 (x_+ - b_1)\ldots(x_+ - b_k), \end{aligned} \tag{10.43}$$

where, as before, $x_\pm = x_1 \pm ix_2$, $a_i, b_i, i = 1,\ldots,k$ are the (mutually different) positions of the 'quark' and 'antiquark' constituents [21] of the instantons in the complex plane, and one of the parameters λ_i can, by the choice of gauge, be set equal to 1.

Suppose now that all the parameters in this field configuration (i.e. a_i, b_i and λ_i) depend on our time parameter t. Then, for a fixed t, (10.42) is still a stationary point of S with respect to space-dependent variations, but not a stationary point (i.e. a classical solution) with respect to general variations (however it approximates such a solution in the limit of weak t-dependence).

If we now insert our z, given by (10.42), with this time dependence into S, we find that

$$S = \int_0^T \mathrm{d}t \int \mathrm{d}^2x \left| \frac{\mathrm{d}z}{\mathrm{d}t} - \left(z^\dagger \frac{\mathrm{d}z}{\mathrm{d}t} \right) z \right|^2 - 2\pi k T. \tag{10.44}$$

The last term can be thought of as representing a potential energy and so is not very interesting; the first one, on the other hand, contains all the dynamics as it represents the total kinetic energy of the system:

$$K(z(t)) = \int \mathrm{d}^2x \left| \frac{\mathrm{d}z}{\mathrm{d}t} - \left(z^\dagger \frac{\mathrm{d}z}{\mathrm{d}t} \right) z \right|^2. \tag{10.45}$$

To study this term we have, first of all, to ensure the finiteness of the total kinetic energy. This condition, as is easy to check, prevents us from allowing the λ_i's to depend on t (as otherwise the kinetic energy would diverge quadratically). In what follows, for simplicity, we choose both $\lambda_i = 1$. In addition we have to impose the constraint that

$$\sum_i^k (a_i - b_i) = \text{constant} \qquad \text{for all t.} \tag{10.46}$$

The above statement shows that the overall relative orientation of the skyrmion quarks and antiquarks cannot change with time.

Next we substitute our expression for the vector f (10.43), and find that the kinetic energy integrand, κ, in (10.45), is given by

$$\begin{aligned}\kappa &= \frac{1}{|f|^4}\,|f_2\,\partial_t f_1 \;-\; f_1\,\partial_t f_2|^2 \\ &= \frac{1}{|f|^4}\,\left|\dot{a}_i\frac{\partial}{\partial a_i} - \dot{b}_i\frac{\partial}{\partial b_i}\right|^2 |f_1|^2|f_2|^2,\end{aligned} \tag{10.47}$$

where $|f|^2 = |f_1|^2 + |f_2|^2$ and $\dot{a}_i = \mathrm{d}a_i/\mathrm{d}t$, $\dot{b}_i = \mathrm{d}b_i/\mathrm{d}t$.

If we now introduce the compact notation

$$\{A_\alpha\} \;=\; \{a_1, \ldots, a_k, b_1 \ldots, b_k\}, \tag{10.48}$$

we find that the second equality above implies that

$$\kappa \;=\; g_{\alpha\bar{\beta}}\,\dot{A}^\alpha \dot{A}^{\bar{\beta}}, \tag{10.49}$$

where the metric $g_{\alpha\bar{\beta}}$ can be written compactly as

$$g_{\alpha\bar{\beta}} \;=\; \frac{\partial}{\partial A^\alpha}\,\frac{\partial}{\partial A^{\bar{\beta}}}\,\log\,|f|^2. \tag{10.50}$$

Thus we see that the kinetic energy defines a Kähler-like metric on the manifold of skyrmion quark and anitiquark parameters.

The total kinetic energy is obtained by integrating the metric over the space coordinates:

$$K \;=\; \int \mathrm{d}^2x\, g_{\alpha\bar{\beta}}\,\dot{A}^\alpha \dot{A}^{\bar{\beta}}, \tag{10.51}$$

and this integral is well defined on the submanifold given by the constraint (10.46).

10.3 $\mathbb{C}P^{N-1}$ Skyrmions

The generalisation of the construction of the previous section is completely straightforward. The vector z now has N components (constrained to satisfy $|z|^2 = 1$) and the k-instanton configuration, discussed in chapter 3, is again given by (10.42), where the N components of the vector f can be taken in the form:

$$f_i \;=\; (x_+ - A_1^i)\,(x_+ - A_2^i)\ldots(x_+ - A_k^i), \tag{10.52}$$

where $i = 1, \ldots, N$, and where we have also chosen all the parameters λ_i equal to 1.

The expressions for the kinetic energy and its density take the same form as in the $\mathbb{C}P^1$ case, except that, when written in terms of f_i, the expression for κ now takes the form:

$$\begin{aligned}\kappa &= \frac{1}{|f|^4}\sum_{i>j}|f_i\,\partial_t f_j - f_j\,\partial_t f_i|^2 \\ &= \frac{1}{|f|^4}\sum_{i>j}\left(\left|\sum_{l=1}^{k}\left\{\dot{A}_l^i\frac{\partial}{\partial A_l^j} - \dot{A}_l^i\frac{\partial}{\partial A_l^i}\right\}\right|^2 |f_i|^2|f_j|^2\right),\end{aligned} \tag{10.53}$$

where $|f|^2 = \sum_{i=1}^N |f_i|^2$ and $\dot{A}_l^j = \mathrm{d}A_l^j/\mathrm{d}t$. This result is a direct consequence of the instanton equation $D_- z = 0$ and the fact that all parameters in f_i are mutually independent.

If we now generalise the compact notation of the previous section, and introduce

$$\{A_\alpha\} = \{A_1^1,\ldots,A_N^1,A_1^2,\ldots,A_N^2,\ldots,A_1^k,\ldots,A_N^k\}, \tag{10.54}$$

we find that the previous expressions for κ (10.49) and the metric $g_{\alpha\bar{\beta}}$ (10.50) still hold.

For the global metric, in order that it does not diverge, we have to impose the constraints

$$\frac{\mathrm{d}}{\mathrm{d}t}\left(\sum_{l=1}^{k}(A_l^i - A_l^j)\right) = 0 \quad \text{for all } i \neq j, \tag{10.55}$$

which are the obvious generalisation of (10.46).

If we now introduce the integrated metric (taking proper care to avoid divergent integrals),

$$G_{\alpha\bar{\beta}} = \int \mathrm{d}^2x\; g_{\alpha\bar{\beta}}, \tag{10.56}$$

we can write down the equations of motion governing the evolution of the quark and antiquark parameters. They are given by

$$\ddot{A}^\alpha + G^{\alpha\bar{\beta}}\,\frac{\partial G_{\epsilon\bar{\beta}}}{\partial A^\gamma}\,\dot{A}^\epsilon\,\dot{A}^\gamma = 0. \tag{10.57}$$

Solutions to these equations, with proper account taken of the constraints, are precisely geodesic motions on the manifold of skyrmions parameters.

What can we say about these geodesics? Clearly, from (10.57) we can show, in a standard way, that there is always one constant of motion:

$$G_{\alpha\bar{\beta}}\,\dot{A}^\alpha\,\dot{\bar{A}}^\beta = \text{constant}. \tag{10.58}$$

Moreover, we can also show that the equation of motion of the centre of mass separates as a free motion. To show this we introduce the vector

$$P_{\bar{\beta}} = \sum_{\alpha} G_{\alpha\bar{\beta}}, \tag{10.59}$$

which can be expressed in terms of the centre of mass position

$$R = \frac{1}{kN} \sum_{\alpha} A^{\alpha} \tag{10.60}$$

as

$$P_{\bar{\beta}} = \frac{1}{kN} \frac{1}{\partial R} \frac{\partial}{\partial \bar{A}^{\beta}} \int \mathrm{d}^2x \log |f|^2. \tag{10.61}$$

The change of variables of integration in (10.61)

$$x_+ \to x_+ - R \tag{10.62}$$

would, naively, seem to make all R-dependence disappear, and one might be tempted to conclude that $P_{\bar{\beta}} = 0$. However, this conclusion would be incorrect, as the integral for $P_{\bar{\beta}}$ diverges and so requires regularisation. If we regularise it by introducing a cut-off $\Lambda > |x|^2$, then our change of variables (10.62) gives rise to an R-dependence through the integration range ($|x_+ - R|^2 < \Lambda^2$). However, as

$$\begin{aligned} \frac{\partial}{\partial R} \int_{|x|^2<\Lambda^2} \mathrm{d}^2x \log |f|^2 &= - \int_{|x|^2<\Lambda^2} \mathrm{d}^2x \frac{\partial}{\partial x_+} \log |f|^2 \\ &= -\frac{i}{2} \oint_{|x|=\Lambda} \mathrm{d}x_- \log |f|^2, \end{aligned} \tag{10.63}$$

it is clear that we need only know the behaviour of the integrand on the circle $x_+ = \Lambda\, e^{i\phi}$ for large Λ. As for large $|x_+|^2$ we have

$$|f|^2 = N|x_+|^{2k} + \dots, \tag{10.64}$$

we see that

$$\begin{aligned} &\lim_{\Lambda\to\infty} -\frac{i}{2} \oint_{|x_+|^2=\Lambda^2} \mathrm{d}x_- \log |f|^2 \\ &= -\tfrac{1}{2} \lim_{\Lambda\to\infty} \int_0^{2\pi} \mathrm{d}\phi\, e^{-i\phi}\, \Lambda \log \left(2\Lambda^2 - \Lambda e^{i\phi} \sum_{\alpha} \bar{A}^{\alpha} - \Lambda e^{-i\phi} \sum_{\alpha} A^{\alpha} + \dots \right) \\ &= \tfrac{1}{2} 2\pi \sum_{\alpha} \bar{A}^{\alpha} + O\left(\frac{1}{\Lambda}\right), \end{aligned} \tag{10.65}$$

and so we see that the regularised expression for $P_{\bar{\beta}}$ is

$$P_{\bar{\beta}} = \frac{\pi}{kN} \frac{\partial}{\partial \bar{A}^{\beta}} \sum_{\alpha} \bar{A}^{\alpha} = \frac{\pi}{kN}. \tag{10.66}$$

Now we can use the fact that $\sum_{\alpha} G_{\alpha\bar{\beta}}$ is a constant independent of $\bar{\beta}$, and that $\sum_{\bar{\beta}} G_{\alpha\bar{\beta}}$ is independent of α. Putting this into the equation of motion, we see that, as

$$\sum_{\beta} \frac{\partial G_{\epsilon\bar{\beta}}}{\partial A^{\alpha}} = 0, \tag{10.67}$$

the second term of the left-hand side of the equations of motion vanishes and we are left with

$$\sum_{\alpha} \ddot{A}^{\alpha} = 0, \tag{10.68}$$

which is precisely the free centre of mass motion equation $\ddot{R} = 0$.

Apart from this result not much is known about the motion in general, although some partial results have been obtained by Ward [22].

10.4 Noninstanton Solutions

Next we consider noninstanton solutions. Such field configurations were discussed in chapter 3, where we showed that they can be interpreted as describing a system of instantons and anti-instantons, which is, however, unstable, as there exist directions in the function space along which the action decreases. The number of such directions is quite small, and it would seem reasonable to consider these field configurations as a model of a system of skyrmions and antiskyrmions.

The simplest solution of this type is given by

$$z_{\alpha} = \frac{P_{+} f_{\alpha}}{|P_{+} f_{\alpha}|}. \tag{10.69}$$

A few lines of algebra then show that the kinetic energy density κ is given by

$$\begin{aligned} \kappa = & \frac{1}{|P_{+} f|^4} \left[(P_{+} f^{\dagger} \cdot P_{+} f)(\partial_t P_{+} f^{\dagger} \cdot \partial_t P_{+} f) \right. \\ & \left. - (\partial_t P_{+} f^{\dagger} \cdot P_{+} f)(P_{+} f^{\dagger} \cdot \partial_t P_{+} f) \right]. \end{aligned} \tag{10.70}$$

If we now make use of the chain rule, and use various identities involving $P_{+} f$, $P_{+} f^{\dagger}$, and their derivatives, and introduce the compact notation $\{A_{\alpha}\}$ for the parameters of f_{α}, as in the previous sections, we find that κ is given by

$$\kappa = g_{\alpha\bar{\beta}} \dot{A}^{\alpha} \dot{\bar{A}}^{\beta}, \tag{10.71}$$

where

$$\begin{aligned} g_{\alpha\bar{\beta}} =& \frac{\partial}{\partial A^\alpha}\frac{\partial}{\partial \bar{A}^\alpha}\log|P_+f|^2 \\ &+ \frac{2}{|P_+f|^2}\Big(\frac{\partial}{\partial A^\alpha}P_+f^\dagger\cdot\frac{\partial}{\partial \bar{A}^\beta}P_+f\Big). \end{aligned} \tag{10.72}$$

We see that our metric is certainly hermitian, and, when written in this form, the first term is also obviously Kähler-like; but what about the second term? Is $g_{\alpha\bar{\beta}}$ Kähler-like and if so can it be written in an obviously Kähler form?

Thinking about this naively, if the metric is to be Kähler-like, then the second term in the expression for $g_{\alpha\bar{\beta}}$, which we will call $h_{\alpha\bar{\beta}}$, should be given by

$$h_{\alpha\bar{\beta}} = \frac{\partial}{A^\alpha}\frac{\partial}{\partial\bar{A}^\beta}V \tag{10.73}$$

for some V. If such a potential V exists, then, by symmetry of partial derivations, we have

$$\frac{\partial}{\partial A^\gamma}h_{\alpha\bar{\beta}} = \frac{\partial}{\partial A^\alpha}h_{\gamma\bar{\beta}}. \tag{10.74}$$

On the other hand, if this result does not hold, it would seem that $h_{\alpha\bar{\beta}}$ is not Kähler-like.

It is easy, though tedious, to show that the condition above is not satisfied. In fact it is not satisfied even for the configurations which are known to correspond to a Kähler-like metric [20]. As an example where the result should hold, we take the specific case of one instanton in $\mathbb{C}P^2$, given by

$$f = \begin{pmatrix} x_+ - a \\ x_+ - b \\ x_+ - c \end{pmatrix}. \tag{10.75}$$

Applying the P_+ operator to this solution, we get

$$P_+f = \frac{1}{|f|^2}\begin{pmatrix} (a-b)(x_- - \bar{b}) + (a-c)(x_- - \bar{c}) \\ (b-a)(x_- - \bar{a}) + (b-c)(x_- - \bar{c}) \\ (c-a)(x_- - \bar{a}) + (c-b)(x_- - \bar{b}) \end{pmatrix}, \tag{10.76}$$

which is essentially a one anti-instanton solution of the $\mathbb{C}P^2$ model, i.e. is of the form

$$P_+f = \begin{pmatrix} \lambda(x_- - \bar{a'}) \\ \mu(x_- - \bar{b'}) \\ \nu(x_- - \bar{c'}) \end{pmatrix} \tag{10.77}$$

for some choice of parameters a', b' c' and λ, μ and ν. We know that this solution should define a Kähler metric in the way discussed in the previous sections.

It is easy to check that the condition (10.74) is not satisfied when for the variables $\{A^\alpha\}$ we take a, b and c, but it is satisfied when we consider the primed variables a', b' and c'. As the relations between two choices of variables are nonanalytic, we see that they define different analytic structures on our manifold (of parameters of solutions).

On the other hand, there are examples where this prescription works perfectly well, and the P_+ operator does give us a further solution which defines a Kähler metric; this is true, for example, for the very special configurations describing two instantons and two anti-instantons located at the same point. These solutions are real and can be constructed as follows: we start with one instanton of the $\mathbb{C}P^1$ model, i.e.

$$\begin{aligned} f_1 &= \lambda(x_+ - a) \\ f_2 &= \mu(x_+ - b) \end{aligned} \qquad (10.78)$$

and then, as in chapter 3, using the Pauli matrices σ_i we translate it into its $O(3)$ form by

$$\varphi_i = z^\dagger \sigma_i z, \qquad (10.79)$$

where $i = 1 \ldots 3$ and $z = f/|f|$.

Next, as in chapter 3, we treat φ_i as the components of a $\mathbb{C}P^2$ field, and as we discussed in chapter 3, this field is a solution of the $\mathbb{C}P^2$ model. Morover, applying the P_+ and P_- operators to this solutions gives us two further solutions corresponding respectively to the two anti-instanton and two instanton solutions of this model.

It is easy, although tedious, to check that all these configurations describe a Kähler-like metric, with the analytic structure induced from the $\mathbb{C}P^1$ configuration (10.78), i.e. given by a and b.

So what about the more general solutions? Does there exist a complex structure in terms of which the metric is Kähler? How should we go about finding this structure? Unfortunately, we do not know the answers to these questions. Recently, Pope *et al* [23] presented expressions involving the Riemann curvature tensor, Ricci tensor and some covariant derivatives constructed from a metric which may, in principle, be used to test whether the metrics we have been considering are Kähler or not. They derived expressions which must vanish if a given metric is Kähler. Unfortunately, the expressions found by Pope *et al* are quite complicated, even when the metric is Einsteinian, and it is difficult to see how these tests can be performed in practice (see however [24]).

10.5 The Hopf Term

As is well known, a purely bosonic classical field theory that admits topological solitons may have fermionic characteristics; in fact, already in 1968

Finkelstein and Rubinstein showed both the spin and the Fermi–Dirac statistics of such a theory can arise via the multi-valuedness of the configuration space.

Similar possibilities exist also for the $\mathbb{C}P^1$ model. As we discussed in chapter 3, the field manifold of this model is $\mathbb{C}P^1$, which is isomorphic to S^2, and the configuration space is the space of continuous maps from the base space (which we know is the compactified Euclidean two-dimensional space, that is S^2) to the field manifold. Next we introduce the time parameter t (in the same way as we did this in the last three sections) and consider the time evolution of the approximate time-dependent structures evolving over the time interval $[0,T]$. If the field configurations are identical at $t = 0$ and at $t = T$, then the base space of our model can be thought of as being given by $S^2 \times S^1$, which is locally isomorphic to S^3. Thus our skyrmions can be considered as defining maps $S^3 \rightarrow \mathbb{C}P^1 \sim S^2$. However, as $\Pi_3(S^2) = Z$, we see that our configuration space in this (2+1)-dimensional model is *infinitely* connected, which allows the existence of *fractional* spin and related statistics. In fact, it was Wilczek and Zee [19], who observed that in this case we can add an additional topological term to the usual expression for the action. This extra term, known in the mathematics literature as the Hopf term of the theory, is locally a total divergence, and so does not alter the classical equations of motion. However, as we will see later, it does affect the quantum properties of the model; it turns out that the Hopf term describes the 'spin' properties of the extended structures of the $\mathbb{C}P^1$ model thus showing that, in a model with this term, the skyrmions we have been discussing so far can have *any* spin, whose value is determined by the coefficient in the Hopf term.

The first expression for the Hopf term was given by Wilczek and Zee [19]; it involved the $O(3)$ parametrisation of the model. Here we will follow the work of [20], which uses the equivalent $\mathbb{C}P^1$ notation.

We begin by introducing a topological current J^μ:

$$J^\mu = -\frac{i}{2\pi}\,\epsilon^{\mu\nu\lambda}\,(D_\nu z^\dagger \cdot D_\lambda z), \tag{10.80}$$

where, as in the previous sections, all the Greek indices run over 0, 1 and 2 and $\epsilon^{012} = 1$.

Observe that this current is conserved (due to the antisymmetry of $\epsilon^{\mu\nu\lambda}$), and that its zero component (J^0) is proportional to the topological charge density introduced in chapter 3.

The conservation of the current enables us to define a 'gauge potential' A_μ through the curl equation:

$$J^\mu = \epsilon^{\mu\nu\lambda}\,\partial_\nu\,A_\lambda. \tag{10.81}$$

However, it is clear that for A_μ we can take the composite gauge field discussed in chapter 3 (except that this time the base space is three dimensional), i.e.

$$A_\mu = i\, z^\dagger \cdot \partial_\mu z. \tag{10.82}$$

For the complete action of the model we can now take

$$S = \int \left(D^\mu z^\dagger \cdot D_\mu + \frac{\theta}{2\pi} A_\mu J^\mu \right) \mathrm{d}t\, \mathrm{d}^2 x, \tag{10.83}$$

where θ is an arbitrary constant coefficient.

The θ-term is the Hopf term of the theory

$$H = \int A_\mu J^\mu \,\mathrm{d}t\, \mathrm{d}^2 x, \tag{10.84}$$

and is formally analogous to the Chern–Simons term of gauge theories [25], as

$$\begin{aligned} A_\mu J^\mu &= -\frac{1}{2\pi} A_\mu \epsilon^{\mu\nu\lambda} \partial_\nu A_\lambda \\ &= -\frac{1}{4\pi} \epsilon^{\mu\nu\lambda} A_\mu F_{\nu\lambda}, \end{aligned} \tag{10.85}$$

where $F_{\mu\nu} = \partial_\mu A_\nu - \partial_\nu A_\mu$. Observe that, due to the antisymmetry properties of $\epsilon^{\alpha\beta\delta}$, the Hopf term involves all three derivatives and so its contribution vanishes for static skyrmions.

There are various ways of showing that the integrand of the Hopf term is locally a total derivative. Perhaps the simplest method consists of writing the integrand of H as

$$h = 2\pi\, A_\mu J^\mu = \epsilon^{\mu\nu\lambda} (z^\dagger \cdot \partial_\mu z)(\partial_\nu z^\dagger \cdot \partial_\lambda z), \tag{10.86}$$

and then writing

$$z_1 = y_1 + i y_2 \qquad z_2 = y_3 + i y_4, \tag{10.87}$$

where $\sum_{i=1,\ldots 4} y_i^2 = 1$. Then, as is easy to check, h is in fact given by the expression for the topological charge density of the $O(4)$ σ model in three dimensions, which, as is well known, is locally given by a total divergence. Thus,

$$h = \epsilon^{abcd} \epsilon^{\mu\nu\lambda} y_a \partial_\mu y_b \partial_\nu y_c \partial_\lambda y_d, \tag{10.88}$$

which, in fact, is the $O(4)$ model Wess–Zumino term studied in chapter 8. There we showed explicitly that this term is locally a total divergence.

It is worth pointing out that, in principle, we can consider the contribution (10.86), not only in the $\mathbb{C}P^1$ case, but also for other values of N. However, it is easy to see that, for $N > 1$, (10.86) is not locally a total divergence.

We can now evaluate the contribution of the Hopf term for our $\mathbb{C}P^1$ skyrmions studied in the previous sections. We write $z = f/|f|$, where f has only two components given by polynomials of degree k in x_+ (see (10.43)). Then our integrand becomes

$$h = \frac{i}{|f|^4}\left[(\partial_t f_1 \partial_+ f_2 - \partial_t f_2 \partial_+ f_1)(f_1 \partial_+ f_2 - f_2 \partial_+ f_1)^\dagger - \text{c.c.}\right]. \qquad (10.89)$$

If we now make use of the identity

$$\partial_+\left(\frac{f_1}{\bar{f}_2 |f|^2}\right) = \frac{1}{|f|^4}(f_2 \partial_+ f_1 - f_1 \partial_+ f_2), \qquad (10.90)$$

we can rewrite h as

$$h = i\left[\partial_-\left((\partial_t f_2 \partial_+ f_1 - \partial_t f_1 \partial_+ f_2)\frac{\bar{f}_1}{f_2 |f|^2}\right) - \text{c.c.}\right]. \qquad (10.91)$$

This form of h allows us to use the divergence theorem in the complex plane to rewrite H as a line integral:

$$H = \frac{1}{2\pi}\int_0^T \frac{1}{2}\left(\oint (\partial_t f_2 \partial_+ f_1 - \partial_t f_1 \partial_+ f_2)\frac{\bar{f}_1}{f_2 |f|^2}\, \mathrm{d}x_+ + \text{c.c.}\right) \mathrm{d}t, \qquad (10.92)$$

where the x_+ contour in the line integral is given by the circle at infinity, and also around all singularities of the integrand.

Inserting the explicit forms of f_1 and f_2 (i.e. given by (10.43)), in which, following our discussion of the kinetic term, we have set $\lambda_2 = 1$ and λ_1=constant, we see that the integrand becomes

$$\left(\sum_{i,j=1}^{k} \frac{\dot{a}_i - \dot{b}_j}{(x_+ - a_i)(x_+ - b_j)}\right)\frac{|f_1|^2}{|f|^2}, \qquad (10.93)$$

where as before the dots denote time derivatives. We can now use the calculus of residues to evaluate our integral (10.92); from the first term of (10.93) we have contributions from the poles at $x_+ = b_j$, from the second term the only contribution comes from the contour at infinity. The final result is therefore given by

$$H = \frac{i}{2}\int_0^T \left(-\sum_{i,j=1}^{k} \frac{\dot{b}_j - \dot{a}_i}{b_j - a_i} - \text{c.c.}\right) \mathrm{d}t. \qquad (10.94)$$

Note that this expression involves line integrals in the complex plane of the relative vectors $b_j - a_i$. For the case of the initial skyrmion configuration at $t = 0$ coinciding with the final configuration at $t = T$ (i.e. the set $\{a_i(T)\}$ is some permutation of the set $\{a_i(0)\}$ and similarly for b_j), the t integral reduces to a closed contour integral, and the value of H is therefore always an integer multiple of 2π. To be specific, let us investigate the case of a one-skyrmion configuration $f = (x_+ - a), \lambda(x_+ - b)$, and consider a 2π rotation

$$a(t) = a\,e^{it}, \qquad b(t) = b\,e^{it}, \quad t \in [0, 2\pi]. \tag{10.95}$$

Then

$$H = -\frac{i}{2}\int_0^{2\pi} (i + i)\,\mathrm{d}t = 2\pi, \tag{10.96}$$

since the relative vector $b(t) - a(t)$ makes one revolution around the origin.

Of course remembering our constraint (10.46), we see that we cannot consider just one skyrmion (which is infinitely heavy), but should consider the relative motion of the constituents of multi-skyrmion systems. However, the discussion given above easily generalises to more complicated systems. Thus we see that, in general, the value of H simply depends on how many times the closed contours of the relative position vectors of the quark and anti-quark constituents of our skyrmions revolve around the origin. As a further example, let us consider a two-skyrmion configuration given by

$$f = ((x_+ - a_1)(x_+ - a_2), \lambda(x_+ - b_1)(x_+ - b_2)), \tag{10.97}$$

in which we consider the exchange of a_1 and a_2.

The expression for H is now given by

$$H = -\frac{i}{4}\int_0^T \mathrm{d}t\left(\frac{\dot a_1}{a_1 - b_1} + \frac{\dot a_2}{a_2 - b_1} + \frac{\dot a_1}{a_1 - b_2} + \frac{\dot a_2}{a_2 - b_2} - \text{c.c.}\right). \tag{10.98}$$

The terms $\dot a_1/(a_1 - b_1)$ and $\dot a_2/(a_2 - b_2)$ combine to give a single closed contour integral and, similarly, the other two terms. The value of H now depends on whether b_1 and b_2 lie within the closed contour traced out by a_1 and a_2, and, if this is the case, how many times this contour revolves. In the case when, say, only b_1 lies inside the contour which revolves once, we find $H = 2\pi$. This value is the same as for the rotation of a single skyrmion considered before, and we see that, as may be anticipated from a discussion of statistics, the exchange operation and the 2π rotation belong to the same connectivity class. This is an explicit realisation of the ideas of Wilczek and Zee [19].

The effect of the Hopf term corresponding to a certain transformation on the parameters of a k-skyrmion wavefunction is simply, as can be seen from the expression of the action, to multiply it by a factor $\exp(i\theta H/2\pi)$. Thus, if we perform a 2π rotation or an exchange as we have just done, this factor is just $\exp(i\theta)$, which, for the choice of the parameter $\theta = 0$, becomes

+1, and so the skyrmion is quantised as a boson, and which, for $\theta = \pi$, becomes -1, i.e. the skyrmion behaves like a fermion. For other values of θ the skyrmions behave as if they had fractional spin; in consequence their statistics properties are also quite complicated.

The idea that the Hopf term can be responsible for the noninteger value of spin of skyrmions has been also studied by Bowick *et al* [26], who looked at this problem within the context of canonical quantisation. In this interesting paper the authors managed to show that, due to the Hopf term, solitons characterised by the topological charge Q have spin $(\theta/2\pi)Q^2$, where θ is the coefficient of the Hopf term. To demonstrate this, the authors first showed that the usual definition of the angular momentum operator

$$J = \epsilon_{ij} \int \mathrm{d}^2x\, x^i\, T^{0j}, \tag{10.99}$$

where $T^{\mu\nu}$ is the symmetric energy–momentum tensor, agrees with the definition of angular momentum as the commutator of the Lorentz boost generators. Even though $T^{\mu\nu}$ is unmodified by the inclusion of the Hopf term, the canonical momentum conjugate to ϕ^a, where ϕ^a is the $O(3)$ model field, is, and in consequence,

$$J = \epsilon_{ij} \int \mathrm{d}^2x\, x^i\, \pi^a(x)\, \partial^j \phi^a(x) + \frac{\theta}{2\pi} Q^2. \tag{10.100}$$

Returning to our previous discussion, we observe that the total Lagrangian that describes the dynamics of the skyrmion consituents can be written as

$$L = G_{\alpha\bar{\beta}}\, \dot{A}^\alpha \dot{\bar{A}}^\beta - \frac{i\theta'}{2} \frac{\mathrm{d}}{\mathrm{d}t} \sum_{i,j} \log \frac{a_i - b_j}{\bar{a}_i - \bar{b}_j}, \tag{10.101}$$

where, as before, the additional constraint $\sum_i (a_i - b_i) =$constant must be imposed, and where θ' is an angular variable. This system has been studied by many people. In particular Arovas *et al* [27] looked in detail at the quantum statistical properties of a system of particles which have fractional spin and so obey nonstandard statistics. Among many interesting results, they found that, treated as a function of θ', the second virial coefficient has cusps at the Bose statistics points. In addition, Wu [28] presented an interesting discussion of the quantisation of systems with fractional statistics, and showed that one can reformulate the interactions of systems with fractional statistics and multi-valuedness of the wavefunction as systems with single-valued wavefunctions which however exhibit very complicated long range interactions. Moreover, it has been even suggested [29] that sigma models with the Hopf term can provide a description of the fractional Hall effect. Clearly, further applications may still be found.

Let us finish by pointing out that, in a recent paper, Polyakov [30] presented arguments which suggest that, for $\theta = \pi$, the Hopf term is not only responsible for the fractional spin of skyrmions but that, in fact, because of its presence, even the basic fields, the z fields of the $\mathbb{C}P^1$ model, have dramatically altered properties. Their propagators are so modified, that they behave as bosons at large momenta but as fermions at small momenta. Moreover, he suggests that, had we had charged fermions in our basic Lagrangian, they would transmute (at large momenta) into pairs of bosons with spins one and zero.

The suggestions of Polyakov are very exciting, but as his paper [30] is very succinct, and only contains a sketch of the proof of his claim, more work will be required to establish the domain of validity of his suggestion. In fact, in a recent paper, Tze [31] discusses Polyakov's claim in detail and shows what assumptions he has to make to rederive Polyakov's result. Moreover, he also discusses comparable effects in higher dimensional theories.

In addition, Polyakov, in collaboration with Dzyaloshinskii and Wiegmann [32], suggested that neutral fermions which result from such a transmutation of bosonic degrees of freedom may be useful in the description of paramagnetic insulators, and may even help to understand high-temperature superconductivity. Clearly, much more will be said on these, and other related topics, in the next few years.

References

[1] Cheng Ta-Pej and Li Ling-Fong 1984 *Gauge Theory of Elementary Particle Physics* (Oxford: Oxford University Press)

[2] 't Hooft G 1974 A planar diagram theory for strong interactions *Nucl. Phys.* B **72** 461; 1974 A two dimensional model for mesons *Nucl. Phys.* B **75** 461

[3] Witten E 1979 Baryons in the 1/N expansion *Nucl. Phys.* B **160** 57

[4] Skyrme T H R 1961 A non-linear field theory *Proc. R. Soc.* A **260** 127

[5] Finkelstein D and Rubinstein J 1968 Connection between spin, statistics, and kinks *J. Math. Phys.* **9** 1762

[6] Balachandran A P, Nair V P, Rajeev S G and Stern A 1982 Exotic levels from topology in the quantum-chromodynamic effective Lagrangian *Phys. Rev. Lett.* **49** 1124; 1983 Soliton states in the QCD effective Lagrangian *Phys. Rev.* D **27** 1153

Witten E 1983 Current algebra, baryons and quark confinement *Nucl. Phys.* B **223** 433

Pak N K and Tze H C 1979 Chiral solitons and current algebra *Ann. Phys., NY* **117** 164

[7] Adkins G S, Nappi C R and Witten E 1983 Static properties of nucleons in the skyrme model *Nucl. Phys.* B **228** 552

[8] Karliner M 1987 Baryon resonances without quarks: a chiral soliton perspective *Skyrmions and Anomalies* ed M Jezabek and M Praszalowicz (Singapore: World Scientific) p 164

Tze H C 1987 Skyrmions and current algebra *Chiral Soliton* ed K F Liu (Singapore: World Scientific)

[9] Brodsky S J, Ellis J and Karliner M 1988 Chiral symmetry and the spin of the proton *Phys. Lett.* **206B** 309

[10] Adkins G S and Nappi C R 1985 Model independent relations for baryons as solitons in mesonic theories *Nucl. Phys.* B **249** 507

[11] Witten E 1983 Global aspects of current algebra *Nucl. Phys.* B **223** 422

[12] Rabinovici E, Schwimmer A and Yankielowicz S 1984 Quantization in the presence of Wess–Zumino terms *Nucl. Phys.* B **248** 523

Balachandran A P, Lizzi F, Rodgers V G J and Stern A 1985 Dibaryons as chiral solitons *Nucl. Phys.* B **256** 525

[13] Date C D, Frishman Y and Sonnenschein J 1987 The spectrum of multi-flavour QCD in two dimensions *Nucl. Phys.* B **283** 365

Frishman Y and Sonnenschein J 1987 Bosonisation of colored flavour fermions and QCD *Nucl. Phys.* B **294** 801

[14] Brihaye Y, Pak N K and Rossi P 1985 Vector mesons within the effective Lagrangian approach *Nucl. Phys.* B **254** 71

[15] Abud M, Maiella G, Nicodemi F, Pettorino R and Yoshida K 1985 Hidden symmetry and skyrmions in the effective Lagrangian approach to QCD *Phys. Lett.* **159B** 155

[16] Aitchison I R, Fraser C and Piron P J 1986 The effective Lagrangian for skyrmion physics *Phys. Rev.* D **33** 1994

[17] Manton N S 1982 A remark on the scattering of BPS monopoles *Phys. Lett.* **110B** 54

[18] Atiyah M F and Hitchin N J 1985 Low energy scattering of non-Abelian monopoles *Phys. Lett.* **107A** 21

Gibbons G W and Manton N S 1986 Classical and quantum dynamics of BPS monopoles *Nucl. Phys.* B **274** 183

[19] Wilczek F and Zee A 1983 Linking numbers, spin, and statistics of solitons *Phys. Rev. Lett.* **51** 2250

[20] Din A M and Zakrzewski W J 1984 Spin and statistics of $\mathbb{C}P^1$ skyrmions *Phys. Lett.* **146B** 341; 1985 Skyrmion dynamics in (2+1) dimensions *Nucl. Phys.* B **259** 667

Wu Y-S and Zee A 1984 Comment of the Hopf Lagrangian and fractional statistics of solitons *Phys. Lett.* **147B** 325

Stokoe I and Zakrzewski W J 1987 Dynamics of solutions of the $\mathbb{C}P^{N-1}$ model in (2+1) dimensions *Z. Phys.* C **34** 491

[21] Fateev V, Frolov I and Schwartz A 1979 Quantum fluctuations of instantons in the non-linear σ model *Nucl. Phys.* B **154** 1

[22] Ward R S 1985 Slowly moving lumps in the $\mathbb{C}P^1$ model in (2+1) dimensions *Phys. Lett.* **158B** 424

[23] Pope C N, Sohnius M F and Stelle K S 1987 Counterterm counterexamples *Nucl. Phys.* B **283** 192

[24] Stokoe I 1987 *PhD thesis* University of Durham

[25] Deser S, Jackiw R and Templeton S 1982 Topologically massive gauge theories *Ann. Phys., NY* **140** 374

[26] Bowick M J, Karabali D and Wijewardhana L C R 1986 Fractional spin via canonical quantisation of the $O(3)$ nonlinear sigma model *Nucl. Phys.* B **271** 417

[27] Arovas D P, Schrieffer J R, Wilczek F and Zee A 1985 *Nucl. Phys.* B **251** 117

Arovas D P, Schrieffer J R and Wilczek F 1984 Fractional statistics and the quantum Hall effect *Phys. Rev. Lett.* **53** 722

[28] Wu Y-S 1984 General theory for quantum statistics in two dimensions *Phys. Rev. Lett.* **52** 2103

[29] Levine H, Libby S B and Pruisken A M M 194 Theory of the quantized Hall effect I, II and III *Nucl. Phys.* B **240** 30, 49, 71

Tao R and Wu Y-S 1984 Fractional statistics and fractional quantized Hall effect *Seattle Preprint*

Affleck I 1986 Exact critical exponents for quantum spin chains, non-linear σ models at $\theta = \pi$ and the quantum Hall effect *Nucl. Phys.* B **265** 409

[30] Polyakov A M 1988 Fermi–Bose transmutations induced by gauge fields *Mod. Phys. Lett.* A **3** 325

[31] Tze C H 1988 Manifold splitting regularisation, self linking, twisting, writhing numbers of spacetime ribbons and Polyakov's Fermi–Bose transmutations *Int. J. Mod. Phys.* A **3** 1959

[32] Dzyaloshinskii I E, Polyakov A M and Wiegmann P B 1988 Neutral fermions in paramagnetic insulators *Landau Institute preprint*

Wiegmann P B 1988 Superconductivity in strongly correlated electronic systems and confinement versus deconfinement phenomena *Phys. Rev. Lett.* **60** 821

11 Bosonisation and Conserved Currents

11.1 Brief Introduction

One of the most amazing aspects of (1+1)-dimensional mathematical physics is the existence of a nonlocal transformation which allows one to reexpress all the observable quantities calculated in a purely bosonic theory in terms of quantities calculated in a purely fermionic theory, thus showing that the two theories are equivalent. This equivalence is sometimes referred to as the 'bosonisation' or the 'fermionisation' of the corresponding theory, and a standard discussion usually provides 'a dictionary'; i.e. a set of rules that specify which operatorial expressions in one theory are equivalent to which expressions in the other theory.

In this chapter we return to two dimensions and discuss some further properties of two-dimensional σ models formulated in Minkowski space. We begin with a discussion of the above mentioned equivalence.

The first example of such an equivalence was found more than twenty years ago, when Coleman and Mandelstam [1] showed how to reexpress all the variables of the massive Thirring model in terms of quantities constructed out of the sine-Gordon model; they also established their quantum equivalence. At first it appeared that this equivalence might hold only for these two models, but then in 1984 Witten [2] generalised this construction and showed that such an equivalence is not an isolated phenomenon but holds for large classes of theories. In particular, he suggested that the equivalence found by Coleman and Mandelstam can be interpreted as an abelian bosonisation, and in his paper [2] he showed how to generalise this equivalence to the nonabelian case.

The discussion of the nonabelian case is, in fact, quite complicated and it brings back the Wess–Zumino term discussed in the previous chapters. Witten's discussion showed that the natural setting for the process of nonabelian bosonisation involves the sigma fields of our chiral models with the Wess–Zumino term in which the coefficient of the Wess–Zumino term takes a special value. Then these sigma fields are equivalent to the free fermion fields associated with the same symmetry group as the original sigma fields. Since the appearance of the paper by Witten, the concept of bosonisation has been the subject of many papers and the literature on this topic is already quite extensive [3].

In this chapter we want to recall some aspects of the construction of Witten following closely Witten's original paper [2] and the excellent review of Goddard and Olive [4]. We finish this chapter with a brief discussion of the existence of conserved local and nonlocal currents in some sigma models.

11.2 σ Models with the Wess–Zumino Term

We now turn to the two-dimensional Minkowski space where our variables are t and x, which we will also call x_0 and x_1 respectively. By analogy with the Euclidean case, we introduce x_+ and x_-, which are now light-cone variables and so are given by $x_+ = x_0 + x_1$ and $x_- = x_0 - x_1$. The action of the $O(N)$ sigma model with the Wess–Zumino term is given by

$$S = S_0 + S_{\mathrm{WZ}}, \tag{11.1}$$

where S_0 is given by

$$S_0 = \frac{1}{4\alpha^2}\int \mathrm{d}^2x \,\mathrm{Tr}\,(\partial^\mu g^\dagger\, \partial_\mu g) \tag{11.2}$$

and

$$S_{\mathrm{WZ}} = \frac{\lambda}{24\pi}\int \mathrm{d}^3x\, \epsilon^{\mu\nu\alpha}\,\mathrm{Tr}\,(g^\dagger \partial_\mu g\, g^\dagger \partial_\nu g\, g^\dagger \partial_\alpha g), \tag{11.3}$$

where $g^\dagger g = gg^\dagger = 1$, and where λ is real.

Strictly speaking the Wess–Zumino term is properly defined only in Euclidean space; we define it in Minkowski space by analytical continuation. The precise definition of the extension of g from being a function of two variables to three variables was given in chapter 8. Moreover, we showed there that, if we want the quantum version of the theory to be independent of the choice of the integration region in (11.3), the coefficient λ must be an integer.

By analogy with the discussion in chapter 8, we calculate the equation of motion corresponding to our Lagrangian, and find that they do not depend on the extension of g and are given by

$$\partial_\mu\left(g^\dagger \partial^\mu g - \frac{\lambda\alpha^2}{4\pi}\,\epsilon^{\mu\nu}\, g^\dagger \partial_\nu g\right) = 0, \tag{11.4}$$

where $\epsilon^{\mu\nu} = -\epsilon^{\nu\mu}$, $\epsilon^{01} = 1$.

Introducing $\partial_\pm = \frac{1}{2}(\partial_0 \pm \partial_1)$ we rewrite (11.4) as

$$\left(\frac{1}{2\alpha^2} + \frac{\lambda}{8\pi}\right) \partial_-(g^\dagger \partial_+ g) + \left(\frac{1}{2\alpha^2} - \frac{\lambda}{8\pi}\right) \partial_+(g^\dagger \partial_- g) = 0 \tag{11.5}$$

and observe that, if $\alpha^2 = 4\pi/\lambda$, the equations of motion reduce to

$$\partial_- (g^\dagger \partial_+ g) = 0, \tag{11.6}$$

while for $\alpha^2 = -4\pi/\lambda$ they give

$$\partial_+ (g^\dagger \partial_- g) = 0. \tag{11.7}$$

Thus, for the first choice, we can define

$$J_+ = g^\dagger \partial_+ g \quad \text{and} \quad J_- = \partial_- g g^\dagger \tag{11.8}$$

and find that they are conserved, i.e. that they satisfy

$$\partial_- J_+ = \partial_+ J_- = 0, \tag{11.9}$$

while for the second choice the conserved currents are given by

$$J_+ = \partial_+ g g^\dagger \quad \text{and} \quad J_- = g^\dagger \partial_- g. \tag{11.10}$$

It is easy to find solutions to the equations of motion for these special values of α^2. Thus, as Witten pointed out [2], when $\alpha^2 = 4\pi/\lambda$, the general solution is given by

$$g(x_+, x_-) = G(x_-)H(x_+), \tag{11.11}$$

where $G(x_-)$ and $H(x_+)$ are arbitrary $O(N)$ valued functions of one variable. For the other choice of α^2, the solution is again given in a factorised form with the order of matrices G and H reversed.

The factorisation property (11.11) of the solution shows that for these special values of α^2, i.e. for $\alpha^2 = \pm 4\pi/\lambda$, left-moving and right-moving waves are completely independent, and so pass through each other without any interference. As we will show later, this property is reminiscent of the free fermion theory in which the left- and right-moving waves are also independent, since they are eigenstates of the γ_5 matrix. Of course, this property is a necessary condition for the boson–fermion equivalence to hold.

Thus far our arguments have been purely classical, so what about quantum corrections? The only parameter altered by such corrections is α, which requires renormalisation (λ being an integer should not and does not undergo any renormalisation). The renormalisation of α depends on both λ and α. In his paper [2] Witten calculated the one-loop beta function for α, and showed that it vanished at $\alpha^2 = \pm 4\pi/\lambda$. He then argued that the complete beta function also had a zero at this point.

To prove the equivalence of our bosonic theory to a free fermion theory, we will look at our conserved currents $J_\pm$ and check what algebra they satisfy. To do this, we calculate their Poisson brackets, as then, in quantum theory, these Poisson brackets are replaced by commutators. At first sight, due to the complexity of the Wess–Zumino term, this calculation appears to be very complicated but, as Witten [2] pointed out, it can be carried out with relative ease at the critical coupling $\alpha^2 = \pm 4\pi/\lambda$.

First of all, due to the current conservation, J_+ is a function only of x_+ and J_- only of x_-. Hence their Poisson bracket vanishes. Thus, it is sufficient to calculate the Poisson bracket of J_+ with itself; the remaining Poisson bracket can be deduced from the symmetry under $x \leftrightarrow -x$ and $g \leftrightarrow g^\dagger$. Following Witten [2], we observe that to calculate this Poisson bracket we can use our light-cone variables and treat x_- and x_+ as new 'space' and 'time' variables respectively. Then, rewriting the complete action S in terms of these variables, we observe that, although the resultant expression is complicated, the integrands in S_0 and in S_{WZ} are first order in 'time' derivatives. This, as Witten has pointed out [2], allows us to compute Poisson brackets without making an explicit choice of the dynamical coordinates and momenta, which, in our case, would be very awkward, due to the nonlinearity of the phase space.

In fact, Witten showed that, in a general case of a theory with dynamical variables ϕ^i, whose action is first order in time derivatives and so is given by

$$S = \int \mathrm{d}t\, A_i(\phi)\, \frac{\mathrm{d}\phi^i}{\mathrm{d}t}, \tag{11.12}$$

the Poisson bracket of any two functions X and Y defined on a phase space is given by

$$[X\,, Y]_{\text{PB}} = \sum_{i,j} F^{ij} \frac{\partial X}{\partial \phi^i} \frac{\partial Y}{\partial \phi^j}, \tag{11.13}$$

where F^{ij} is the inverse matrix of $F_{ij} = \partial_i A_j - \partial_j A_i$. Hence, the problem reduces to the calculation of the matrix F^{ij}. However, as F_{ij} is the coefficient of $\delta\phi^i \mathrm{d}\phi^j/\mathrm{d}t$ in the calculation of the change in S under an arbitrary infinitesimal variation $\phi^i \to \phi^i + \delta\phi^i$, it is sufficient to know the matrices F^{ij} and F_{ik} relative to any basis of tangent vectors to the phase space and in our case a convenient basis is provided by the matrices $g^\dagger\, \delta g(x_-)$. Of course, F and F^{-1} act both on the Lie algebra index of these matrices and on x_-.

Choosing $\alpha^2 = 4\pi/\lambda$, we find that the variation of the action S is given by

$$\delta S = \frac{\lambda}{4\pi}\int \mathrm{d}x_-\mathrm{d}x_+ \,\mathrm{Tr}\left\{g^\dagger\delta g\,\frac{\mathrm{d}g^\dagger}{\mathrm{d}x_-}\frac{\mathrm{d}g}{\mathrm{d}x_+}\right\}. \tag{11.14}$$

Hence, we see that our matrix F is given by a tensor product of an identity which acts on the Lie algebra index and of $(\lambda/4\pi)(\mathrm{d}/\mathrm{d}x_-)$, which acts on the spatial variable x_-. For the inverse matrix, F^{-1}, the operator acting on the spatial part is, of course, given by $(4\pi/\lambda)(\mathrm{d}/\mathrm{d}x_-)^{-1}$.

To calculate the Poisson brackets of our currents, first of all, we consider X and Y given by $X = \mathrm{Tr}\{A(\mathrm{d}g/\mathrm{d}x_-)g^\dagger(x_-)\}$ and $Y = \mathrm{Tr}\{B(\mathrm{d}g/\mathrm{d}x_-)g^\dagger(x'_-)\}$, where A and B are Lie-valued constant matrices. Then, as was shown by Witten [2], to calculate the Poisson bracket of X and Y it suffices to consider the product of variations of X and Y, i.e. $\delta X\delta Y$ and then within it replace $\delta\phi^i\delta\phi^j$ by F^{ij}. In our case, the role of $\delta\phi^i$ is played by $(g^\dagger\delta g(x_-))^a$, where a denotes the Lie algebra index. Thus, we see that

$$\begin{aligned}\delta X\delta Y &= \mathrm{Tr}\left\{g^\dagger(x_-)Ag(x_-)\right.\\ &\quad\left.\times\frac{\mathrm{d}}{\mathrm{d}x_-}(g^\dagger\delta g(x_-))\right\}\cdot\mathrm{Tr}\left\{g^\dagger(x'_-)Bg(x'_-)\frac{\mathrm{d}}{\mathrm{d}x_-}(g^\dagger\delta g(x'_-))\right\}\end{aligned} \tag{11.15}$$

and so, given our form of the F matrix, to calculate the Poisson bracket, we replace $(g^\dagger\delta g(x_-))^a\cdot(g^\dagger\delta g(x'_-))^b$ by $(4\pi/\lambda)\delta^{ab}G(x_-,x'_-)$, where $G(x_-,x'_-)$ is an inverse of $\mathrm{d}/\mathrm{d}x_-$. In this way we find that the Poisson bracket of our X and Y is given by

$$\begin{aligned}[X,Y]_{\mathrm{PB}} &= -\frac{4\pi}{\lambda}\delta'(x_- - x'_-)\,\mathrm{Tr}\{g^\dagger(x_-)Ag(x_-)g^\dagger(x'_-)Bg(x'_-)\}\\ &= -\frac{4\pi}{\lambda}\delta(x_- - x'_-)\,\mathrm{Tr}\,[A,B]\frac{\mathrm{d}g}{\mathrm{d}x_-}g^\dagger - \frac{4\pi}{\lambda}\delta'(x_- - x'_-)\,\mathrm{Tr}\,AB.\end{aligned} \tag{11.16}$$

Recalling that J_- is given by $(\lambda/2\pi)(\mathrm{d}g/\mathrm{d}x_-)g^\dagger$ we see that, after quantisation, we have

$$\begin{aligned}[\mathrm{Tr}\,AJ_-(x_-),\mathrm{Tr}\,BJ_-(x'_-)] &= 2i\,\mathrm{Tr}\,([A,B]J_-(x_-))\delta(x_- - x'_-)\\ &\quad+\frac{\lambda}{\pi}i\delta'(x_- - x'_-)\,\mathrm{Tr}\,AB.\end{aligned} \tag{11.17}$$

If we now choose to take for A and B the generators of the Lie algebra of our group we find that we obtain an expression corresponding to a current algebra with a Schwinger term [5].

11.3 Kac–Moody Algebra

Taking for the matrices A and B of the previous section the generators of the group, we rewrite the current algebra derived in the previous section as

$$[J^a(\xi), J^b(\eta)] = if^{abc}J^c(\xi)\delta(\xi-\eta) + \frac{i\lambda}{2\pi}\delta^{ab}\delta'(\xi-\eta), \tag{11.18}$$

where f^{abc} are the structure constants of the group. This algebra of the currents is, in fact, equivalent to one of the Kac–Moody algebras [6], which have recently attracted considerable interest [7]. To introduce Kac–Moody algebras and to establish this equivalence, we impose periodic boundary conditions on our currents with respect to their variables ξ and η (x_- and x'_- of the previous section). Thus, we assume that we can set

$$J^a(\xi) = \frac{1}{L}T^a(z), \tag{11.19}$$

where $z = \exp(2\pi i\xi/L)$, and where L is some unspecified constant which sets the period of T^a.

Next, we perform the Laurent expansion of T^a:

$$T^a(z) = \sum_{n=-\infty}^{\infty} T^a_{-n}\, z^n, \tag{11.20}$$

where the coefficients of the expansion are given by

$$T^a_n = \int_0^L J^a(\xi)\, e^{2\pi i\xi n/L}\, \mathrm{d}\xi, \tag{11.21}$$

or, using the periodicity conditions, by the Cauchy contour integrals:

$$T^a_m = \oint \frac{T^a(\rho)\rho^m}{2\pi i\rho}\, \mathrm{d}\rho. \tag{11.22}$$

If we now multiply both sides of (11.18) by $\exp(2\pi i\xi n/L)$ and $\exp(2\pi i\eta m/L)$, and then integrate over ξ and η, we obtain

$$[T^a_n, T^b_m] = if^{abc}T^c_{m+n} + \lambda n\delta^{ab}\delta_{n+m,0}, \tag{11.23}$$

which is the standard form of the Kac–Moody algebra, as discussed, say, in [4].

Of course, the two expressions for the algebra, (11.23) and (11.24), are completely equivalent. Thus to derive (11.18) from (11.23) we multiply both sides of (11.23) by z^{-n} and ω^{-m}, where $z = \exp(2\pi i\xi/L)$ and $\omega =$

$\exp(2\pi i\eta/L)$, and then sum over all the integer values of n and m. Then, using the property that

$$\sum_{n=-\infty}^{\infty} e^{2\pi i(\xi-\eta)n/L} = L\delta(\xi-\eta), \tag{11.24}$$

and

$$\sum_{m=-\infty}^{\infty} m\, e^{2\pi i(\xi-\eta)m/L} = \frac{L^2}{2\pi i}\delta'(\xi-\eta), \tag{11.25}$$

we recover (11.18).

The Kac–Moody algebra we have briefly discussed here can be put in the deeper mathematical setting of infinite dimensional group theory. All interested readers can find many interesting details in the review article of Goddard and Olive [4], parts of which we follow quite closely in this chapter.

So far we have discussed the Kac–Moody algebras; recently considerable attention has been paid also to the so-called 'Virasoro algebras' [8]. These algebras, which were first studied extensively in the early 1970's, are naturally associated with the Kac–Moody algebras; for each Kac–Moody algebra we can introduce a Virasoro algebra in such a way that they form a semi-direct product. To see how a Virasoro algebra arises in the context of our σ models with the Wess–Zumino term, we look at the canonical classical energy–momentum tensor of our model. As we argued in the last section, if we choose the coupling constant α at one of its special values $\alpha^2 = \pm 4\pi/\lambda$, then the renormalisation beta function vanishes and the model preserves its classical conformal invariance. Thus the energy–momentum tensor, which is traceless at the classical level, remains traceless also after quantisation.

To construct the associated Virasoro algebra we look at the energy–momentum tensor of the theory. We calculate it in the usual way [9], and find that, when $\alpha^2 = 4\pi/\lambda$, it has the current–current, or Sugawara, form [10]:

$$\theta_{\mu\nu} = \sum_a \frac{1}{\lambda}(J^a_\mu J^a_\nu - \tfrac{1}{2} g_{\mu\nu} J^a_\lambda J^{a\lambda}). \tag{11.26}$$

The only nonzero components of this tensor are

$$\theta_{++} = \sum_a \frac{1}{2\lambda} J^a_+ J^a_+, \tag{11.27}$$

and also θ_{--}, which is described by a similar expression involving J^a_-'s. As J^a_+ is a function of only x_+ and J^a_- of only x_-, we see that θ_{++} depends only on x_+ and θ_{--} only on x_-.

In quantum theory both these expressions can be divergent, and so they have to be regularised by some sort of normal ordering procedure. Such a regularisation is possible, as, at our special value of α^2, the currents become free fields which satisfy the commuting Kac–Moody algebras. Following Goddard and Olive [4], we introduce the normal ordering in terms of T^a_m defined in (11.21):

$$\begin{aligned} {}^{\times}_{\times} T^a_m T^a_n {}^{\times}_{\times} &= T^a_m T^a_n, \quad m < 0, \\ &= T^a_n T^a_m, \quad m > 0. \end{aligned} \tag{11.28}$$

Next, we take the moments of θ_{++}, i.e.

$$\frac{1}{2\lambda} \int_0^L J^a_+ J^a_+ \exp\left(\frac{2\pi i \xi p}{L}\right) \mathrm{d}\xi, \tag{11.29}$$

and define $\tilde{L}_p$ operators, which are given by these moments, up to overall $\hbar^2/L$ factors, and so take the form:

$$\tilde{L}_p = \frac{1}{2\lambda} \sum_{-\infty}^{\infty} {}^{\times}_{\times} T^a_n T^a_{p-n} {}^{\times}_{\times}. \tag{11.30}$$

If we now calculate the commutation relations of $\tilde{L}_p$ with T^a_m, we find, using (11.23) that, if we ignore the normal ordering operation, we obtain

$$[\tilde{L}_p, T^a_n] = -nT^a_{p+n}. \tag{11.31}$$

However, the normal ordering spoils this result, and the correct calculation produces an extra term given by

$$\begin{aligned} &\frac{i}{2\lambda} f^{bac}[-T^b_{n-1}T^c_{p+1} - T^b_{n-2}T^c_{p+2} - \cdots - T^b_0 T^c_{n-1} \\ &\qquad + T^c_{n+p}T^b_0 + T^c_{n+p-1}T^b_1 + \cdots + T^c_{p+1}T^b_{p-1}] \\ &= -\frac{i}{2\lambda} n\, f^{bac} f^{cbd}\, T^d_{p+n} = -\frac{i}{2\lambda} \kappa n\, T^a_{p+n}, \end{aligned} \tag{11.32}$$

where κ is defined by $f^{bac} f^{cbd} = \kappa \delta^{ad}$. This shows that, if we take

$$L_p = \frac{\beta}{2\lambda} {}^{\times}_{\times} T^b_n T^b_{p-n} {}^{\times}_{\times}, \tag{11.33}$$

where β is given by

$$\beta = \frac{1}{1 + \kappa/2\lambda}, \tag{11.34}$$

the commutator of this operator with T^a_m is again given by (11.31), i.e. is given by

$$[L_p, T^a_m] = -mT^a_{p+m}. \tag{11.35}$$

Let us now consider the commutator of two L_i operators, i.e. $[L_p, L_m]$. Using $[A, BC] = B[A, C] + [A, B]C$ and (11.35), we find

$$\begin{aligned}[L_p, L_m] &= \left[L_p, \alpha \sum_n {}^{\times}_{\times} T^a_n T^a_{m-n} {}^{\times}_{\times}\right] \\ &= -\alpha \sum_{n<0} (m-n) T^a_n T^a_{p+m-n} - \alpha \sum_{n<0} n T^a_{p+n} T^a_{m-n} \\ &\quad -\alpha \sum_{n\geq 0} n T^a_{m-n} T^a_{p+n} - \alpha \sum_{n\geq 0} (m-n) T^a_{m+p-n} T^a_n ,\end{aligned} \tag{11.36}$$

where $\alpha = \beta/2\lambda$.

Writing out explicitly the terms on the right-hand side of (11.36), we find that they correspond to $(p-m)L_{p+m}$ and some further terms, which are given by

$$\begin{aligned}&\alpha[T^a_{p-1}T^a_{m+1} + 2T^a_{p-2}T^a_{m+2} + \cdots + pT^a_0 T^a_{m+p} \\ &-pT^a_{m+p}T^a_0 - (p-1)T^a_{m+p-1}T^a_1 - \cdots - T^a_{m+1}T^a_{p-1}].\end{aligned} \tag{11.37}$$

However, using (11.23), we see that

$$T^a_{p-i}T^a_{m+i} - T^a_{m+i}T^a_{p-i} = \lambda(p-i)G\delta_{p+m\ 0}, \tag{11.38}$$

where $G = \delta^{aa} = \dim g$. Thus, all the terms in (11.37) are proportional to $G\delta_{p+m\ 0}$, and so, performing their sum, we see that (11.37) is given by $\frac{1}{6}p(p^2-1)\alpha\lambda G$. Putting everything together, we find that the L_p operators satisfy

$$[L_p, L_m] = (p-m)L_{p+m} + \tfrac{1}{12}cp(p^2-1)\delta_{p+m\ 0}, \tag{11.39}$$

where $c = \beta G = 2\lambda \dim g/(2\lambda + \kappa)$.

The derived expression (11.39) is the celebrated Virasoro algebra [8]. We have just shown that the moments of the energy–momentum tensor, when properly normalised, satisfy the Virasoro algebra, and together with the moments of the currents, which satisfy a Kac–Moody algebra, form a semi-direct product and so are related to each other by

$$[L_m, T^a_n] = -nT^a_{m+n}. \tag{11.40}$$

Let us finish this section by pointing out that the L_p operators are, in fact, generators of the conformal symmetry in two dimensions which, as we mentioned in chapter 3, consists of holomorphic transformations

$$x_+ \to \zeta = \zeta(x_+). \tag{11.41}$$

The Virasoro algebra (11.39) is then the centrally extended algebra of this conformal group. To see this we observe that we can define

$$T_\zeta f(x_+) = f(\zeta^{-1}(x_+)), \tag{11.42}$$

for any function f of x_+ into some vector space, and then consider the effects of (11.41) close to the identity

$$\zeta(x_+) = x_+ \exp(-i\epsilon(x_+)), \quad \zeta^{-1}(x_+) \sim x_+ + ix_+\epsilon(x_+). \tag{11.43}$$

Then

$$T_\zeta f(x_+) \sim f(x_+) + ix_+\epsilon(x_+)\frac{\mathrm{d}}{\mathrm{d}x_+}f(x_+), \tag{11.44}$$

and so performing the Laurent expansion of $\epsilon(x_+)$

$$\epsilon(x_+) = \sum_{n=-\infty}^{\infty} \epsilon_{-n}x_+^n, \tag{11.45}$$

we find that the generators of (11.41) are given by

$$L_n = -x_+^{n+1}\frac{\mathrm{d}}{\mathrm{d}x_+}, \tag{11.46}$$

where n can take any integer value. The algebra of these generators satisfies (11.39) without the central term. However, as shown by Goddard and Olive [4], the central extension is then essentially unique (the only arbitrariness is the value of the constant c in (11.39)).

11.4 Free Massless Fermions

Consider now free massless fermions, whose theory, as was shown by Witten [2] is equivalent to our σ model with the Wess–Zumino term. Let us take N noninteracting copies of a free real massless fermion field. Then, each of these fields is described by an anticommuting spinor field ψ. If we now choose a real representation of γ matrices in two dimensions:

$$\gamma_0 = \begin{pmatrix} 0 & 1 \\ 1 & 0 \end{pmatrix}, \quad \gamma_1 = \begin{pmatrix} 0 & -1 \\ 1 & 0 \end{pmatrix}, \quad \gamma_5 = \gamma_0\gamma_1 = \begin{pmatrix} 1 & 0 \\ 0 & -1 \end{pmatrix} \tag{11.47}$$

then, following Witten [2] and Goddard and Olive [4], we can put

$$\psi = \begin{pmatrix} \psi^- \\ \psi^+ \end{pmatrix}, \tag{11.48}$$

where $\psi^{\pm}$ are clearly the eigenvectors of our γ_5 matrix. Notice that our choice of γ matrices is not the Minkowski version of the Euclidean γ matrices used before, and, in consequence, our $\psi_{\pm}$ fields are also different. However, each choice gives eigenvectors which are linear combinations of the eigenvectors of the other choice, so the translation of any results derived using one set to the description of the other set is straightforward. The set we use here is perhaps slightly more convenient when working in Minkowski space and, as we said earlier, is used by Witten and Goddard and Olive. As we follow these papers very closely, and given that these papers contain more details and develop this subject in much more depth, we try to keep the same notation as Goddard and Olive, so as to minimise the additional work for any reader who, having read this chapter, will want to broaden his knowledge by reading [4].

The action of our free fermion fields is given by

$$S_{\mathrm{F}} = \tfrac{1}{2}i \int \mathrm{d}^2x\, \bar{\psi}\gamma^{\mu}\partial_{\mu}\psi, \tag{11.49}$$

where, as usual, $\bar{\psi} = \psi^{\dagger}\gamma_0$. The equation of motion corresponding to this Lagrangian is the massless Dirac equation, which, when written in terms of $\psi^{\pm}$, takes the form

$$\partial_{+}\psi^{-} = \partial_{-}\psi^{+} = 0. \tag{11.50}$$

This means that for the solutions of these equations of motion, ψ^{+} and ψ^{-} are functions only of x_{+} and x_{-} respectively.

As $\psi^{\pm}$ are independent fields, after quantisation they anticommute with each other, while each of them satisfies the canonical anticommutation relations:

$$\{\psi^{\pm}(x), \psi^{\pm}(y)\} = \hbar\,\delta(x-y). \tag{11.51}$$

Thus, restricting ourselves to only one component, say, ψ^{+}, and recalling that we have N noninteracting copies of our field, we see that the anticommutation relations are given by

$$\{\psi_i(x), \psi_j(y)\} = \hbar\delta(x-y)\delta_{ij}. \tag{11.52}$$

To construct a current algebra, and so (as shown in the previous sections) a Kac–Moody algebra, we consider ψ_i to lie in a real representation of a symmetry group g, say $O(N)$, the generators of which are given by $T^a = iM^a$, where M^a are $N \times N$ real antisymmetric matrices satisfying

$$[M^a, M^b] = f^{abc}M^c. \tag{11.53}$$

The group g is clearly a symmetry group of our fermionic model and the associated conserved currents, when written in terms of our light-cone components, are given by

$$J^a_{\pm} = \tfrac{1}{2}i\,\psi^{\dagger}_{\pm}\,M^a\,\psi\pm, \tag{11.54}$$

where now by $\psi_\pm$ we denote a column vector whose components are given by $\psi_i^\pm$. We notice that J_+^a is a function of only x_+ while J_-^a is a function of only x_- and as ψ_+ and ψ_- anticommute, we can study all properties of $J_+^a(x_+)$ and $J_-^a(x_-)$ independently.

Denoting either component by $J^a(x)$, where x stands for the appropriate argument (x_+ for J_+ and x_- for J_-), we will now show that our anticommutation relations (11.52) imply the current algebra (11.18) with a specific value of λ.

To do this, we have to repeat some of the steps followed in the previous sections. First of all, we assume that our currents are periodic:

$$J^a(x) = J^a(x+L). \tag{11.55}$$

This will be achieved if we require that our fermion field $\psi(x)$ is either periodic or antiperiodic. The usual approach corresponds to the requirement of antiperiodicity, but, in fact, the periodic case can be used too. In string theory, where these fields were first defined, the periodic case corresponds to a Ramond field [11] and the antiperiodic one to a Neveu–Schwarz field [12]. We will follow the conventional path and use the antiperiodic fields:

$$\psi(x+L) = -\psi(x). \tag{11.56}$$

As in the case of the σ models with the Wess–Zumino term discussed in the previous sections, where we replaced $J^a(\xi)$ by $T^a(z)$, we also replace $\psi_i(x)$ by a dimensionless field $H_i(z)$, which is related to it by

$$\psi_i(x) = \left(\frac{\hbar}{L}\right)^{1/2} H_i(z), \quad \text{where} \quad z = \exp\left(\frac{2\pi i x}{L}\right). \tag{11.57}$$

Due to our requirement of antiperiodicity (11.56) we see that, if we expand $H_i(z)$ in the Laurent series:

$$H_i(z) = \sum_r b_{-r}^i \, z^r, \tag{11.58}$$

r belongs to the set of all (positive and negative) half-integer values, which we denote by $r \in Z + \frac{1}{2}$.

The canonical anticommutation relations (11.52) are now equivalent to

$$\{b_r^i, b_s^j\} = \delta^{ij}\delta_{r,-s}, \tag{11.59}$$

where r and $s \in Z + \frac{1}{2}$. The b_r^i operators are, of course, the fermionic annihilation and creation operators, and they satisfy the hermiticity condition $b_r^{i\,\dagger} = b_{-r}^i$. We use them to define a Fock space; we apply b_r^i to a vacuum state $|0\rangle$, which is required to satisfy

$$b_r^i|0\rangle = 0, \quad \text{for all} \quad r > 0. \tag{11.60}$$

Next, to avoid divergences, we introduce a normal ordering relative to this vacuum:

$$
\begin{aligned}
{}^\circ_\circ b^i_r b^j_s {}^\circ_\circ &= -b^j_s b^i_r \quad \text{if} \quad r > 0, \\
&= b^i_r b^j_s \quad \text{if} \quad r < 0.
\end{aligned} \tag{11.61}
$$

Of course we have to introduce this normal ordering also in (11.54), and we see that, to study our currents, we have to consider expressions such as ${}^\circ_\circ H_i(z) H_j(y) {}^\circ_\circ$.

Following Goddard and Olive [4], we introduce the Wick [13] contraction function $\Delta(z,\zeta)$ by

$$
H^i(z)H^j(\zeta) = {}^\circ_\circ H^i(z)H^j(\zeta) {}^\circ_\circ + \delta^{ij}\Delta(z,\zeta), \quad |z| > |\zeta|, \tag{11.62}
$$

and evaluate it finding that, as

$$
\delta^{ij}\Delta(z,\zeta) = \sum_k \sum_r \theta(r) \{b^j_k b^i_r\} z^{-r}\zeta^{-k} = \sum_{r>0} \delta^{ij} \left(\frac{\zeta}{z}\right)^r, \tag{11.63}
$$

$\Delta(z,\zeta)$ is given by

$$
\Delta(z,\zeta) = \frac{\sqrt{z\zeta}}{z-\zeta}. \tag{11.64}
$$

Now we are ready to calculate the commutator of the two currents $J^a(z)$ and $J^b(\zeta)$, and check whether it leads to the current algebra (11.18). Clearly, the problem is reduced to the calculation of

$$
\begin{aligned}
&\left[{}^\circ_\circ H_i(z) M^a_{ij} H_j(z) {}^\circ_\circ , {}^\circ_\circ H_k(\zeta) M_{kl} H_l(\zeta) {}^\circ_\circ \right] \\
&= \sum_{rs} \sum_{tu} \left[{}^\circ_\circ b^i_{-r} M^a_{ij} b^j_{-s} {}^\circ_\circ , {}^\circ_\circ b^k_{-t} M^a_{kl} b^l_{-u} {}^\circ_\circ \right] z^{r+s} \zeta^{t+u}.
\end{aligned} \tag{11.65}
$$

However, lengthy calculations involving the normal orderings show that

$$
\begin{aligned}
{}^\circ_\circ b^i_r b^j_s b^k_t b^l_u {}^\circ_\circ &= {}^\circ_\circ b^i_r b^j_s {}^\circ_\circ \, {}^\circ_\circ b^k_t b^l_u {}^\circ_\circ \\
&+ {}^\circ_\circ b^i_r b^k_t {}^\circ_\circ \delta^{lj}\delta_{u+s,0} - {}^\circ_\circ b^i_r b^l_u {}^\circ_\circ \delta^{kj}\delta_{t+u,0} + {}^\circ_\circ b^j_s b^l_u {}^\circ_\circ \delta^{ki}\delta_{r+t,0} \\
&- {}^\circ_\circ b^j_s b^k_t {}^\circ_\circ \delta^{li}\delta_{u+r,0} + \delta^{ik}\delta^{lj}\delta_{u+s,0}\delta_{r+t,0} - \delta^{kj}\delta^{il}\delta_{t+s,0}\delta_{r+u,0}.
\end{aligned} \tag{11.66}
$$

Using this result and using also our definition of $\Delta(z,\zeta)$ (11.62), we can now perform summations over r, z, t and u (with the factors z^{-r-s} and ζ^{-t-u} reinstituted), and find that, in the region $|z| > |\zeta|$, we have

$$
\begin{aligned}
&{}^\circ_\circ H^i(z)H^j(z) {}^\circ_\circ \, {}^\circ_\circ H^k(\zeta)H^l(\zeta) {}^\circ_\circ = {}^\circ_\circ H^i(z)H^j(z)H^k(\zeta)H^l(\zeta) {}^\circ_\circ \\
&+ \Delta(z,\zeta) {}^\circ_\circ \{ H^i(z)H^l(\zeta)\delta^{jk} + H^j(z)H^k(\zeta)\delta^{il} - H^i(z)H^k(\zeta)\delta^{jl} \\
&- H^j(z)H^l(\zeta)\delta^{ik} \} + \Delta(z,\zeta)^2 (\delta^{jk}\delta^{il} - \delta^{jl}\delta^{ik}).
\end{aligned} \tag{11.67}
$$

Although this expression was derived under the assumption that $|z| > |\zeta|$, we can use the right-hand side of (11.67) to extend the range of the definition of our expression by analytical continuation to other values of z and ζ. If we denote our expression by $h^{ijkl}(z,\zeta)$, then we see that it is symmetric under the interchange of pairs of indices ij and kl, provided that we interchange also z and ζ. Hence, if we calculate the moments of our currents (11.22), we find that both $T^a_m T^b_n$ and $T^b_n T^a_m$ can be written as double Cauchy integrals of the same integrand:

$$-\frac{1}{4}\oint \frac{d\zeta\, \zeta^n}{2\pi i\zeta}\oint \frac{dz\, z^m}{2\pi i z} M^a_{ij} M^b_{kl}\, h^{ijkl}(z,\zeta). \tag{11.68}$$

The difference between the two expressions resides in the orientation of the contours of integration; in the first case these contours are such that $|z| > |\zeta|$, and in the second this inequality is reversed. As the only singularities of the integrand are at z and $\zeta = 0$ or ∞ and also at $z = \zeta$, we see that the commutator $[T^a_m, T^b_n]$ is given by the same integrand integrated over the difference of the two contours, which, by Cauchy's theorem, will give a nonvanishing expression only for the terms which are singular at $\zeta = z$. Looking at h^{ijkl}, we see that the first term in (11.67) is regular at $\zeta = z$, and so does not contribute. The terms involving one contraction function Δ have a simple pole at $\zeta = z$, and yield $if^{abc}T^c_{m+n}$, and the Schwinger term comes from the terms containing double poles (i.e. Δ^2).

Let us calculate first the contribution due to the double poles. To do this we calculate

$$I = \oint dz\, z^{m-1} \oint d\zeta\, \zeta^{n-1} \frac{z\zeta}{(z-\zeta)^2}, \tag{11.69}$$

in which we choose the integration contours to be given by $z = \rho e^{i\phi}$ and $\zeta = \eta e^{i\psi}$, where $\rho > \eta$. As $\rho > \eta$, we can expand

$$\frac{1}{(1-\zeta/z)} = \sum_{j=0}^{\infty}(j+1)\left(\frac{\eta}{\rho}\right)^j e^{ij(\psi-\phi)}, \tag{11.70}$$

and so, given that

$$\int_0^\infty d\kappa\, e^{i\kappa l} = 2\pi\, \delta_{l0}, \tag{11.71}$$

we find that $I = -4\pi^2 m\theta(m)\, \delta_{n+m,0}$. The contribution of the other orientation of the contours gives a similar expression (with m and n interchanged and so also effectively with the complementary θ functions), and so, using the fact that $(\delta^{jk}\delta^{il} - \delta^{jl}\delta^{ik})M^a_{ij}M^b_{kl} = 2\,\mathrm{Tr}\, M^a M^b$, we find that our Schwinger term contribution is given by $-\frac{1}{2}m\delta_{m+n,0}\,\mathrm{Tr}\, M^a M^b$.

To see that the terms involving a single pole at $\zeta = z$ reproduce $if^{abc}T^c_{m+n}$, we observe that, due to the δ functions in (11.67), their contribution to (11.68) involves $4\,{}^\circ_\circ\, H(z)M^aM^bH(\zeta)\,{}^\circ_\circ\,\Delta(z,\zeta)$. However, by Cauchy's theorem, instead of integrating our expression over the difference of the two integration contours, we can integrate it over a z contour enclosing $z = \zeta$ once, followed by a ζ contour enclosing the origin once. As the residue of the pole at $z = \zeta$ is 1, we find that the integration over z essentially eliminates the Δ function, and sets in all other expressions $z = \zeta$. Then, using the commutation relation of the M's (i.e. $[M^a, M^b] = f^{abc}M^c$), and remembering that the $H^i(\zeta)$ anticommute, we obtain the required result after the integration over ζ. Thus we have showed that the moments of our free fermion model currents, like those of the σ model with the Wess–Zumino term, satisfy the Kac–Moody algebra (11.23). Thus the currents themselves satisfy a current algebra with a Schwinger term. In chapter 9 we mentioned the possibility of the appearance of an anomalous term in the commutator of the generators of time-independent gauge transformations in some theories with fermions. In fact, the discussion given here can be adapted to these theories, establishing the existence of this term in theories where quantum effects break gauge invariance.

Let us look next at a Virasoro algebra associated with this Kac–Moody algebra. To do this, by analogy with the sigma model case discussed before, we consider the symmetric energy–momentum tensor of the theory. This tensor is given by

$$\theta^{\mu\nu} = \tfrac{1}{4}\{\bar\psi\gamma^\mu\overleftrightarrow{\partial}^\nu\psi + \bar\psi\gamma^\nu\overleftrightarrow{\partial}^\mu\psi\}, \tag{11.72}$$

with an implicit summation over the internal index of ψ. As in the sigma model case, this tensor is traceless and, when written in terms of light-cone components, its only nonvanishing components are θ_{++} and θ_{--}.

Let us consider one of these components, say θ_{++}, which is given by $\psi_+\overleftrightarrow{\partial}_+\psi_+$, and so is proportional to $\frac{1}{4}[z\mathrm{d}H/\mathrm{d}z, H]$. Thus, recalling that we have to normal order all our quantities, we see that θ_{++} is proportional to

$$L(z) = \tfrac{1}{2}\,{}^\circ_\circ\, z\frac{\mathrm{d}H}{\mathrm{d}z}H\,{}^\circ_\circ. \tag{11.73}$$

Performing the Laurent series expansion $L(z) = \sum_n L_{-n}\,z^n$, and using the expansion of $H(z)$ in terms of b^i_r, we find that

$$L_n = -\tfrac{1}{2}\sum_r {}^\circ_\circ\, rb^i_rb^i_{n-r}\,{}^\circ_\circ, \tag{11.74}$$

where n and r take integer and half-integer values respectively. Now, using the anticommutation relations satisfied by b^i_r (11.59), it is easy to show that

$$[L_n, b^i_k] = -(\tfrac{1}{2}n + k)b^i_{n+k}. \tag{11.75}$$

The calculation of the commutator of two L_k is much more tedious, but relatively straightforward. We observe that

$$[b^i_\alpha b^j_\beta, b^k_\gamma b^l_\delta] = \{b^k_\gamma b^k_\gamma b^i_\alpha\}b^l_\delta b^j_\beta - b^k_\gamma\{b^l_\delta b^i_\alpha\}b^j_\beta - b^i_\alpha b^k_\gamma\{b^l_\delta b^j_\beta\} + b^i_\alpha\{b^k_\gamma b^j_\beta\}b^l_\delta, \tag{11.76}$$

and then we calculate all the anticommutators using our basic anticommutation relations (11.53). As $\{b^i_\alpha, b^i_\beta\} = N\delta_{\alpha,\beta}$ we finally obtain

$$[L_m, L_n] = (m-n)L_{m+n} + \tfrac{1}{24}Nm(m^2-1)\delta_{m,-n}, \tag{11.77}$$

which is just (11.39) with $c = N/2$.

We have thus established that one can choose our two models, one fermionic and one bosonic, in such a way that they are characterised by the same Kac–Moody algebras and their energy–momentum tensors satisfy the same Virasoro algebras. In this sense both theories are equivalent. One may ask to what extent these results generalise, if we take complex spinor representations and/or consider different groups of symmetries. Of course the necessary condition for this equivalence is the equality of the c numbers $N/2$ of (11.78) and c of (11.39); however, this requirement is not sufficient and any interested reader can consult the review article of Goddard and Olive [4] or the original paper by Goddard *et al* [14], where these conditions are spelled out in detail.

Of course the appearance of the Schwinger term in the algebra of currents can also be shown more directly by applying the procedure of point splitting as discussed by Affleck [15]. As this is a very useful technique with many applications we reproduce here Affleck's derivation of the Schwinger term.

Consider, for simplicity, the $U(1)$ case and let us look at the bilinear in the fermi fields $P = \psi^{+\dagger}\psi^+$, which is a function of only x_-. To define the current we have to normal order P and point split it; i.e. we define

$$J(x_-) = \lim_{\epsilon\to 0} :\psi^{+\dagger}(x_- - \epsilon)\psi^+(x_- + \epsilon):. \tag{11.78}$$

The expression for the current–current commutator $[J(x_-), J(y_-)]$ then follows from the fermion field anticommutator (11.52), i.e.

$$\{\psi^{+\dagger}(x_-), \psi^+(y_-)\} = \delta(x_- - y_-). \tag{11.79}$$

Thus we obtain

$$\begin{aligned}[J(x_-), J(y_-)] &= \lim_{\epsilon\to 0}[J(x_-), :\psi^{+\dagger}(y_- - \epsilon)\psi^+(y_- + \epsilon):] \\ &= \lim_{\epsilon\to 0}[\delta(x_- - y_- + \epsilon) - \delta(x_- - y_- - \epsilon)]\psi^{+\dagger}(y_- - \epsilon)\psi^+(y_- + \epsilon).\end{aligned} \tag{11.80}$$

This expression may, naively, appear to vanish but is, in fact nonzero due to the singularity that occurs in bringing the two fields to the same point. To see this we write

$$\begin{aligned} \psi^{+\dagger}(y_- - \epsilon)\psi^+(y_- + \epsilon) = &: \psi^{+\dagger}(y_- - \epsilon)\psi^+(y_- + \epsilon) : \\ &+ \langle 0|\, \psi^{+\dagger}(y_- - \epsilon)\psi^+(y_- + \epsilon)\, |0\rangle\,, \end{aligned} \tag{11.81}$$

and then we observe that we can take the limit $\epsilon \to 0$ inside the normal ordering without encountering a singularity. On the other hand

$$\langle 0|\, \psi^{+\dagger}(y_- - \epsilon)\psi^+(y_- + \epsilon)\, |0\rangle = \frac{i}{4\pi\epsilon}. \tag{11.82}$$

Thus

$$\begin{aligned} [J(x_-),\, J(y_-)] &= \lim_{\epsilon \to 0}[\delta(x_- - y_- + \epsilon) - \delta(x_- - y_- - \epsilon)]\frac{i}{4\pi\epsilon} \\ &= \frac{i}{2\pi}\,\delta'(x_- - y_-). \end{aligned} \tag{11.83}$$

The obtained expression represents the required Schwinger term. A similar discussion of the nonabelian case reproduces (11.18).

Let us finish this section by pointing out that our construction, based as it is on the equivalence of the currents, has automatically provided us with a 'dictionary' for bosonisation; it is the currents which provide the translation from one description to the other one. In addition, as discussed by Witten [2], the bilinears of the two different currents (one J_+^a the other J_-^b) are proportional to the bosonic sigma matrices g themselves.

Of course, Witten's nonabelian bosonisation has been the subject of many papers [3]. Let us mention here only one of them, namely a paper of Di Vecchia and Rossi [16]. In this paper the authors prove the rules of bosonisation of Witten by using path integral techniques. They consider $U(N)$ and $O(N)$ sigma models and then, using path integral methods, calculate the generating functional for the currents, which they later show to be equal to the corresponding one in the appropriate free fermionic theory, provided that this theory is regularised in a specific way. In the $U(N)$ case, the theory should be regularised in such a way that only the vector current corresponding to the $U(1)$ factor of the $U(N)$ theory is conserved. More details can be found in the original reference [16].

11.5 Models Without the Wess–Zumino Term

In the last section we studied σ models with the Wess–Zumino term. These models have conserved currents and, as Witten pointed out [2], at a specific value of the coefficient of the Wess–Zumino term, both currents J_+ and

J_- are separately conserved, and this conservation survives quantisation. This allowed Witten to argue that this model is equivalent to a free fermion theory. But what about other values of this coefficient? Then we would expect some renormalisation effects and the quantum currents would be different from the classical ones. This result was also obtained by Braaten *et al* [17], as we discussed in chapter 9.

If, when considering a quantum theory, we want to treat the Wess–Zumino term as a tool for breaking some symmetries and for leading to the equations of motion with the required symmetries, then the three-dimensional space used in the definition of the Wess–Zumino term is an artefact of our construction. Thus, so that the quantum theory will not depend on the precise definition of this space, the coefficient of the Wess–Zumino term must be quantised. Hence λ must be an integer. The discussions of Witten and Braaten *et al* hold for any value of this integer, except $\lambda = 0$. So what is the situation with σ models without the Wess–Zumino term?

The existence of conservation laws in pure sigma models (i.e. without the Wess–Zumino term) and of their survival after quantisation, have, of course, been studied much longer. First Pohlmeyer and Polyakov [18] showed that the $O(N)$ σ model possesses an infinite number of classical conservation laws represented by local currents; then Lüscher and Pohlmeyer [19] showed that, in this same model, there exists also an infinite number of classical nonlocal currents which are also conserved. Then various people showed that such currents can be found in many other models.

11.5.1 Nonlocal Currents

An interesting general discussion about the existence of conserved nonlocal currents was given by Brezin *et al* [20], so let us repeat it here. The discussion of Brezin *et al* [20] is based on the assumption that we can introduce a set of matrices $A_\mu(x,t)$, which satisfy two basic conditions:

1. the matrices $A_\mu(x,t)$ can be considered as describing pure gauge field configurations, i.e. there exists a nonsingular matrix g such that
$$A_\mu(x,t) = g^{-1}(x,t)\partial_\mu g(x,t), \qquad (11.84)$$
2. the matrices $A_\mu(x,t)$ are conserved as a consequence of the equations of motion of the model, i.e.
$$\partial^\mu A_\mu(x,t) = 0. \qquad (11.85)$$

Before we show that these assumptions lead to the conservation of nonlocal currents, let us point out that the conditions given above are automatically satisfied in the sigma models (without the Wess–Zumino term), in which the Lagrangian density is provided by the standard expression:

$$L = \frac{1}{2\alpha}\,\mathrm{Tr}\,(\partial^\mu g\,\partial_\mu g^{-1}). \qquad (11.86)$$

In fact, as we said in previous chapters, the various models differ from each other by the conditions we impose on the matrix g, which we can choose to be orthogonal or unitary, and which we can demand to be also hermitian.

The construction of Brezin *et al* is inductive. First they use the matrix $A_\mu(x,t)$ to construct a covariant derivative D_μ defined by

$$D_\mu^{ab} = \delta^{ab}\partial_\mu + A_\mu^{ab}. \tag{11.87}$$

Then, from our first condition it follows that this covariant derivative has a vanishing curvature

$$[D_\mu, D_\nu] = 0, \tag{11.88}$$

while the second condition tells us that ∂^μ commutes with D_μ:

$$\partial^\mu D_\mu = D_\mu \partial^\mu. \tag{11.89}$$

The construction of the currents is inductive. Suppose that we have constructed the nth conserved current $J_\mu^{(n)}$. Then there exists a function $\chi^{(n)}(x,t)$ such that

$$J_\mu^{(n)} = \epsilon_{\mu\nu}\partial^\nu \chi^{(n)}, \quad n \geq 0. \tag{11.90}$$

The $(n{+}1)$th current is then defined by

$$J_\mu^{(n+1)} = D_\mu \chi^{(n)}, \quad n \geq 0. \tag{11.91}$$

The induction starts with $\chi^{(0)}$, which we choose to be given by $\chi^{(0)} = 1$, and so we see that our first current is A_μ, which is conserved by the equation of motion. To prove the conservation of the $(n{+}1)$th current, given the conservation of the nth one, we observe that from (11.90) we have

$$\partial^\mu J_\mu^{(n+1)} = D_\mu \partial^\mu \chi^{(n)}, \quad n \geq 1. \tag{11.92}$$

However, from (11.91) we have

$$\partial^\mu \chi^{(n)} = -\epsilon^{\mu\nu} J_\nu^{(n)}. \tag{11.93}$$

Thus, we see that

$$\partial^\mu J_\mu^{(n+1)} = -D_\mu \epsilon^{\mu\nu} J_\nu^{(n)} = -\epsilon^{\mu\nu} D_\mu D_\nu \chi^{(n-1)} = 0, \tag{11.94}$$

due to the fact that $\epsilon^{\mu\nu} D_\mu D_\nu$ is proportional to the commutator of D_0 and D_1, which vanishes.

The conserved currents give us the conserved charges:

$$Q^n(t) = \int_{-\infty}^{\infty} dx\, J_0^{(n)}(x,t), \quad \frac{dQ^n(t)}{dt} = 0. \tag{11.95}$$

They are nonlocal, since, in order to determine their form, we have to know $\chi^{(n)}$, which can be found only after the integration of (11.91). Their explicit form is rather complicated. For example, the explicit form of the first nontrivial charge is given by

$$Q^{(2)}(t) = -\int_{-\infty}^{\infty} \mathrm{d}x\, A_1(t,x) + \int_{-\infty}^{\infty} \mathrm{d}x\, A_0(t,x)\chi^{(1)}(t,x), \tag{11.96}$$

where $\chi^{(1)}(t,x)$ can be found from the integration of (11.91) for $n = 1$ and so can be chosen to be given by

$$\chi^{(1)}(t,x) = \int_{-\infty}^{x} \mathrm{d}x'\, A_0(t,x'). \tag{11.97}$$

Observe that the integration constant that we can add to $\chi^{(1)}(t,x)$ is equivalent to the addition to $Q^{(2)}(t)$ of a piece proportional to $Q^{(1)}(t)$, i.e. the integral over x of $A_0(t,x)$.

11.5.2 Local Currents

The existence of classical local conservation laws is associated with the conformal invariance of our theory, which we have already exploited in our discussion of σ models with the Wess–Zumino term given in the previous sections. In fact the appearance of the Virasoro algebra was associated with this symmetry and its persistence, in an unmodified form, after quantisation. For this, however, we needed a specific value of the coefficient of the Wess–Zumino term. Now we are considering the theory without this term and so we expect the quantum effects at least to alter the form of the currents, if not to spoil the invariance completely. An interesting study of this problem is given by Goldschmidt and Witten [21] and so we repeat it here.

To do this, we, once again, introduce light-cone variables $x_+ = x_0 + x_1$ and $x_- = x_0 - x_1$ and observe that as a result of the conservation of the energy–momentum, the energy–momentum tensor $T_{\mu\nu}$, when written in terms of these variables, satisfies

$$\partial_+ T_{--} = \partial_- T_{++} = 0. \tag{11.98}$$

Here we have exploited the fact that in a conformal theory the energy–momentum tensor is traceless, and in light-cone variables its trace is given by T_{-+}, which must therefore vanish.

From (11.99) we can immediately derive an infinite number of classical conservation laws,

$$\partial_+[(T_{--})^n] = 0, \quad \partial_-[(T_{++})^n] = 0, \tag{11.99}$$

where n is any integer.

When we quantise the theory we expect, in general, conformal invariance to be broken and so the trace T_{+-} of the energy–momentum will not vanish. Thus our conservation laws (11.100) will not be valid in the form as written. However, it may happen that the extra expressions, which will appear on the the right-hand side of (11.100) as a result of quantisation, can be written as total divergences:

$$\partial_+[(T_{--})^n] = \partial_+ J + \partial_- K, \quad \partial_-[(T_{++})^n] = \partial_+ J' + \partial_- K' \tag{11.100}$$

for some functions J, K, J' and K'. Then we can interpret the additional terms as quantum corrections to our currents, and we see that the conservations laws do survive although in a modified form.

To see whether quantum corrections to (11.100) can, indeed, be rewritten as total divergences we have to look at specific models and, for each of them, investigate what expressions can appear on the right-hand sides of (11.100).

Following Goldschmidt and Witten [21] we consider the $O(N)$ σ model, for which, as we recall from the previous chapters, the Lagrangian density is given by

$$L = \frac{1}{2\alpha}\, \partial_\mu \vec{\phi} \cdot \partial_\mu \vec{\phi}, \tag{11.101}$$

where $\vec{\phi}$ denotes an N-component real field which satisfies $\vec{\phi} \cdot \vec{\phi} = 1$. In this theory $T_{++} = \partial_+ \vec{\phi} \cdot \partial_+ \vec{\phi}$, and so classically

$$\partial_-[(\partial_+ \vec{\phi} \cdot \partial_+ \vec{\phi})^n] = 0. \tag{11.102}$$

What are the possible quantum modifications of (11.103) for, say, $n = 2$?

Clearly, (11.103) will, in general, be modified to

$$\partial_-[(\partial_+ \vec{\phi} \cdot \partial_+ \vec{\phi})^2] = \sum_i \gamma_i A_i, \tag{11.103}$$

where the γ_i are c-numbers, which may depend on the coupling constant and the A_i are local operators, polynomials in the field $\vec{\phi}$, which must have the same dimension and quantum numbers as the left-hand side of (11.103). Goldschmidt and Witten [21] then construct a full list of operators which can appear on the right-hand side of (11.104), realising that operators which vanish due to the equations of motion do not contribute. They then point out that there are only three such operators, which are given by

$$\begin{gathered} A_1 = \partial_- \vec{\phi} \cdot \partial_+^4 \vec{\phi}, \quad A_2 = (\partial_- \vec{\phi} \cdot \partial_+ \vec{\phi})(\partial_+ \vec{\phi} \cdot \partial_+^2 \vec{\phi}), \\ A_3 = (\partial_- \vec{\phi} \cdot \partial_+^2 \vec{\phi})(\partial_+ \vec{\phi} \cdot \partial_+ \vec{\phi}). \end{gathered} \tag{11.104}$$

Then they make a list of operators, which they call B's, which have the same quantum numbers as the A's but which are total divergences. They are given by

$$B_1 = \partial_-[\partial_+^2\vec{\phi}\cdot\partial_+^2\vec{\phi}], \quad B_2 = \partial_+[(\partial_-\vec{\phi}\cdot\partial_+\vec{\phi})(\partial_+\vec{\phi}\cdot\partial_+\vec{\phi})],$$
$$B_3 = \partial_+[\partial_+^3\vec{\phi}\cdot\partial_-\vec{\phi}]. \tag{11.105}$$

Each of the B's can be written as a certain linear combination of A's, and, as the number of B's is the same as the number of A's, it is possible to rewrite any of the A's as a combination of B's. Thus we see that (11.104) is modified to

$$\partial_-[(\partial_+\vec{\phi}\cdot\partial_+\vec{\phi})^2] = \sum_i c_i B_i, \tag{11.106}$$

for some choice of the c_i which are c-numbers. Since the B's are explicitly total divergences our expression (11.100) corresponds to a conservation law. These conservation laws were also discovered independently by Polyakov [22].

Repeating this discussion for other σ models, Goldschmidt and Witten observe that in the case of $U(N)$ or $O(N)$ chiral models, the classical conservation laws are given by

$$\partial_-\{[\mathrm{Tr}\,(g^{-1}\partial_+g)^m]^n\} = 0, \tag{11.107}$$

for any choice of m and n, and where g denotes our chiral field which satisfies $g^{-1}g = 1$. Then they look at the conserved currents of dimension four, and point out that there are two of them; they are given by the expression above with $m = 4$, $n = 0$ and by the expression with $m = n = 2$. Repeating then the discussion of the $O(N)$ case, they find five possible anomaly terms, i.e. A's. As there are five A's and only four B's this means that one of the two classical conservation laws does not survive quantisation; however, there is one combination of the two classical laws for which the quantum corrections are given by a total divergence. We see that the $U(N)$ and $O(N)$ models have, each, at least one local current conservation law at the quantum level:

$$\partial_-\{a[\mathrm{Tr}\,(g^{-1}\partial_+g)^2]^2 + b[\mathrm{Tr}\,(g^{-1}\partial_+g)^4]\} = \sum_i c_i B_i, \tag{11.108}$$

for some choice of a, b and c_i.

Finally, Goldschmidt and Witten consider also the $\mathbb{C}P^{N-1}$ model (for $N > 2$). They find that all classical conservation laws are destroyed by quantum corrections. This should not surprise us. In chapter 6 we studied these models in the $1/N$ expansion. We found that the quantum corrections profoundly altered some properties of the theory. In particular, we found that the quantum effects induced the kinetic term for the composite gauge field $A_\mu = z^\dagger \partial_\mu z$, which in turn provides a confining potential for the quanta of the z field. Hence, quantum effects are responsible for the confinement of the z field, and so the classical and quantum spectra of the theory are very different. Of course, our discussion in chapter 6 relied on the $1/N$ expansion; the lack of the local conservation law after quantisation, and the fact that the $\mathbb{C}P^1$ model is equivalent to the $O(3)$ model, whose fields are bilinears in the $\mathbb{C}P^1$ fields, all suggest that these results are true in general.

11.5.3 S matrices

The existence of locally conserved currents which survive quantisation has profound implications for the properties of the theory. In fact, as shown by Zamolodchikov and Zamolodchikov [23], there is no particle production in such theories and, as a consequence, the S matrices of these theories must factorise into products of two-particle matrices. The conditions of factorisability are, in fact, quite restrictive. In their paper [23] the Zamolodchikovs studied the effects of these restrictions together with further constraints such as unitarity and crossing, which any relativistic theory should satisfy. They found that, if one further combines these restrictions with some notions about the symmetry of the spectrum of particles in the theory, the equations which follow from the factorisation restrictions can be solved and the explicit form of the S matrices determined. In particular, in the case of the $O(N)$ model they assume that the spectrum of the model exhibits $O(N)$ symmetry, and consists of massive particles grouped into multiplets which transform as vectors under this symmetry. This seems a very reasonable assumption, and is supported by the $1/N$ calculations [24]. Proceeding in this way, and making similar assumptions about the spectrum of each model, the Zamolodchikovs determine the S matrix for the sine-Gordon model (or, equivalently, the massive Thirring model) and find two more solutions, of which one is the S matrix for the $O(N)$ model and the other describes the scattering of elementary particles of the Gross–Neveu model [25]. We do not report here any details of their calculations as these calculations are rather technical and involved; the interested reader can find all the details in the original reference [23].

Since the original Zamolodchikov paper many other papers [26–29] have appeared in which the S matrices of various models are determined either by solving the corresponding factorisation equations or by relating the bosonic model to a fermionic model, which can then be solved by the so-called 'Bethe ansatze'. The latter approach was invented by Polyakov and Wiegmann [30], who applied it to the case of the $U(N)$ chiral and $O(4)$ and $O(3)$ sigma models. In particular, studying the $SU(2)$ chiral model, Polyakov and Wiegmann showed that this model is quantum mechanically equivalent to a special fermionic model, which can be 'solved' exactly (i.e.

for which we can find all its S-matrix elements). Thus they show that the models based on the Lagrangian densities

$$L = \frac{1}{2f_0} \operatorname{Tr}(\partial_\mu g^{-1} \partial_\mu g), \tag{11.109}$$

where $g \in SU(2)$ and

$$L_1 = \sum_{\alpha=1}^{N} i\psi_\alpha^\dagger \partial\!\!\!/ \psi_\alpha + \lambda_0 \left(\sum_{\alpha=1}^{N} \psi_\alpha^\dagger \gamma_\mu \tau^i \psi_\alpha \right)^2, \tag{11.110}$$

where τ^i are the familiar Pauli matrices, are equivalent in the limit $N \to \infty$. To prove this they introduce an auxiliary field

$$A_\mu^i = \sum_{\alpha=1}^{N} \psi_\alpha^\dagger \gamma_\mu \tau^i \psi_\alpha, \tag{11.111}$$

and then integrate out the ψ fields and obtain

$$S_1 = \int \mathrm{d}^2x \left(\frac{1}{2\lambda_0} A_\mu^2 \right) + N \ln \det[(i\partial\!\!\!/ + \gamma_\mu A_\mu)]. \tag{11.112}$$

Then they evaluate the second term in the expression above by observing that, since it represents the fermionic determinant in a background field, its effects can be represented by a corresponding σ model with the Wess–Zumino term. The σ fields of this model are so chosen that their auxiliary field A_μ^i can be constructed out of them. Next they take the limit of $N \to \infty$, and show that in this limit the N factor which appears in (11.113) quenches the fluctuations of the A_μ^i field forcing it to become pure gauge; i.e. to satisfy $F_{\mu\nu} = 0$. But the Lagrangian density

$$L = \left(\frac{1}{2\lambda_0} A_\mu^2 \right), \tag{11.113}$$

taken together with the constraint

$$F_{\mu\nu} = 0, \tag{11.114}$$

is equivalent to the chiral field g, whose dynamics is governed by (11.110).

Having established this equivalence they are now ready to use the various techniques which exist for looking at the spectrum and other properties of a fermionic theory. In particular, the standard procedure [31] uses the diagonalisation of the transfer matrix which, when coupled with the periodicity requirements of the fermionic wavefunctions, leads to various Bethe-ansatz equations. We will not discuss these equations, as they are rather involved (the interested reader can consult [30]), but only say that the solutions of these equations provide information on the spectrum and other properties of the theory, and lead to S-matrix elements which can then be compared with those found by the Zamolodchikovs [23]. It is reassuring that, where they overlap, both approaches give the same results.

Actually, in their calculation, for some technical reasons, Polyakov and Wiegmann use a fermionic model different from (11.111), for which they establish their equivalence. The model they use corresponds to the completely integrable theory of a single fermion field with an isospin S. They argue that this new theory is equivalent to (11.111) for $N = 2S$.

The results of Polyakov and Wiegmann that we have just reported were then generalised to other models in a series of papers by Wiegmann [26] and by other people [27]. Moreover these ideas were further generalised by Kirillov and Smirnov [28], who used them to calculate the form factors in the $O(3)$ model.

At the same time Abdalla and co-workers [29] studied the effects of nonlocal charges which we discussed at the beginning of this section. They found that in many theories (e.g. $\mathbb{C}P^{N-1}$ models with fermions coupled minimally or supersymmetrically) these charges also survive quantisation in a modified form (all quantum corrections turn out to be given by total divergences). Then they also produce restrictions on the spectrum of the theory and on the form of the scattering of the particles from this spectrum. This has allowed Abdalla *et al* to determine various S-matrix elements of these models.

11.6 Conclusions and Omissions

In this book we have studied various σ models, These models, the first of which was introduced by Gell-Mann and Levy [32] almost thirty years ago to describe the theory of the beta decay, have since become a useful laboratory for testing various theoretical ideas. As such, they are normally considered in two dimensions, where the calculations are simpler and the results easier to obtain. At the same time it is usually hoped that these results are not misleading as to what to expect in dimensions more relevant from the physics point of view.

In this book we have concentrated on treating sigma models as a laboratory in which we can check various ideas from particle physics theory and, in particular, the relevance of classical solutions to quantisation, the topological properties of gauge theories and the skyrmion models of elementary particles. Thinking about quantisation in terms of path integrals, we have performed most of our studies in Euclidean space paying particular attention to the classical solutions of the equations of motion (or, strictly speaking, to the Euler–Lagrange equations) and their solutions. We have found that the spectrum of our solutions is quite complicated but, in the grassmannian case, all the solutions fall into various classes characterised by two integer numbers, each of which corresponds to a degree of some holomorphic map. Moreover, all of them can be related to some basic holomorphic solutions, from which they can be obtained via a simple procedure of taking derivatives and the Gramm–Schmidt orthogonalisation of the resultant vectors. As we showed in chapters 3 and 4, the two integers which characterise each of our solutions are simply related to the integer which describes the topological properties of our fields, and so appear in the Atiyah–Singer index theorem. They are also related to the index of the fluctuation operator around these solutions (although this relation is less straightforward). To see both these relations it is convenient to solve the associated background Dirac problem. Moreover, all noninstantonic (or nonanti-instantonic) solutions are unstable; they correspond to the saddle points of the action.

In the chiral-model case the spectrum of the solutions is richer. All the solutions of grassmannian models are automatically also solutions of these models; in addition there exist further solutions which depend on additional parameters, and this dependence is such that these solutions can be thought of as interpolating between various grassmannian solutions. As in the grassmannian models case, the action of all of our solutions of the chiral models is still quantised, but this time all our solutions are unstable.

We have also looked at the case of σ models with the Wess–Zumino term, and have found that the solutions to these models can be be found by solving the auxiliary linear problem, the consistency condition for which constitutes the integrability condition for the original σ model (without the Wess–Zumino term). Moreover, the solutions to this linear problem can be found quite easily (given the solutions of the chiral problem) and, as we have shown, they correspond to replacing the uniton factors $(1-2R)$ of the chiral model solution by $[1-2R/(1+\lambda)]$. Like the solutions of the chiral models, the solutions of the model with the Wess–Zumino term are also unstable.

The fact that most of our solutions are not the minima of the action, but only its saddle points, leads to problems with the implementation of our programme of quantisation via the functional integration method, as discussed in the first chapter. Clearly, some sort of analytical continuation of these negative modes will be required in order to carry it out. However, for this we have to know the spectrum of the fluctuation operator, or use some indirect methods. In chapter 4 we discussed such indirect methods in the case of instantonic solutions of the $\mathbb{C}P^1$ model, and commented on the problem of studying the effects of unstable solutions. Unfortunately, the problem of determining the complete spectrum of the fluctuation operator around a general solution has turned out to be a very difficult task; in general, not even the spectrum of the negative modes is known.

We have also discussed other approaches to studying the quantum properties of our sigma models, namely, those based on the $1/N$ expansions and, in this chapter, those based on proving the equivalence of a given sigma model to a suitable fermionic model and then solving the latter by the Bethe ansatz, or by determining their S-matrix elements from factorisation, unitarity and crossing. In the $O(N)$ model case both approaches show that the spectrum of the quantum states consists of massive particles (thus their mass is generated by quantum corrections), and that the original $O(N)$ symmetry, broken at the classical level, is restored. In the $\mathbb{C}P^{N-1}$ case the quantum corrections generate the confining potential for the quanta of the z_α field, and the spectrum consists of their massive bound states. The situation in the other models is similar.

Most of our discussion has been concerned with maps only between compact spaces; however, various studies have been performed when this restriction has been lifted. In particular, Abbott [33] investigated solutions of the $\mathbb{C}P^{N-1}$ models without the requirement of finiteness of the action. In addition, there exist various studies of the classical solutions of the compact sigma models in Minkowski space (both 1+1 and 1+2 dimensions) — see, e.g. [34–36]. Moreover, noncompact sigma models have been also studied. If we restrict ourselves to the simplest noncompact generalisations of the $\mathbb{C}P^{N-1}$ model, such as those discussed in [37] or [38], we find that the solutions are, again, given by holomorphic maps, but this time it is much harder to impose the conditions of finiteness of the action. For the simplest of these models, namely, the model defined on the $U(1,1)/(U(1)\times U(1))$ coset space, we can adapt the original discussion of Woo [39], and show that this model does not possess any nontrivial finite action solutions. For larger models of this class it is easy to see that some finite action solutions can be found by various embeddings of compact models.

Noncompact models were also investigated by Davis *et al* [37] from the point of view of the $1/N$ expansions. They found that, in contradistinction to the compact case, quantum corrections do not generate kinetic terms for the composite fields.

We have also looked at the effects of fermions which can be added to these models; we have studied the cases when they are introduced in the simplest possible way, and when they are introduced in such a way that the resultant theory is supersymmetric. We have found that, depending on how the fermions are coupled in, their effects can alter the spectrum of the theory. We have not discussed a further generalisation of such models corresponding to the inclusion of supersymmetry with central charges as suggested by Gates [40]. The interested reader can consult [41] for a detailed investigation of such models in the $1/N$ limit.

We have also studied more general sigma models (with the Wess–Zumino term) and presented the Braaten *et al* interpretation of such models in terms of models in which the target space is endowed with nonvanishing torsion.

In chapter 10 we discussed briefly the Skyrmion model of baryons, and then studied in some detail the (2+1)-dimensional skyrmions based on our sigma models. We looked at the possibility of adding the Hopf terms to our Lagrangian in the $O(3)$ sigma model case, and discussed the resultant fractionalisation of spin.

Finally, in this last chapter, we have discussed various concepts associated with Witten's non-abelian bosonisation, the conservation of local and nonlocal currents and Kac–Moody and Virasoro algebras.

There are, however, whole areas of the world of two-dimensional σ models that we have not managed to discuss in this book. Thus we have not said anything about either the standard perturbative expansions of our sigma models involving the renormalisation of various composite operators in these models nor the recently proposed new perturbation theory which, from the onset, includes some nonperturbative effects. The interested reader can consult the original literature — for recent papers see [42–44].

Moreover, sigma models have been the subject of many numerical investigations, involving various lattice approximations, etc; for a recent paper presenting such calculations of the mass gap see [45].

The original ideas of the equivalence of some bosonic and some fermionic theories, which we discussed in the first sections of this chapter, have been pursued further (see, e.g. [46]; however [47] has a warning that this equivalence may not hold for some groups) and has led to a whole new area of study and applications associated with the so-called conformal field theories. Here we send the interested reader to the paper of Knizhnik and Zamolodchikov [48] and to the papers by Belavin *et al* [49] and Friedan *et al* [50], who formulated the basic rules of such field theories. For recent studies of various properties of these theories see [51] and for applications of these ideas in the area of critical phenomena see the review by Cardy [52].

The last few years have seen a growing interest in the so-called string models of elementary particle physics. These models are seen by many as the basic building blocks on which the final theory will be constructed. For a recent book on this topic see [53]. However, it has been realised that, although string Lagrangians describe the whole spectrum of these models, the lowest lying states (and in our current understanding, the only states relevant from a physical point of view) are described by an effective theory represented by sigma models (this time, of course, not in two dimensions). This had led to an increased interest in these models — for recent papers on this equivalence see [54–57]. Finally, we would like to mention recent recent papers by Witten [58–60], in which he studies some more mathematical aspects of the Virasoro representation theory and its relation to quantisation, and in [60] he presents and discusses purely topological field theories, in which there are no particle interpretable fluctuations but, which, nevertheless, are nontrivial. Clearly, we will hear more on these topics in the near future.

Classical solutions of various σ models have also been of interest to pure mathematicians. They consider maps between various manifolds and call harmonic those maps which satisfy the Euler–Lagrange equations of the appropriate variational problem. These maps, very often, correspond exactly to what we have called the classical equations of motion. Moreover, as mathematicians like to consider maps from compact spaces, and often choose this space to be S^2, this corresponds, in our language, to considering only finite action solutions, which is, in fact, the condition we want to impose too; in our case, the motivation for this restriction comes from a consideration of their contribution in various path integrals. These mathematicians also consider ±holomorphic maps; they correspond to our instantons and anti-instantons.

Let me finish by presenting a list of some references from pure mathematical literature. The classic older papers are:

Eells J and Sampson J H 1964 Harmonic mappings of Riemannian manifolds *Am. J. Math.* **86** 109

Calabi E 1976 Minimal immersions of surfaces in Euclidean spheres *J. Diff. Geom.* **1** 111

Eells J and Lemaire L 1979 A report on harmonic maps *Bull. Lond. Math. Soc.* **10** 1

The more recent papers include

Burns D 1982 Harmonic maps from $\mathbb{C}P^1$ to $\mathbb{C}P^N$ *Lecture Notes in Mathematics* **949** (Berlin: Springer)

Leung P F 1982 On the stability of harmonic maps *Lecture Notes in Mathematics* **949** (Berlin: Springer) p 122

Eells J and Lemaire L 1983*Selected Topics in Harmonic Maps. CBMS Regional Conf. in Mathematics 50* (Providence, RI: Am. Math. Soc.)

Eells J and Wood J C 1983 Harmonic maps from surfaces to complex projective spaces *Adv. Math.* **49** 217

Erdem S and Glasbrook J F 1983 Harmonic maps of Riemann surfaces to indefinite complex hyperbolic and projective Spaces *Proc. Lond. Math. Soc.* **47** 547

Erdem S and Wood J C 1983 On the construction of harmonic maps into a Grassmannian *J. Lond. Math. Soc.* **28** 161

Ramantham J 1984 Harmonic maps from S^2 to $G_{2,4}$ sl J. Diff. Geom. **19** 207

Wolfson J 1984 Harmonic maps of S^2 into $G(2, n, \mathbb{C})$ *Complex Differential Geometry and Non-Linear Differential Equations* (Brunswick: College Press)

The most recent papers include

Chern S-S and Wolfson J G 1985 Harmonic maps of S^2 into a complex Grassmannian Manifold *Proc. Natl Acad. Sci., USA* **81** 2217

Burstall F E, Rawnsley J H and Salamon S M 1987 Stable harmonic 2-spheres in symmetric spaces *Bull. Am. Math. Soc.* **16** 249

Burstall F E and Wood J C 1986 The construction of harmonic maps into complex Grassmannians *J. Diff. Geom.* **23** 255

Wolfson J G 1986 Harmonic maps of the two-sphere into the complex hyperquadratic *J. Diff. Geom.* **24** 141

Burstall F E and Rawnsley J H 1987 Stability of classical solutions of two dimensional Grassmannians *Commun. Math. Phys.* **110** 311

The papers by Uhlenbeck and Wood quoted in chapters 7 and 8.

Clearly, the interaction between pure mathematicians and theoretical physicists will continue, and will lead to an even better understanding of the classical solutions of the equations of motion/harmonic maps.

References

[1] Coleman S 1975 Quantum sine-Gordon equation as the massive Thirring model *Phys. Rev.* D **11** 2088

Mandelstam S 1975 Soliton operators for the quantised sine-Gordon equation *Phys. Rev.* D **11** 3026

[2] Witten E 1984 Non-Abelian bosonisation in two dimensions *Commun. Math. Phys.* **92** *455*

[3] Date C D, Frishman Y and Sonnenschein J 1987 The spectrum of multi-flavour QCD in two dimensions *Nucl. Phys.* B **283** 365

Frishman Y and Sonnenschein J 1987 Bosonisation of colored flavour fermions and QCD *Nucl. Phys.* B **294** 801

[4] Goddard P and Olive D I 1986 Kac–Moody and Virasoro algebras in relation to quantum physics *Int. J. Mod. Phys.* A **1** 303

[5] Adler S and Dashen R 1968 *Current Algebra and Applications to Particle Physics* (Reading, MA: Benjamin)

[6] Moody R V 1967 Lie algebras associated with generalised Cartan matrices *Bull. Am. Math. Soc.* **73** 217

Kac V G 1985 *Infinite Dimensional Lie Algebras* (Camabridge: Cambridge University Press)

[7] Bowcock P and Goddard P 1987 Virasoro algebras with central charge $c < 1$ *Nucl. Phys.* B **285** 651

Gepner D 1987 On the spectrum of 2D conformal field theories *Nucl. Phys.* B **287** 111

Di Francesco P, Saleur H and Zuber J B 1987 Modular invariance in non-minimal two dimensional conformal theories *Nucl. Phys.* B **285** 454

Guadaguini E 1987 Current algebra in sigma models on homogenous spaces *Nucl. Phys.* B **290** 430

[8] Virasoro M 1970 Subsidiary conditions and ghosts in dual-resonance models *Phys. Rev.* D **1** 2933

[9] Ramond P 1981 *Field Theory: a Modern Primer* (Reading, MA: Benjamin-Cummings)

[10] Sugawara H 1968 A field theory of currents *Phys. Rev.* **170** 1659

[11] Ramond P 1971 Dual theory of free fermions *Phys. Rev.* D **3** 2415

[12] Neveu A and Schwarz J H 1971 Factorisable dual model of pions *Nucl. Phys.* B **31** 86; 1971 Quark model of dual pions *Phys. Rev.* D **4** 1109

[13] Wick G C 1950 The evaluation of the collision matrix *Phys. Rev.* **80** 268

[14] Goddard P, Nahm W and Olive D 1985 Symmetric spaces, Sugawara's energy momentum tensor in two dimensions and free fermions *Phys. Lett.* **160B** 111

[15] Affleck I 1986 Exact critical exponents for quantum spin chains, non-linear σ Models at $\theta = \pi$ and the quantum Hall effect *Nucl. Phys.* B **265** 409

[16] Di Vecchia P and Rossi P 1984 On the equivalence between the Wess–Zumino action and the free Fermi theory in two dimensions *Phys. Lett.* **140B** 344

[17] Braaten E, Curtright T L and Zachos C K 1985 Torsion and geometrostatis in nonlinear sigma models *Nucl. Phys.* B **260** 630

[18] Pohlmeyer K 1976 Integrable Hamiltonian systems and interactions through quadratic constraints *Commun. Math. Phys.* **46** 207

Polyakov A M 1977 Hidden symmetry of the two dimensional chiral fields *Phys. Lett.* **72B** 224

[19] Lüscher M and Pohlmeyer K 1978 Scattering of massless lumps and nonlocal charges in the 2 dimensional nonlinear σ model *Nucl. Phys.* B **137** 46

[20] Brezin E, Itzykson C, Zinn-Justin J and Zuber J-B 1979 Remarks about the existence of non-local charges in two dimensional models *Phys. Lett.* **82B** 442

[21] Goldschmidt Y Y and Witten E 1980 Conservation laws in some two-dimensional models *Phys. Lett.* **91B** 392

[22] Polyakov A M 1975 Interaction of Goldstone particles in two dimensions. Applications to ferromagnets and massive Yang–Mills fields *Phys. Lett.* **59B** 79

[23] Zamolodchikov A B and Zamolodchikov Al B 1979 Factorised S-matrices in two dimensions as the exact solutions of certain relativistic quantum field theories *Ann. Phys., NY* **120** 253

[24] Brezin E, Le Guillou J C and Zinn-Justin J 1976 Renormalisation of the nonlinear σ model in $2+\epsilon$ dimensions *Phys. Rev.* D **14** 2615

Bardeen W A, Lee B W and Shrock R E 1976 Phase transition in the non-linear σ model in a $(2+\epsilon)$ dimensional continuum *Phys. Rev.* D **14** 985

[25] Gross D and Neveu A 1974 Dynamical symmetry breaking in asymptotically free field theories *Phys. Rev.* D **10** 3235

[26] Wiegmann P B 1984 On the theory of nonabelian Goldstone bosons in two dimensions; exact solution of the $SU(N) \times S(N)$ nonlinear σ model *Phys. Lett.* **141 B** 217; 1984 Exact factorised S-matrix of the chiral field in two dimensions *Phys. Lett.* **142B** 173; 1985 Exact solution of the $O(3)$ nonlinear σ model *Phys. Lett.* **152B** 209

Ogievetsky E, Reshetkhin N and Wiegmann P 1987 The principal chiral field in two dimensions on classical Lie algebras: the Bethe ansatz solutions and factorised theory of scattering *Nucl. Phys.* B **280** 45

[27] de Vega H J and Karowski M 1987 Exact Bethe ansatz solution of the $O(N)$ symmetric theories *Nucl. Phys.* B **280** 225

[28] Kirillov A N and Smirnov F A 1988 Form factors in $O(3)$ nonlinear σ model *Int. J. Mod. Phys.* A **3** 731

[29] Abdalla E and Abdalla M C B 1980 On the exact S-matrix for $\mathbb{C}P^{N-1}$ and $SU(N)$ chiral Thirring model *Nuovo Cimento* A **57** 334

Abdalla E, Abdalla M C B and Gomes M 1982 Anomaly cancellation in the supersymmetric $\mathbb{C}P^{N-1}$ model *Phys. Rev.* D **25** 452; 1983 Nonlocal charge of the $\mathbb{C}P^{N-1}$ model and its supersymmetric extension to all orders *Phys. Rev.* D **27** 825

Abdalla E and Lima-Santos A 1988 Factorisable S-matrix for $SO(D)/(SO(2) \times SO(D-2))$ nonlinear σ models with fermions *Mod. Phys. Lett.* A **3** 311

Abdalla E, Abdalla M C B and Forger M 1988 Exact S-matrices for anomaly free nonlinear σ models on symmetric spaces *Nucl. Phys.* B **297** 374

[30] Polyakov A and Wiegmann P 1983 Theory of nonabelian Goldstone bosons in two dimensions *Phys. Lett.* **131B** 121; 1984 Goldstone fields in two demensions with multivalued actions *Phys. Lett.* **141B** 223

[31] Belavin A A 1979 Exact solution of the two dimensional model with asymptotic freedom *Phys. Lett.* **87B** 117

Andrei N and Lowenstein I H 1979 Diagonalisation of the chiral invariant Gross–Neveu Hamiltonian *Phys. Rev. Lett.* **43** 1698

[32] Gell-Mann M and Levy M 1960 The axial vector current in beta decay *Nuovo Cimento* **16** 705

[33] Abbott R B 1982 Elliptic solutions of $\mathbb{C}P^{N-1}$ models and their properties *Z. Phys.* C **15** 51

[34] Harnad J, Saint-Aubin Y and Shnider S 1984 Backlund transformation for non-linear sigma models with values in Riemannian symmetric spaces *Commun. Math. Phys.* **92** 329

[35] Ward R S 1988 Soliton solutions in an integrable chiral model in 2+1 dimensions *J. Math. Phys.* **29** 386

[36] Forgacs P, Horvath Z and Zakrzewski W J 1984 Classical solutions of $\mathbb{C}P^{N-1}$ models in 2+1 dimensions *Nucl. Phys.* B **248** 187

[37] Davis A C, Macfarlane A J and van Holten J W 1983 Composite gauge bosons in sigma models and supergravity *Phys. Lett.* **125B** 151

Davis A C, Freeman M D and Macfarlane A J 1985 Dynamical generation in non-compact σ models: a no-go theorem *Nucl. Phys.* B **258** 393

[38] Erdem S and Glasbrook J F 1983 Harmonic maps of Riemann surfaces to Indefinite Complex hyperbolic and projective spaces *Proc. Lond. Math. Soc.* **47** 547

[39] Woo G 1977 Pseudoparticle configurations in two dimensional ferromagnets *J. Math. Phys.* **18** 1264

[40] Gates S J 1984 Superspace formulation of new non-linear sigma models *Nucl. Phys.* B **238** 349

[41] Davis A C, Freeman M D and Macfarlane A J 1985 Quantum theory of supersymmetric sigma models with central charges *Nucl. Phys.* B **256** 299

[42] Wegner F 1987 Anomalous dimensions for the nonlinear sigma model in $2+\epsilon$ dimensions, I and II *Nucl. Phys.* B **280** 193, 211

[43] Osborn H 1987 Renormalisation and composite operators in nonlinear σ models *Nucl. Phys.* B **294** 595

[44] Davis A C and Nahm W 1985 A perturbative approach to mass-gap generation — the nonlinear σ model *Phys. Lett.* **155B** 404; 1985 Nonperturbative effects in the perturbation theory — the $\mathbb{C}P^{N-1}$ model *Phys. Lett.* **159B** 294

[45] Patrascioiu A 1987 Mass gap in non-Abelian σ models and gauge theories *Phys. Rev. Lett.* **58** 2285

[46] Goddard P, Kent A and Olive D 1985 Virasoro algebras and coset space models *Phys. Lett.* **152B** 88

[47] Ashworth R M 1987 The inequivalence of fermionic and bosonic theories sharing the same energy–momentum tensor *Nucl. Phys.* B **280** 321

[48] Knizhnik V G and Zamolodchikov A B 1984 Current algebra and Wess–Zumino model in two dimensions *Nucl. Phys.* B **247** 83

[49] Belavin A A, Polyakov A M and Zamolodchikov A B 1984 Infinite conformal symmetry in two dimensional quantum field theory *Nucl. Phys.* B **241** 333

[50] Friedan D, Qiu Z and Shenker S H 1984 Conformal invariance, unitarity and two dimensional exponents *Vertex Operators in Mathematics and Physics* ed J Lepowsky *et al* (Berlin: Springer) p 419

[51] Christe P and Flume R 1987 The four-point correlations of all primary operators of the d=2 conformally invariant $SU(2)$ σ-model with Wess–Zumino term *Nucl. Phys.* B **282** 466

[52] Cardy J 1986 Conformal invariance *Phase Transitions and Critical Phenomena* vol 11, ed C Domb and J L Lebowitz (New York: Academic)

[53] Green M B, Schwarz J H and Witten E 1987 *Superstring Theory* vols I and II (Cambridge: Cambridge University Press)

[54] Gross D J and Sloan J H 1987 The quartic effective action for the heterotic string *Nucl. Phys.* B **291** 41

[55] Callan C, Friedan D, Martinec E and Perry M J 1985 Strings in the fields background *Nucl. Phys.* B **262** 593

[56] Friedling B E and Jevicki A 1986 Nonlinear σ models as S-matrix generating functional on strings *Phys. Lett.* **174B** 75

[57] Metsaev R R and Tseytlin A A 1987 Order α (two loop) equivalence of the string equations of motion and the σ model Weyl Invariance conditions: dependence on the dilaton and the antisymmetric tensor *Nucl. Phys.* B **293** 385

[58] Witten E 1988 Quantum field theory Grassmannians and algebraic curves *Commun. Math. Phys.* **113** 529

[59] Witten E 1988 Coadjoint orbits of the Virasoro group *Commun. Math. Phys.* **114** 1

[60] Witten E 1988 Topological quantum field theory *Commun. Math. Phys.* **117** 353; 1988 Topological sigma models *Commun. Math. Phys.* **117** 411

Appendices

Appendix A

Here we discuss a possible way of dealing with the zero modes of a fluctuation operator. We base this method on the example of the double well discussed in section 4 of chapter 3. There we found that

$$\eta_0(\tau - \tau_o) = \frac{1}{\sqrt{S_0}} \frac{\mathrm{d}q_{\mathrm{cl}}(\tau, \tau_0)}{\mathrm{d}\tau} \tag{A1}$$

was a zero mode of the fluctuation operator $\hat{O}$. As q_{cl} does not depend individually on τ and τ_0, but only on their difference, we exhibit this fact explicitly in our definition of η_0. To eliminate the direction η_0 from the functional space of fluctuations we recall that

$$\int_{-\infty}^{\infty} \delta[f(x) - a]\,\mathrm{d}x = \int_{-\infty}^{\infty} \delta(x - b)\frac{1}{f'}\Big|_{x=b}\,\mathrm{d}x = \frac{1}{f'}\Big|_{x=b}, \tag{A2}$$

where we assumed that $x = b$ is the solution of $f(x) = a$. Thus we have a convenient 'resolution' of 1:

$$f'|_{x=b} \int_{-\infty}^{\infty} \delta[f(x) - a]\,\mathrm{d}x = 1. \tag{A3}$$

Next we consider a general fluctuation $\eta(\tau) = q(\tau) - q_{\mathrm{cl}}(\tau, \tau_0)$ and require that it is orthogonal to $\eta_0(\tau - \tau_0)$. We can impose this orthogonality by inserting into the functional measure the factor:

$$\delta\left(\int_{-\infty}^{\infty} \mathrm{d}\tau' \eta_0(\tau' - \tau_0)\eta(\tau')\right). \tag{A4}$$

At the same time we notice that

$$\int_{-\infty}^{\infty} \mathrm{d}\tau' \eta_0(\tau' - \tau_0) q_{\mathrm{cl}}(\tau' - \tau_0) = \frac{1}{s\sqrt{S_0}} q_{\mathrm{cl}}^2(\tau' - \tau_0)\Big|_{\tau'=-\infty}^{\tau'=\infty}, \tag{A5}$$

which is clearly independent of τ_0 (actually it is 0 in our case). Calling this quantity R, we observe that the correct procedure corresponds not to the insertion of (A4) but to the whole of (A3), in which we have set

$$a = R \qquad f(\tau_0) = \int_{-\infty}^{\infty} \mathrm{d}\tau' \eta_0(\tau' - \tau_0) q(\tau'). \tag{A6}$$

So we insert into the integration measure

$$1 = \int_{-\infty}^{\infty} \mathrm{d}\tau_0 \, \delta\left(\int_{-\infty}^{\infty} \mathrm{d}\tau' \eta_0(\tau' - \tau_0)\eta(\tau')\right) \Delta(q), \tag{A7}$$

where

$$\begin{aligned} \Delta(q) &= \frac{\mathrm{d}}{\mathrm{d}\tau_0} \int_{-\infty}^{\infty} \mathrm{d}\tau' \eta_0(\tau' - \tau_0) q(\tau') = \int_{-\infty}^{\infty} \mathrm{d}\tau' q(\tau') \frac{\mathrm{d}}{\mathrm{d}\tau_0} \eta_0(\tau' - \tau_0) \\ &= -\int_{-\infty}^{\infty} \mathrm{d}\tau' \frac{\mathrm{d}}{\mathrm{d}\tau'} \eta_0(\tau' - \tau_0) q(\tau') = \int_{-\infty}^{\infty} \eta_0(\tau' - \tau_0) \frac{\mathrm{d}q(\tau')}{\mathrm{d}\tau'} \mathrm{d}\tau'. \end{aligned} \tag{A8}$$

We can now make a further *approximation* (often not stated very clearly):

$$\Delta(q) \cong \Delta(q_{\mathrm{cl}}) = \int_{-\infty}^{\infty} \eta_0(\tau' - \tau_0) \frac{\mathrm{d}q_{\mathrm{cl}}(\tau' - \tau_0)}{\mathrm{d}\tau'} \mathrm{d}\tau' = \sqrt{S_0}. \tag{A9}$$

Thus the total functional integral can be rewritten as

$$\begin{aligned} &\int \mathrm{D}[\eta] \exp\left(-\frac{S_0}{\hbar} - \frac{1}{2\hbar}\int \mathrm{d}\tau' \eta \hat{O} \eta\right) \\ &= \int_{-\infty}^{\infty} \mathrm{d}\tau_0 \int \mathrm{D}[\eta] \delta\left(\int_{-\infty}^{\infty} \eta_0(\tau' - \tau_0)\eta(\tau')\mathrm{d}\tau'\right) \\ &\quad \times \Delta(q) \exp\left(-\frac{S_0}{\hbar} - \frac{1}{2\hbar}\int \mathrm{d}\tau' \eta \hat{O} \eta\right) \\ &\simeq T\sqrt{S_0} \exp\left(-\frac{S_0}{\hbar}\right) \left(\det{}' \hat{O}\right)^{-1/2}, \end{aligned} \tag{A10}$$

where the $'$ over det denotes that the zero mode is to be ignored. Note that our approximation (A9) is justified if, in our calculations, we are interested only in the functional determinant. It is easy to convince oneself that the terms dropped by making this approximation are of higher order in the coupling constant g; hence they have to be included if, in our perturbative calculations, we want to go beyond the calculation of only the functional determinant.

Actually, in field theory we often scale out a coupling constant g. This corresponds to $S \mapsto S/g$. Then each zero mode brings in a factor $1/\sqrt{g}$.

Appendix B

In this appendix we give some very useful results on the use of the ζ-function technique of regularisation of functional determinants. To specify this technique we first of all consider a $N \times N$ hermitian matrix A with eigenvalues $\lambda_1, \lambda_2, \ldots \lambda_N$ (not necessarily distinct). Then we define the ζ function of this matrix as

$$\zeta_A(s) = \sum_{i=1}^{N} \lambda_i^{-s}. \tag{B1}$$

Then clearly

$$\zeta_A(0) = \sum_{i=1}^{N} 1 = N, \tag{B2}$$

i.e. equals the dimension of A and

$$\zeta_A'(0) = -\sum_{i=1}^{N} \lambda_i^{-s} \ln \lambda_i|_{s=0} = -\sum_{i=1}^{N} \ln \lambda_i = -\ln \det A. \tag{B3}$$

These two properties form the basis of our definition of functional determinants. To see how the regularisation comes about, we recall the definition of the Riemann zeta function:

$$\zeta(s) = \sum_{n=1}^{\infty} \frac{1}{n^s} = \frac{1}{\Gamma(s)} \int_0^{\infty} \mathrm{d}t\, t^{s-1} \frac{1}{e^t - 1}. \tag{B4}$$

The expression on the right-hand side is suitable for the analytical continuation of $\zeta(s)$ to all s, and it reveals a pole at $s = 1$. Performing this continuation we can show that

$$\zeta(0) = -\tfrac{1}{2}. \tag{B5}$$

To do this we expand

$$\frac{1}{e^t - 1} = \frac{e^{-t}}{t}(1 + \tfrac{1}{2}t \pm \ldots), \tag{B6}$$

and so performing the integration over t, we find

$$\zeta(s) = \frac{1}{\Gamma(s)}[\Gamma(s-1) + \tfrac{1}{2}\Gamma(s) + \ldots] = \frac{1}{s-1} + \tfrac{1}{2} + \ldots, \tag{B7}$$

where the terms denoted by $(\ldots)$ go to 0 as $s \to 0$. Thus $\zeta(0) = -\frac{1}{2}$. In a similar way, though with more effort, we can show that

$$\zeta'(0) = -\tfrac{1}{2}\ln(2\pi). \tag{B8}$$

The analogous definition, suitable for analytical continuation, for a finite-dimensional matrix A is

$$\zeta_A(s) = \frac{1}{\Gamma(s)} \int_0^\infty \mathrm{d}t\, t^{s-1}\, \mathrm{Tr}\,(e^{-At}). \tag{B9}$$

Notice that in the Riemann zeta function case we can write

$$\frac{1}{e^t - 1} = e^{-t} + e^{-2t} + \cdots = \mathrm{Tr}\, e^{-At}, \tag{B10}$$

where

$$A = \begin{pmatrix} 1 & 0 & \ldots \\ 0 & 2 & 0 \\ \ldots & \ldots & \ldots \\ 0 & 0 & \ddots \end{pmatrix}.$$

In the functional case we have an infinite-dimensional matrix A, and so we generalise further and replace the matrix A by a differential operator D^2 and e^{-At} by the associated Green function $G(x, y, t)$, which satisfies

$$\frac{\partial}{\partial t} G(x, y, t) = D^2\, G(x, y, t). \tag{B11}$$

On this Green function we impose the boundary condition that

$$G(x, y, t) = \delta(x - y) + O(t) \qquad \text{as } t \to 0. \tag{B12}$$

Thus

$$\zeta_{-D^2}(s) = \frac{1}{\Gamma(s)} \int_0^\infty \mathrm{d}t\, t^{s-1} \int \mathrm{d}^d x\, G(x, x, t). \tag{B13}$$

This expression has a pole at $s = 1$ but can be shown to be regular at $s = 0$, so that both $\zeta_{-D^2}(0)$ and $\zeta'_{-D^2}(0)$ are well defined and calculable.

A further comment can be made about the scaling properties of the determinants. As is easy to check,

$$\zeta_{A/\lambda}(s) = \lambda^s\, \zeta_A(s), \tag{B14}$$

and

$$\zeta'_{A/\lambda}(0) = \ln \lambda\, \zeta_A(0) + \zeta'_A(0). \tag{B15}$$

This result generalises to the case of functional determinants.

Appendix C

In this appendix we calculate the coefficients $a_1(x,x)$ and $\hat{a}_1(x,x)$, which are used in the calculation of the effects of fluctuations around the k-instanton solution in the $\mathbb{C}P^1$ model. These coefficients appear in the expansion of the Green functions $G(z,y,t)$ and $\hat{G}(z,y,t)$, which are given by

$$G(z,y,t) = \frac{e^{-|x-y|^2/t}}{\pi t}[a_0(x,y) + a_1(x,y)t + \ldots] \tag{C1}$$

and

$$\hat{G}(z,y,t) = \frac{e^{-|x-y|^2/t}}{\pi t}[\hat{a}_0(x,y) + \hat{a}_1(x,y)t + \ldots]. \tag{C2}$$

These expressions are Green functions of the operators

$$Q = \varrho\partial_+\varrho^{-2}\partial_-\varrho \qquad \text{and} \qquad \hat{Q} = \varrho^{-1}\partial_-\varrho^2\partial_+\varrho^{-1} \tag{C3}$$

respectively. Let us give all the details of the calculation of $a_1(x,x)$. To do this we set

$$G(z,y,t) = G_0(z,y,t)[a_0(z,y) + a_1(z,y)t + \ldots], \tag{C4}$$

where G_0 is the Green function corresponding to $\varrho = 1$, and then observe that

$$\begin{aligned} QG &= \varrho\partial_+\varrho^{-2}\partial_-\varrho[G_0(a_0 + a_1t + \ldots)] \\ &= \partial_+\partial_-G_0(a_0 + a_1t + \ldots) + (\partial_-G_0)\varrho\partial_+[\varrho^{-1}(a_0 + a_1t + \ldots)] \\ &\quad +\partial_+(G_0)\varrho^{-1}\partial_-[\varrho(a_0 + a_1t + \ldots)] + G_0Q(a_0 + a_1t + \ldots). \end{aligned} \tag{C5}$$

However, as

$$QG = \frac{\partial}{\partial t}G = \frac{\partial G_0}{\partial t}(a_0 + a_1t + \ldots) + G_0a_1 + \ldots, \tag{C6}$$

and as

$$\partial_+\partial_-G_0 = \frac{\partial G_0}{\partial t}, \tag{C7}$$

we find that

$$\begin{aligned} G_0a_1\ldots &= \partial_-G_0\varrho\partial_+[\varrho^{-1}(a_0 + a_1t + \ldots)] \\ &\quad +\partial_+G_0\varrho^{-1}\partial_-[\varrho(a_0 + a_1t + \ldots)] + G_0Q(a_0 + a_1t + \ldots). \end{aligned} \tag{C8}$$

On the other hand G_0 satisfies

$$\begin{aligned} \partial_-G_0 &= -\frac{(z-y)}{t}G_0, \\ \partial_+G_0 &= -\frac{(z-y)^*}{t}G_0, \end{aligned} \tag{C9}$$

and so we see that

$$\partial_- G_0|_{z=y} = \partial_+ G_0|_{z=y} = 0, \tag{C10}$$

giving us

$$a_1|_{z=y} = (Qa_0)|_{z=y}, \tag{C11}$$

and showing that the problem has been reduced to the calculation of $a_0(x,y)$.

However, from the previous expressions we see that a_0 satisfies

$$(\partial_- G_0)\varrho\partial_+(\varrho^{-1}a_0) + (\partial_+ G_0)\varrho^{-1}\partial_-(\varrho a_0) = 0, \tag{C12}$$

and so that

$$(x-y)\varrho\partial_+(\varrho^{-1}a_0) + (x^* - y^*)\varrho^{-1}\partial_-(\varrho a_0) = 0. \tag{C13}$$

If we now take ∂_- of this expression, we obtain

$$(x-y)\partial_-[\varrho\partial_+(\varrho^{-1}a_0)] + \varrho^{-1}\partial_-(\varrho a_0) + (x^* - y^*)\partial_-[\varrho^{-1}\partial_-(\varrho a_0)] = 0. \tag{C14}$$

Next we take $\varrho\partial_+(\varrho^{-1}\cdot)$ of this expression and set $x = y$, obtaining

$$\partial_-[\varrho\partial_+(\varrho^{-1}a_0)]\Big|_{x=y} + \varrho\partial_+[\varrho^{-2}\partial_-(\varrho a_0)]\Big|_{x=y} = 0, \tag{C15}$$

as all other terms $\sim (x-y)$ or $(x^* - y^*)$. Thus we obtain

$$Q\,a_0\Big|_{x=y} + \partial_-[\varrho\partial_+(\varrho^{-1}a_0)]\Big|_{x=y} = 0, \tag{C16}$$

and so, expanding the last term,

$$Qa_0\Big|_{x=y} = -\partial_+\partial_- a_0\Big|_{x=y} + (\partial_+ \ln\varrho)\,\partial_- a_0\Big|_{x=y} + (\partial_+\partial_- \ln\varrho)a_0\Big|_{x=y} \tag{C17}$$

On the other hand, setting $x = y$ in (C14) gives us

$$\partial_- a_0\Big|_{x=y} = -a_0\Big|_{x=y}\,\partial_- \ln\varrho. \tag{C18}$$

In a similar way we find that

$$\partial_+ a_0\Big|_{x=y} = a_0\Big|_{x=y}\,\partial_+ \ln\varrho. \tag{C19}$$

Next we take ∂_+ of (C14) and set $x = y$, obtaining

$$2\,\partial_+\partial_- a_0 \Big|_{x=y} = -2a \Big|_{x=y} \partial_+ \ln \varrho\, \partial_- \ln \varrho. \tag{C20}$$

Combining all the terms together we find that

$$Q a_0 \Big|_{x=y} = -\partial_+\partial_- \ln \varrho, \tag{C21}$$

and so finally that

$$a_1 \Big|_{x=y} = -\partial_+\partial_- \ln \varrho. \tag{C22}$$

In a similar way we demonstrate that

$$\hat{a}_1(x,y) \Big|_{x=y} = \partial_-\partial_+ \ln \varrho, \tag{C23}$$

as $\hat{a}_1$ is obtained from a_1 by the replacement $\varrho \leftrightarrow \varrho^{-1}$ and $\partial_+ \leftrightarrow \partial_-$.

Appendix D

In this appendix we compute $\check{F}^{(\alpha)}(p)$ and $\check{F}^{(\lambda)}_{\mu\nu}(p)$, which arise in the $1/N$ expansion of the $\mathbb{C}P^{N-1}$ model discussed in chapter 6. To calculate $\check{F}^{(\alpha)}(p)$, which is given by

$$\check{F}^{(\alpha)}(p) = \int \frac{\mathrm{d}^2 q}{(2\pi)^2} \frac{1}{m^2 + q^2} \frac{1}{m^2 + (p+q)^2}, \tag{D1}$$

we first of all rewrite the propagators, using the Feynman parametrisation trick,

$$\frac{1}{D_1\, D_2} = \int_0^1 \mathrm{d}x \int_0^1 \mathrm{d}y\, \frac{\delta(1-x-y)}{(D_1 x + D_2 y)^2}, \tag{D2}$$

and obtain

$$\begin{aligned}
&\int \mathrm{d}^2 q \frac{1}{m^2+q^2}\, \frac{1}{m^2+(p+q)^2} \\
&\quad = \int_0^1 \mathrm{d}x \int \mathrm{d}^2 q \frac{1}{[(m^2+q^2)x+(m^2+(q+p)^2)(1-x)]^2} \\
&\quad = \int \mathrm{d}^2 q \int_0^1 \mathrm{d}x\, \frac{1}{[m^2+q^2+p^2 x(1-x)]^2} = \pi \int_0^1 \mathrm{d}x\, \frac{1}{[m^2+p^2 x(1-x)]},
\end{aligned} \tag{D3}$$

where we have changed the integration variables and then performed the new q integration. Next we change the variables further by setting $x(1-x) = u$ and then, using

$$\int_0^1 \frac{\mathrm{d}v}{\sqrt{v}} \frac{1}{b - av} = \frac{1}{\sqrt{ab}} \ln \frac{\sqrt{b} + \sqrt{a}}{\sqrt{b} - \sqrt{a}},$$

we find

$$\check{F}^{(\alpha)}(p) = \frac{1}{2\pi} \frac{1}{\sqrt{p^2(p^2+4m^2)}} \ln \frac{\sqrt{p^2+4m^2} + \sqrt{p^2}}{\sqrt{p^2+4m^2} - \sqrt{p^2}}. \tag{D4}$$

The calculation of $\check{F}^{(\lambda)}_{\mu\nu}(p)$ is more involved. To perform it we need some formulae which often arise in dimensional regularisation calculations. We need two such formulae:

$$\int \frac{\mathrm{d}^{2\omega} l}{(2\pi)^{2\omega}} \frac{1}{[l^2 + M^2 + 2l \cdot p]^A} = \frac{\Gamma(A-\omega)}{(4\pi)^\omega \Gamma(A)} \frac{1}{(M^2 - p^2)^{A-\omega}}, \tag{D5}$$

and

$$\int \frac{\mathrm{d}^{2\omega} l}{(2\pi)^{2\omega}} \frac{l_\mu l_\nu}{[l^2 + M^2 + 2l \cdot p]^A}$$
$$= \frac{1}{(4\pi)^\omega \Gamma(A)} \left(p_\mu p_\nu \frac{\Gamma(A-\omega)}{(M^2-p^2)^{A-\omega}} + \frac{\delta_{\mu\nu}}{2} \frac{\Gamma(A-\omega-1)}{(M^2-p^2)^{A-\omega-1}} \right).$$

In our case we want to evaluate

$$\begin{aligned} \check{F}^{(\lambda)}_{\mu\nu}(p) = {} & 2\delta_{\mu\nu} \int \frac{\mathrm{d}^2 q}{(2\pi)^2} \frac{1}{q^2+m^2} \\ & - \int \frac{\mathrm{d}^2 q'}{(2\pi)^2} \frac{4q'_\mu q'_\nu}{[(q'-\frac{1}{2}p)^2+m^2][(q'+\frac{1}{2}p)^2+m^2]}, \end{aligned} \tag{D6}$$

and so we want to use the formulae above with $\omega \to 1$. Using them we find

$$\check{F}^{(\lambda)}_{\mu\nu}(p) = 2\delta_{\mu\nu} \frac{\Gamma(1-\omega)}{(4\pi)^\omega (m^2)^{1-\omega}} - \frac{4}{(4\pi)^\omega} \int_0^1 \mathrm{d}x \left(\frac{\Gamma(2-\omega)}{4} \right.$$
$$\left. \frac{p_\mu(1-2x)p_\nu(1-2x)}{[m^2+\frac{1}{4}p^2-\frac{1}{4}p^2(1-2x)^2]^{2-\omega}} + \frac{1}{2}\delta_{\mu\nu} \frac{\Gamma(1-\omega)}{[m^2+\frac{1}{4}p^2-\frac{1}{4}p^2(1-2x)^2]^{1-\omega}} \right).$$

The $\delta_{\mu\nu}$ terms are individually divergent as $\omega \to 1$ so we consider them first. We expand the denominator

$$\left(1 + \frac{p^2}{4m^2} - \frac{p^2(1-2x)^2}{4m^2}\right)^{\omega-1}$$

$$= 1 + (\omega - 1)\ln\left(1 + \frac{p^2}{4m^2} - \frac{p^2(1-2x)^2}{4m^2}\right) + O((\omega-1)^2),$$

and find that in the limit $\omega \to 1$ the $\delta_{\mu\nu}$ terms are given by

$$\frac{2\delta_{\mu\nu}}{4\pi}\int_0^1 \mathrm{d}x \ln\left(1 + \frac{p^2}{4m^2} - \frac{p^2(1-2x)^2}{4m^2}\right)$$

$$= -\frac{\delta_{\mu\nu}}{\pi} + \frac{\delta_{\mu\nu}}{2\pi}\sqrt{\frac{4m^2+p^2}{p^2}}\ln\frac{\sqrt{p^2+4m^2}+\sqrt{p^2}}{\sqrt{p^2+4m^2}-\sqrt{p^2}}, \tag{D7}$$

where we have used

$$\int \mathrm{d}x \ln(a^2 - x^2) = x\ln(a^2-x^2) - 2x + a\ln\frac{x+a}{a-x}.$$

Next we look at the $p_\mu p_\nu$ terms. They are given by

$$\begin{aligned} -\frac{p_\mu p_\nu}{4\pi}\int_0^1 \mathrm{d}x \frac{(1-2x)^2}{[m^2 + \frac{1}{4}p^2 - \frac{1}{4}p^2(1-2x)^2]} \\ = -\frac{p_\mu p_\nu}{4\pi}\int_0^1 \mathrm{d}v \frac{v^2}{[m^2+\frac{1}{4}p^2 - \frac{1}{4}p^2v^2]}. \end{aligned} \tag{D8}$$

We perform the v integration using

$$\int_0^1 \mathrm{d}v \frac{-v^2}{a - bv^2} = \frac{1}{b} - \frac{a}{b}\frac{1}{2\sqrt{ab}}\ln\frac{\sqrt{a}+\sqrt{b}}{\sqrt{a}-\sqrt{b}},$$

and so determine their contribution as

$$\frac{p_\mu p_\nu}{p^2}\left(\frac{1}{\pi} - \frac{\sqrt{p^2+4m^2}}{2\pi\sqrt{p^2}}\ln\frac{\sqrt{p^2+4m^2}+\sqrt{p^2}}{\sqrt{p^2+4m^2}-\sqrt{p^2}}\right). \tag{D9}$$

Further Reading — General Texts

We present here a list of some books, review articles and longer articles, which we feel the interested reader may want to consult after having read this book, to deepen his knowledge of the topics covered here or to extend it into closely related areas. Some of the titles on this list appear also in the list of references given at the end of each chapter.

Affleck I 1986 Exact critical exponents for quantum spin chains, non-linear σ Models at $\theta = \pi$ and the quantum Hall effect *Nucl. Phys.* B **265** 409

Aitchison I J R 1987 Berry phases, magnetic monopoles, and Wess–Zumino terms or how the skyrmion got its spin *Acta Phys. Polon.* B **18** 207

Coleman S 1985 *Aspects of Symmetry* (Cambridge: Cambridge University Press)

DeWitt B 1983 *Supermanifolds* (Cambridge: Cambridge University Press)

Friedan D and Shenker S 1987 The analytic geometry of two dimensional conformal field theory *Nucl. Phys.* B **281** 509

Goddard P and Mansfield P 1986 Topological structures in field theories *Rep. Prog. Phys.* **49** 725

Goddard P and Olive D I 1980 Kac–Moody and Virasoro algebras in relation to quantum physics *Int. J. Mod. Phys.* A **1** 303

Green M B, Schwarz J H and Witten E 1987 *Superstring Theory* vols I and II (Cambridge: Cambridge University Press)

Howe P 1987 Two-dimensional non-linear σ models *Frontiers of High-Energy Physics* ed I G Halliday (Bristol: Adam Hilger)

Hull C M 1987 Lectures on non-linear sigma models *Super Field Theories* ed H C Lee *et al* (New York: Plenum)

Jackiw R 1977 Quantum meaning of classical field theory *Rev. Mod. Phys.* **49** 681

Martinec E 1987 Conformal field theory *Nucl. Phys.* **281** 157

Percacci R 1986 *Geometry of Nonlinear Field Theories* (Singapore: World Scientific)

Perelomov A M 1987 Chiral models: geometrical aspects *Phys. Rep.* **146** 135

Petersen J L 1985 Non-Abelian chiral anomalies and Wess–Zumino effective anomalies *Acta Phys. Polon.* **16** 271

Rajaraman R 1982 *Solitons and Instantons* (Amsterdam: North-Holland)

Trieman S B, Jackiw R, Zumino B and Witten E (ed) 1985 *Current Algebra and Anomalies* (Singapore: World Scientific)

Tze C-H 1987 Skyrmions and current algebra *Chiral Soliton* ed K F Liu (Singapore: World Scientific)

Index